Biological Aspects of Human Sexuality

Biological Aspects of Human Sexuality

Fourth Edition

Herant A. Katchadourian
Stanford University

HOLT, RINEHART AND WINSTON, INC.

Fort Worth   Chicago   San Francisco   Philadelphia
Montreal   Toronto   London   Sydney   Tokyo

For Stina, Nina, and Kai

Publisher	Ted Buchholz
Acquisitions Editor	Christina Oldham
Senior Project Editor	Christine Caperton
Production Manager	Ken Dunaway
Art & Design Supervisor	Vicki McAlindon Horton
Cover Designer	Rhonda Campbell

Author photo by Chuck Painter, News and Publications, Stanford University
Anatomical Illustrations by Warren Budd/Lee Ames & Zak Studio
Charts and graphs by Vantage Art, Inc.

Library of Congress Cataloging-in-Publication Data

Katchadourian, Herant A.
    Biological aspects of human sexuality/Herant A. Katchadourian. —
    4th ed.
    Includes bibliographical references.
    1. Human reproduction.   2. Sex.   I. Title.
QP251.K36   1990   89-19856
612.6—dc20

ISBN: 0-03-032852-7

NOTE: This work is derived from FUNDAMENTALS OF HUMAN SEXUALITY, Fifth Edition by Herant A. Katchadourian, copyright © 1989, 1985, 1980, 1975, 1972 by Holt, Rinehart and Winston, Inc.

Address Editorial Correspondence To: 301 Commerce Street, Suite 3700, Fort Worth, TX 76102

Address Orders To: 6277 Sea Harbor Drive, Orlando, FL 32887
1-800-782-4479, or 1-800-433-0001 (in Florida)

Printed in the United States of America
0   1   2   3   0 1 6   9   8   7   6   5   4   3   2

Holt, Rinehart and Winston, Inc.
The Dryden Press
Saunders College Publishing

Preface

College courses in human sexuality vary widely in their scope and depth of coverage. While there is no common core for such courses that instructors can readily agree on, the biological aspects of sexuality are clearly central to any meaningful exploration of the subject. It is hardly conceivable that topics like sexual anatomy, physiology, endocrinology, reproduction, contraception, and sexually transmitted diseases could be left out of even the most elementary introduction to sexuality.

Yet the majority of instructors in human sexuality are behavioral scientists with no specialized training in biology or medicine. It is therefore particularly important that they have at their disposal a text that deals with these topics in a clear and authoritative manner. For these reasons, we issue the fourth edition of this brief volume on the biological aspects of human sexuality. The text is based largely on the more comprehensive textbook, *Fundamentals of Human Sexuality* (fifth edition). Chapter 1 draws from various parts of the larger volume to provide a context for the discussion of sexual biology. Chapters 2 to 7 are presented with very few changes. Chapter 8 (which corresponds to Chapter 15 of *Fundamentals*) is likewise only slightly modified.

While retaining its strengths, the fifth edition of *Fundamentals* (which is the source of this text) has been revised extensively in substance, style, and format. Some of these changes have been necessitated by new developments in the field and new problems, such as AIDS, requiring greater attention. Other changes have been in response to the thoughtful comments and criticisms of four sets of reviewers representing a broad spectrum of disciplines and teaching experiences in the field. As a result, we have shifted the level of emphasis in a number of areas; we have made the level of presentation more accessible to a broader range of students; and we have been responsive to the needs of instructors in a greater variety of institutions, while retaining the intellectual vigor and integrity of the text. Equally important changes have been made in the figures and photographs illustrating the text.

The text attempts to be comprehensive without being exhaustive in coverage. Many important aspects of sex are dealt with, but the main emphasis is on topics likely to be of concern to the majority of readers.

Virtually all of the chapters in this edition have been rewritten. The changes from the previous edition thus amount to more than an ordinary revision. Fast-moving areas like AIDS research have been monitored to provide readers with the most up-to-date information possible. The text now provides an extensive coverage of AIDS and other sexually transmitted diseases. In addition to the biomedical and the psychosocial aspects of these problems, special care has been taken to provide students with the necessary information and guidance to help them make safe and sensible sexual decisions. The same is true for contraception and other topics with a bearing on sexual health.

Some of the most important changes pertain to the illustrations. We now have a uniform style for figures and charts, all of which have been redrawn. The purpose of these illustrations is to add substance to the text, not to decorate it. The addition of a second color has provided an important new aspect to the appearance and instructional purposes of the text.

The effectiveness of a text depends first on its content, then on its manner of presentation. If there is no substance to a book, no amount of "pedagogical" gimmickry will make it worthwhile. However, if the material presented does not engage the reader's attention, it will do little good.

We continue to use boxes to highlight special topics and to present literary excerpts. Key terms in the text are italicized. At the end of each chapter there is a set of review questions and thought questions. Suggested readings provide a list of accessible sources for further study. A detailed glossary (with pronunciation guide) defines important terms. An extensive list of references provides documentation of sources and directs the student to additional sources of information.

The appearance of this volume is in no way a relinquishing of our commitment to the broader psychosocial perspective on human sexuality that characterizes *Fundamentals of Human Sexuality*. Biology may be the bedrock on which sex is based, but there is far more to human sexuality than may be contained within the domain of biology. This volume is therefore not a substitute but an adjunct to other texts and sources that provide the necessary psychosocial perspective to understanding human sexuality.

ACKNOWLEDGMENTS

Numerous colleagues and reviewers have contributed greatly to the text's success through the successive editions. Space limitations do not permit the reiteration of acknowledgments to reviewers from previous editions, but that does not diminish my appreciation for their help.

In the preparation of this edition, Julian Davidson was particularly helpful with the updating of materials on physiology and hormones and Sylvia Cerel Bowen on contraception and sexually transmitted diseases.

A special word of thanks is owed to the successive groups of reviewers who provided invaluable advice and criticisms at various phases of working. They are: John M. Allen, The University of Michigan; Wayne Anderson, University of Missouri, Columbia; Ann Auleb, San Francisco State University; Elaine Baker, Marshall University; Janice Baldwin, University of California, Santa Barbara; M. Betsy Bergen, Kansas State University; Ruth Blanche, Montclair State College; Stephen W. Bordi, West Valley College; James E. Cherry, Charles Stewart Mott Community College; Dennis M. Dailey, The University of Kansas; Ronald S. Daniel, California State Polytechnic University, Pomona; Wayne Daugherty, San Diego State University; William A. Fisher, The University of Western Ontario; Susan Fleischer, Queens College; Suzanne G. Frayser, University of Denver; Grace Galliano, Kennesaw College; Frederick P. Gault, Western Michigan University; Brian A. Gladue, North Dakota State University; Barbara Gordon-Lickey, University of Oregon; John T. Haig, Philadelphia College of Textiles and Science; Sandra Hamilton, University of Oregon; Donald E. Herrlein, Northeastern State University; Ray W. Johnson, North Texas State University; Ethel Kamien, University of Lowell; Sander M. Latts, University of Minnesota; Ronald S. Mazer, University of Southern Maine; Roger N. Moss, California State University, Northridge; Daniel P. Murphy, Creighton University; Andrea Parrot, Cornell University; Sara Taubin, Drexel University; Richard M. Tolman, University of Illinois, Chicago; Marlene Tufts, Clackamas Community College; Charles Weichert, San Antonio College; Donald Whitmore, The University of Texas, Arlington; Edward W. Wickersham, Pennsylvania State University; Midge Wilson, DePaul University.

The production of a book is almost as arduous a task as its writing. Laurie Burmeister went through the rigors of typing and preparing the manuscript for publication with her competent, tireless, and cheerful dedication. Jane Knetzger, developmental editor, played a central role throughout the process. Paula Cousin, senior project manager, masterfully guided the complex task of preparing the book for production.

I also wish to thank Susan Driscoll and Ted Buchholz, publishers of Behavioral and Social Sciences; Susan Arellano and Christina Oldham, acquisitions editors; Christine Caperton, senior project editor; Kristin Zimet and Jeanette Johnson, copy editors; Annette May-

eski and Ken Dunaway, production managers; Judy Allan and Vicki McAlindon Horton, design supervisors; Elsa Peterson, photo researcher; Lisa Bossio, who wrote the glossary; Marion Geisinger, who assisted with the photo program; and Karee Galloway, who assisted with permissions.

The members of my family, to whom this book is dedicated, and in particular, my wife, Stina, relieved me of many burdens to make this task easier. To her, and to our children, Nina and Kai, I express my affectionate gratitude.

H.A.K.

CREDITS

The author is indebted to the following for photographs and permissions to reproduce them. Copyright for each photograph belongs to the photographer or agency credited, unless specified otherwise.

Cover, Michelangelo *Madonna and Child*, Florence Casa Buonarrotti, Scala/Art Resource; **7**, Harlow Primate Laboratory, University of Wisconsin; **19**, by permission of the Kinsey Institute for Research in Sex, Gender, and Reproduction; **35**, © Joel Gordon 1972; **52**, Dr. Julian Davidson; **80**, Blackwell Scientific Publications Limited; **80**, Blackwell Scientific Publications Limited; **81**, Blackwell Scientific Publications Limited; **90**, Johns Hopkins University Press; **91**, Johns Hopkins University Press; **99**, Bilder Lexicon; **116**, Dr. Austin I. Dodson, Jr.; **117**, Courtesy of Centers of Disease Control, HHS; **121 (above)**, Courtesy of Centers of Disease Control, HHS; **121 (below)**, Dr. John Wilson/Photo Researchers; **122**, Candy Tedeschi, R.N.C.; **153**, Reprinted by permission of Macmillan Publishing Company from *For Women of All Ages* by Sheldon H. Cherry. Copyright © 1979 by Sheldon H. Cherry; **156 (left)**, © Mariette Pathy Allen/Peter Arnold, Inc.; **156 (right)**, Suzanne Szasz/Photo Researchers; **157**, Erika Stone; **159**, Suzanne Arms Wimberley; **179**, © Joel Gordon 1988; **185**, © Joel Gordon 1981; **186**, © Joel Gordon 1988; **209**, Abraham Menashel/Photo Researchers.

About the Author

Herant Katchadourian, M.D., is Professor of Psychiatry and Behavioral Sciences, Professor of Human Biology, and Professor of Education (by courtesy) at Stanford University, where he has also served as Vice Provost and Dean of Undergraduate Studies.

He received his undergraduate and medical degrees (with Distinction) from the American University of Beirut and his specialty training in psychiatry at the University of Rochester, New York.

In addition to this text (which has been translated into French, Spanish, Portuguese, and Chinese), Dr. Katchadourian is the author of *Human Sexuality: Sense and Nonsense, The Biology of Adolescence*, and *Fifty: Midlife in Perspective*. He is the co-author of *Careerism and Intellectualism Among College Students* and the editor of *Human Sexuality: A Comparative and Developmental Perspective*. His other publications are in the areas of cross-cultural psychiatry and the life cycle.

In 1968, Dr. Katchadourian initiated one of the first college courses in human sexuality at a major American university. Since then, well over 12,000 students have taken his course. He has been selected Outstanding Professor and Class Day speaker six times by Stanford seniors. He received the Richard W. Lyman Award of the Stanford Alumni Association in 1984. He is a member of the Alpha Omega Alpha medical honor society.

Brief Contents

Detailed Contents

Boxed Features

Biological Aspects of Human Sexuality

The Biology of Sex in Perspective

Sex is the source of human life and central to life from birth to death. Whether or not we actively engage in it, sexuality is part of our everyday thoughts and feelings; it is rooted in our dreams and longings, fears and frustrations.

Sex dominates the lives of some people; for others it plays a lesser role, either by choice or by circumstance. For most of us sexual interests wax and wane, depending on a host of internal needs and external circumstances. At each phase of life sexuality unfolds in distinctive ways.

At the biological level, the primary function of sex is reproduction—having children. Biology provides the machinery for sexual functions and behavior. Its components range from genes to genitals. Sexual functions are regulated by your nerves and hormones and sustained by the circulatory, muscular, and other systems of your body.

At the psychological level, sex consists of a variety of behaviors and relationships aimed at erotic pleasure, affection, and other needs.

At the societal level, sex pervades numerous facets of life. It makes your culture unique, flavoring its art, its history, its laws, and its values.

In addition sexuality has many indirect roles in your life. It is a big part of your gender identity (masculinity, femininity) and of what people expect of you in various social roles. Sex can express dominance or hostility. It contributes to your self-esteem and social status, and in many other ways shapes your life from infancy to old age.

The sexual machinery in our bodies makes sexual activity possible, but such activity is only given meaning as human experience in its psychological and social contexts. We see genital organs like the penis and vagina not only for what they are in a *naturalistic* sense but for what they represent in a *symbolic* sense. They may be imbued with feelings of power, dominance, beauty, and pride as well as shame, guilt, and ugliness. These feelings are determined by our culture, which generates and manipulates symbols through language and imagery. A relentlessly biological perspective reduces us to reproductive machines; an insis-

tently sociocultural approach transforms us into disembodied abstractions.

In this book, we are concerned with the biology of sex. This truncated approach to sexuality may be misleading in and of itself. Hence, to provide a somewhat broader context in which to understand the biology of sex, we shall briefly consider, in this chapter, some of the salient psychological and social factors that give meaning to sexuality as a human experience.

DETERMINANTS OF SEXUAL BEHAVIOR

There are certain physiological functions, like eating, whose primary purpose is obvious. Even though food preferences vary widely among individuals and groups, everyone must eat something: fish or fowl, raw or cooked, with or without fork and knife. Similarly, waste products must be eliminated from the body regardless of the social customs attached to the process. No society can interfere with the physiological functions of eating and elimination without jeopardizing life.

What about sex? Can it be compared to either eating or elimination? Both comparisons have been tried, but neither is quite adequate. We can, of course, refer metaphorically to "sexual hunger." In fact, at certain times people and animals may prefer copulating to eating. Nevertheless, sex is not necessary for sustaining life, except in the broad sense of preserving the species. In fact, there is no evidence that abstinence from sexual activity is necessarily detrimental to health.

The comparison to elimination also has some superficial validity, since most men and women experience a buildup of sexual tension; so we can speak of periodic neurophysiological "discharge" through orgasm, with the understanding that sexual activity is not literally the "discharging" of anything that would otherwise be "dammed up" like the waste products of the body.

Sexual behavior is ultimately the interactional outcome of three types of forces—biological, psychological, and social. This should not imply a hierarchy among these factors as to which one is more important: these forces are complementary and integrated rather than mutually exclusive.

Biological Origins of Sexuality

Biological explanations of sexual behavior have traditionally been based on the concept of *instinct,* an innate force that compels people to act a certain way. Yet it has been difficult to define, let alone demonstrate, the existence of such motivating forces in humans.

Sex and reproduction are so intimately linked in most living beings that one explanation of sexual behavior is simply the need to reproduce. Sex, in this sense, is part of a deep-rooted biological incentive for animals to mate and perpetuate their species. But since lower animals cannot know that mating results in reproduction, what mysterious force propels them to mate? Despite the reproductive consequences of copulation, sexual behavior cannot be scientifically explained in teleological terms, as behavior in which animals engage for the purpose of reproduction. People, of course, do engage in sex at an individual level in order to reproduce. Yet a great deal of human sexual activity serves no reproductive function.

A simpler explanation is that humans and animals engage in sex for physical pleasure. The incentive is in the act itself, rather than in its possible consequences. Sexual behavior in this sense arises from a psychological drive, associated with sensory pleasure, and its reproductive consequences are a by-product. We are only beginning to understand the neurophysiological basis of pleasure. It has been demonstrated, for example, that there are "pleasure centers" in the brain that, when electrically stimulated, cause animals to experience intense pleasure. Such pleasure centers may also exist in humans.

What about *sex hormones?* They are a fascinating and problematic subject of study. We know, for instance, that they begin to exert their influence before birth and are vital in sexual development. Although these hormones may also be intimately linked to sexual behavior, the link is not a simple one, and we have yet to discover a substance that might represent a true "sex fuel." We shall have more to say about the relationship of hormones and sexual behavior in Chapter 4.

Psychosocial Determinants of Sexual Behavior

In one sense, psychological or social forces are merely reflections and manifestations of underlying biological processes. For example, Freud argued that the *libido,* or sex drive, is the psychological representation of a biological sex instinct.

In another sense psychological factors are independent, even though they must be mediated through the neurophysiological mechanisms of the brain. But these mechanisms are considered only as the intermediaries through which thought and emotions operate, rather than as their primary determinants. Let us again use hunger as an example. When hunger motivates a person to eat, the behavior of the individual is relatively independent of psychological and social factors. Although such factors may influence the individual's eating behavior, they are not the main determinants of why a person eats. Yet if people dislike pork or are expected to abstain from eating it, they act from personal preference or religious conviction. Their motivation is still mediated through the brain, but it originates in learned patterns of behavior rather than in biological factors. The same may be true for sex, although a biological imperative is less clear in this case.

Theories of the psychological motivation for sexual behavior are therefore fundamentally of two types. In the first, psychological factors are considered to be representations or extensions of biological forces. In the second, it is assumed that patterns of sexual behavior are largely acquired through a variety of psychological and social mechanisms.

If we accept that the primary goals of sexuality are *reproduction* and *erotic pleasure,* we can also discern a number of secondary, though by no means unimportant, goals that are "nonsexual." Paramount among these is the use of sex as a vehicle of expressing and obtaining *love.* This may include deep and genuine affection or its shallow and stereotyped parodies. Adolescent girls may indulge in joyless sexual encounters to maintain popularity and acceptance by their peers and as a defense against loneliness. Teenage boys may be driven to sexual exploits to defend against self-doubts about masculinity and power.

The contribution of sex to *self-esteem* is important. Each of us needs a deep and firmly rooted conviction of personal worth and a reaffirmation of such worth from significant others. An important component of a person's self-esteem is his or her sexual standing.

Sexuality is clearly an important component of an individual's self-concept or *sense of identity*. Awareness of sexual differentiation precedes that of all other social attributes in the child: the child knows the self as a boy or girl long before it learns to associate the self with national, ethnic, religious, and other cultural groupings.

Although developing an awareness of one's biological sex is a relatively simple matter, the acquisition of a sense of sexual identity is a more complex, culturally relative process. Traditionally we have assumed that biological sex (maleness and femaleness) and its psychological attributes (masculinity and femininity) are two sides of the same coin. In traditional and stable societies this assumption may have been (and may still be) valid. But in the technologically advanced and rapidly changing societies such direct correspondence between biological sex and psychological attributes is being challenged more and more vigorously as occupational and social roles for men and women become progressively blurred.

For some people sexual activity is a form of *self-expression* in a creative, or esthetic, sense. Sex also functions as a form of interpersonal *communication*. The very act of giving oneself to another fully, willingly, and joyfully can be an exquisite means of expressing affection. But sex can also be used as a weapon to convey *dominance* and *anger*.

The association of sex and *aggression* is very broad and encompasses biological, psychological, and social considerations. We need only point out here that aggressive impulses may be expressed through sexual behavior and vice versa, and that there is an intimate relation between these two powerful drives in all kinds of sexual liaisons.

Just as psychological functions are intimately linked with biological forces, so also they are tied to social factors. In fact, distinctions between what is primarily psychological and what is social often tend to be arbitrary. As a rule, in referring to social or cultural factors, we emphasize the interpersonal over the intrapsychic and group processes over personal ones.

Sexuality is often considered a *cohesive* force that binds the family together. In this sense it serves a social goal, which is why society facilitates sexual aims by providing opportunities for contacts with sexual partners. Sexuality can also have a *divisive* influence, and this potential may be one reason for the ambivalence with which sex has been viewed in many societies, especially our own.

Sex symbolizes *social status*. Like dominant male primates in a troop, men with most power have often had first choice of the more desirable females. Beauty is naturally pleasing to the eye, but the company of a beautiful woman is also a testimony and a tribute to a man's social standing. A woman's status is more often enhanced by her man's social importance than by his looks. As women gain true social parity with men, these attitudes are likely to undergo profound changes.

Finally, sex figures prominently in an individual's *moral* standing. In Western cultures, sex is used as a moral yardstick more consistently than any other form of behavior, at both the personal and public levels. Many of us feel greater guilt and are often punished more severely for sexual transgressions than for other offenses. As a result sexual behavior or its restraint becomes the means to social acceptability.

EVOLUTIONARY PERSPECTIVES ON SEXUALITY

In their reproductive functions, humans clearly represent evolutionary extensions of primate and mammalian patterns. To what extent could the same be said about sexual behaviors more generally? Have we irrevocably broken off the behavioral patterns of our evolutionary forebears, or, underneath our cultural refinements and complexities, do we still behave in ways that continue to serve the same basic evolutionary purposes?

The basic premise of the evolutionary approach is that our sexual structures and behaviors have evolved through two processes: natural selection and sexual selection. *Natural selection* favors survival and depends on the success of both sexes in staying alive within whatever conditions of life prevail at a given time. *Sexual selection* has to do with all of the physical and behavioral characteristics of an animal that enhance its being chosen by a mate.

Primate reproduction depends on male-female interaction, hence evolutionary approaches to sexual relations focus on male-female differences and the effects of those differences on sexual interactions. Darwin noted in *The Descent of Man and Selection in Relation to Sex* (1871), the striking sex differences in structure and behavior among many species of animals: males were larger, stronger, more aggressive; they took the initiative in courting and fought off their rivals.[1] Darwin attributed these differences to sexual selection, the evolutionary process that favored those individuals that had a reproductive advantage over others of the same sex within the species.

Sexual selection has two components: *intrasexual selection,* which entails the competition of males with males for females whereby some males breed more than others; *female choice,* which results in the female selecting the males who will father her offspring. The purpose of both processes is *reproductive success:* to produce as many offspring with as high a chance of survival as possible, to contribute maximally to the gene pool of the next generation.

The recent formulations of sociobiology

are not exclusively based on the biological determinism of sexual behaviors. Nonetheless they have proven to be quite controversial. As popularly perceived, sociobiology renders human beings into biological puppets. And by linking certain behaviors (like male dominance and sexual promiscuity) to evolutionary antecedents, it appears to lend legitimacy to sexually prejudicial attitudes and practices.

The process of learning is essential for explaining the wide repertory of our sexual behaviors, no matter how one looks at it. Sociobiologists do not deny that. But in their view, learning is *biologically directed,* which predisposes the individual to learn certain things more readily than others and, hence, to behave in certain ways more often than others. As E. O. Wilson states, "Only small parts of the brain resemble a tabula rasa . . . the remainder is more like an exposed negative waiting to be dipped into developer fluid" (Wilson, 1975b, p. 156). In this view, society is the "developer fluid"; it has the crucial task of bringing the image out of the negative, but it is not the inscriber of the image or the author of our sexual "script."

Based on these assumptions and on evidence from various quarters, sociobiologists formulate explanations of why and how we behave sexually. Since these have mainly to do with various *reproductive strategies* used by men and women, we shall next discuss the sociobiological perspective pertaining to sex and interpersonal relations.

Evolutionary Background of Sexual Relationships

As Frank Beach has stated, "To interpret the sexual behavior of men and women in any society it is necessary to first recognize the nature of any fundamental mammalian pattern and then to discover the ways in which some of its parts have been suppressed as a result of social pressures brought to bear upon the individual" (quoted in Diamond, 1965, p. 167).

The evolutionary background of our sexual relationships can be ascertained by studying the behavior of living animals and by the

---

[1]This is by no means a universal pattern. There are many species in which the sexes are similar in size or the females are more aggressive.

study of early human fossils. The sexual behavior of *nonhuman primates* is particularly interesting in this regard as long as one avoids ascribing human motives and meanings to animal behavior.

Among lower mammals, sex and reproduction are fairly standardized behaviors largely regulated by hormones centered around the estrous cycle of the female (Chapter 4). The pattern largely persists among nonhuman primates, consisting of *monkeys* and *apes* (which include baboons, chimpanzees, and gorillas). Female monkeys and apes have monthly estrous cycles during which they are *in heat* for a period of about 10 days that coincides with ovulation. During this time, the females develop swellings and color changes in the genital region (*sex skin*). These visual cues, along with olfactory stimuli from pheromones (Chapter 4) elicit sexual arousal in the male.

Sexual interactions are by and large restricted to these periods. For example, when a female baboon enters estrus, she may sexually solicit the younger males, but as she nears her ovulatory peak it is the dominant male baboon in the troop who takes over, and for several days the couple lives in a *consort relationship* with frequent copulations. The association usually ends as estrus subsides, though some consort relationships extend over longer periods. When another female comes into heat, the dominant male now establishes a new consort relationship with her.

Female chimpanzees exhibit similar behavior. During 35-day monthly cycles, they are in estrus for about 6 days, during which they eagerly and indiscriminately copulate with all males in the troop, except their own sons. When signs of estrus subside, their sexual interest wanes although they may continue to be mounted. Gibbons are exceptional among the primates in their *monogamous* associations. They live in family groups marked by closeness between male and female and a lack of clear dominance patterns.[2]

[2]Accounts of primate sexual behavior can be found in Rowell (1972); Hinde (1974); Lancaster (1979); and Symons (1979).

Emergence of the Hominid Sexual Pattern

Unlike other primate females, women have no estrus; they are potentially sexually responsive during any period including menstruation, pregnancy, and nursing. They can reach orgasm readily, whereas the orgasmic ability of other nonhuman female primates is not firmly established. Women ovulate "silently" with no genital swelling. Instead, they have prominent breasts and buttocks, smooth and hairless skin, forward rotated vagina (which facilitates face-to-face intercourse), a high-pitched voice, and other secondary sexual characteristics to highlight their erotic attraction.

When, why, and how did these transformations come about? Our earliest ancestors (the *protohominids*) lived in the dense forests of East Africa, feeding on fruits and nuts, some 15 to 20 million years ago. They were organized in loosely knit groups whose members more or less fended for themselves, except for mothers who looked after their offspring. About 10 million years ago, woodlands began to replace forests, and early *hominids* were forced to live in more open surroundings. Now vulnerable to predators and in need of new sources of foods, they gradually turned to hunting. In this setting emerged the erect posture, larger brain, and the use of tools that compensated for the relative lack of physical prowess of these puny early humans (Fisher, 1983).

The basic attachment among these early humans was between mother and infant. In their tree-dwelling days, mothers could look after themselves and their offspring with relative ease. As they came into estrus they copulated more or less indiscriminately until they got pregnant; beyond that they had no particular need of the male. With the shift to ground dwelling, life became more difficult for mothers, who were now more dependent on males to protect them and share the meat from their hunt while they themselves continued to provide whatever food could be gathered. These associations were facilitated by more selective and stable patterns of male and female relationships known as *pair-bonding*, with attachments that lasted over longer periods of time and enhanced the chances of survival of the infants in their care.

The mechanism of estrus had worked well to attract the male for insemination. But estrus lasted for too short a period to allow the female to get much additional service out of a male. Through natural selection, the *continuous receptivity* of the human female and her orgasmic capacity evolved to cement the pair bonds between male and female (Morris, 1967). The male did not need to wander around anymore looking for females in heat. All he needed to do was stay close to the females at hand and copulate as circumstances permitted.

Evidence from other sources lends credence to the notion that sexuality (especially increased female sexuality) promoted pair-bonding and paternal investment in offspring (Hrdy, 1981). Much of human affectional behavior can in turn be viewed as arising from and sustained by these sexual rewards whereby our ancestral "naked ape" developed the capacity for falling in love, for becoming sexually "imprinted" on a single partner, and for evolving pairbonds (Morris, 1967, p. 64) By further extension, such human ties resulted in the formation of *primate societies*. As the male's attachment to the female extended to her offspring, and he came to recognize her child as his, the pairbond was further cemented. The male thus became domesticated in the service of the family.

This model of sex as personal and social cement is challenged by some primatologists. For example, Symons questions the very assumption that shifts in the hominid pattern of nonreproductive sexual relations evolved to cement pair-bonding or group cohesiveness (Symons, 1979). The claim that the loss of estrus or constant sexual attraction formed the basis of persistent pairings and social groups is contradicted by the observation that among some primates (such as the Japanese macaque monkey), stable social associations continue in the absence of persistent sexual attractivity or receptivity: some of the shortest and most clearly marked estrous periods occur in animals that also have highly developed social organizations. The seasonality of sexual relations

further contradicts the notion of sexual attraction as the basis for persistent social groupings of primates (Lancaster and Lee, 1965). Therefore, rather than sex, it is perhaps the division of labor and other social and psychological rewards of associating with members of one's species that have bonded individuals and cemented group ties.

Implications of the Evolutionary Perspective

The sexual behavior of primates may look beguilingly human (Figure 1.1). Yet there is no direct way to simply draw inferences from such observations to how humans behave in the contemporary world, let alone to how they behaved during prehistory. Nonetheless, insights gained from evolutionary study may help us trace the continuity of behavioral patterns from the dim past through the last 20,000 years of cultural history.

Sex was neither designed primarily for reproduction (asexual reproduction is more efficient) nor for giving and receiving pleasure

Figure 1.1   Basic adult copulatory posture in monkeys.

(pleasure at best is an inducement to undertake the burdens of reproduction). The main function of sex is the creation of *genetic diversity* in offspring, which allows for greater adaptability and hence higher chances of survival (Wilson, 1978). Reproductive success calls for different strategies for males and females. The investment of a female in her eggs (and resultant offspring) is far greater than that of the male in his sperm, which are produced in the billions. A woman can at most generate several scores of living infants; a man can theoretically father hundreds of children.[3]

This differential investment in offspring by male and female, the sexual dimorphism that favors the male in physical strength, and his greater aggressiveness, are presumably the genetically determined bases of the differences in how male and female approach sexual relationships. As Wilson sums it up, "It pays males to be aggressive, hasty, fickle and indiscriminating. In theory, it is more profitable for females to be coy, to hold back until they can identify the males with the best genes. In species that rear young, it is also important for the females to select males who are more likely to stay with them after insemination" (Wilson, 1978, p. 129).

A variety of calmly reasoned as well as impassioned objections have been raised to these sociobiological assertions. Some question the validity of the observations; others object to the inferences drawn from the data. Many social scientists tend to reject its strong sense of biological determinism. Those concerned with the oppression of women bristle at what they see as tacit vindication for the mistreatment of women by men. Others worry that even if that is not the intent of sociobiologists, there is the potential for the misapplication of their theories to sexist ends.

[3]The *Guiness Book of World Records* credits Mrs. Vassiliyev, a Russian peasant, with 69 children born through 27 pregnancies (16 pairs of twins, 7 sets of triplets and 4 sets of quadruplets) during the 18th century. The paternity record is said to be held by Moulay Ismail, "the Bloodthirsty," former emperor of Morocco (1672–1727) who reputedly fathered 548 sons and 340 daughters (McWhirter, 1987).

Sociobiologists disavow the motives ascribed to them and point out that recognizing the residues of our primate heritage does not necessarily mean endorsing them as adaptive, equitable, or desirable in the modern world. But simply because we find certain conclusions socially unpalatable, does not justify their out-of-hand rejection (Hrdy, 1981).

Mary Jane Sherfey has offered another evolutionary perspective on female sexuality. She claims that, "*To all intents and purposes, the human female is sexually insatiable in the presence of the highest degree of sexual satiation.*"[4] The rise of modern civilization was contingent on the suppression of this female hypersexuality because it would interfere with maternal responsibilities; furthermore, with the rise of agriculture, property rights, and kinship laws, large families of known parentage could not evolve until the inordinate sexual demands of women were curbed.

The overall conclusions one can draw from the evolutionary perspective are as follows. First, animals do not sexually interact in a random and pointless manner. They behave in the way they do because their behavior has proved adaptive—that is why it has survived.

Second, there is no single way in which all living organisms interact. The polygamy of baboons is no better or worse than the monogamy of gibbons; they are different adaptations to different conditions. Neither pattern in itself says anything about how humans should relate to each other.

Third, there are distinct differences in the strategies of males and females in sexual interaction. Their interests may converge or diverge, calling for cooperation or competition, depending on the circumstances.

Fourth, we exercise far greater control over our environment, ourselves, and each other, than do the animals we have descended from. Hence, whatever our evolutionary proclivities, we can largely shape the world that we live in through our culture. But to do this efficiently, and to recognize the biological lim-

[4]Sherfey (1973, p. 112). For a critique of Sherfey's views, see Heiman (1968); Symons (1979); Hrdy (1981).

its within which we must operate, we need to understand where we come from, which is what the study of evolution is all about.

THEORIES OF SEXUAL BEHAVIOR

There are no behavioral *theories* (Greek for "contemplation") that are specific to sexuality; theoretical approaches to sexual behavior are the application of more general theories of human behavior to the sexual realm. Thus our understanding of sexual behavior eventually depends on the solution of the more basic riddle of human behavior in general.

Theories are useful to the extent that they can be tested and verified by empirical evidence; theories that pertain to sexual behavior are handicapped in this regard since naturally occurring sexual behavior is generally concealed from observation, and sexual experimentation is socially unacceptable. Theories of sexual behavior, therefore, tend to remain highly speculative.

Theories that pertain to sexuality are usually of two kinds: *motivational theories* that attempt to explain why people behave sexually the way they do, and *developmental theories* (often called *theories of psychosexual development*) that trace the origins and patterns of sexual behavior.[5]

Theories seek to explain objective reality but they themselves are part of the human intellectual enterprise and thus subject to historical and cultural influences. Just as sexual behavior is profoundly influenced by its social content, theorizing about such behavior also reflects the particular intellectual temper of the time and must be understood in that setting. One of the key conflicts between theories of human behavior has been, and remains, over the respective role of biological and cultural determinants, historically known as the *nature versus nurture* controversy (Box 1.1).

Even though almost everyone now recognizes the importance of both sets of factors, the matter is by no means resolved.

Theory of Instincts

Prior to the theory of evolution, animals were seen as categorically different from humans. Without benefit of reason or soul, animal behavior was ascribed to *instincts* (Latin, *incite*), which were conceived as biological forces that compelled the organism to act in characteristic ways without the necessity of learning those behaviors. Animals copulated, reproduced, and looked after their offspring because that was part of their "animal nature"; they did not need to be taught how to perform any of these functions. In the context of evolutionary theory, the concept of instincts was extended to humans as well. By the turn of the century, this idea had caught on so firmly that psychologists like William McDougall sought to explain human behavior through an ever-expanding list of instincts (McDougall, 1908).

By attempting to explain everything instincts explained nothing; so the term gradually fell into disrepute among social scientists. But the fundamental notion of a biologically determined *predisposition,* or tendency, to act in certain ways persisted. The term "instinct," however, is now used to refer to processes far more complex than previously understood.[6] To what extent instinctual processes, as currently defined, apply to humans, is, however, still not quite clear. Yet efforts to extend insights from animal to human sexual behavior continue through naturalistic observations in the wild as well as experimental work in the laboratory.

The sexual behavior of animals generally represents involuntary responses to external stimuli resulting in predictable and relatively

---

[5]*Motivate* literally means to "cause to move." Psychologists use the concept of motivation to refer to those factors that energize behavior and give it direction. For example, a rat is said to be motivated by hunger to seek food purposefully and energetically.

[6]Tinbergen (1951) has defined instinct as "a hierarchically organized nervous mechanism which is susceptible to certain priming, releasing, and directing impulses of internal as well as external origin, and which responds to these impulses by coordinated movements that contribute to the maintenance of the individual and species."

Box 1.1

NATURE VERSUS NURTURE

Whether behavior is preordained by nature or acquired by nurture has preoccupied thoughtful observers since antiquity. In the 18th century there was a critical examination of these issues. Especially influential were the ideas of the English philosopher John Locke (1632–1704). Locke argued that our minds at birth are like blank tablets (*tabula rasa*); everything that defines us as individuals is learned (Russell, 1945).

In the middle of the 19th century, Darwin's theory of evolution by natural selection was a momentous challenge. Human beings were now seen as but one link in a long chain of organisms. Biological factors shaped by evolution determined the nature of individuals and societies. Darwin himself said little about these matters, but Francis Galton applied the concept of natural selection to all aspects of human character and history. It was Galton who in 1874 coined the phrase "nature and nurture" to separate "under two distinct heads the innumerable elements of which personality is compared. Nature is all that a man brings with him into the world; nurture is every influence from without that affects him after birth" (Freeman, 1983, pp. 3–50).

Galton had no doubt that nature was by far the more important determinative force. Under his influence, there developed a school of thought in biology that discounted the importance of cultural influences. The application of these principles led to the *eugenics* movement, aimed at "race improvement." These ideas and other excesses of *social Darwinism* were applied to justify colonialism, racism, and the dominance of the upper class in Victorian society.

Not all biologists adhered to these views. T. H. Huxley, for instance, insisted there were cultural processes in the "evolution of society," which was "a process essentially different from that which brought about the evolution of species, in the state of nature." However, by the early part of the 20th century, biological determinism had gained the upper hand.

The opposition to the extreme evolutionists was led by Franz Boas (1858–1942), the founder of anthropology in the United States, and his students:

Alfred Kroeber, Robert Lowie, and later on Ruth Benedict and Margaret Mead, among others. The central issue for Boas was the ways in which cultures shaped and sometimes shackled the lives of individuals. He began his career at a time when evolutionary determinism reigned supreme not only in biology but in anthropology. E. B. Tyler represented the view that cultural phenomena were just as subordinate to the laws of evolution as natural phenomena: "Our thoughts, wills, and actions" accorded with "laws as definite as those that govern the motion of waves . . . and the growth of plants and animals." These laws were biologically determined. Against this view, Boas put forth his conviction that culture "is an expression of the achievements of the mind and shows the cumulative effects of the activities of many minds." These cultural activities entailed phenomena to which the laws of biology did not apply.

Discussions between these two intellectual camps proved increasingly futile as the evolutionists and eugenicists (now led by Charles Davenport in the United States) pressed their cause with fervor. In response, the proponents of cultural determinism undertook to go their separate way, declaring their complete independence: an "eternal chasm" (in the words of Kroeber) now separated cultural anthropology from biology. Subsequent conceptions of culture saw it as quite independent of evolutionary antecedents or origins. Culture was "not a link in any chain, not a step in any path, but a leap to another plane." The laws of biology and evolution had nothing to do with it.

The revolt of the anthropologists against biological determinism was reinforced by the work of other social scientists. In sociology, the doctrine of Emile Durkheim (1858–1917) that society is "a thing in itself," and in psychology the behaviorist theory of J. B. Watson (1878–1958) were highly compatible with cultural determinism. These developments, combined with discoveries in genetics and revulsion against Nazi eugenics in the 1930s, ensured the victory of cultural determinism.

The pendulum began to swing back over the past several decades as advances in genetics, the study of social behavior of animals by ethologists, and major fossil discoveries by paleontologists shed new light on human evolution. The application of evolutionary principles to understanding human behavior by sociobiologists is once again generating controversy.

---

fixed behavior patterns. These patterns are inherited, complex, adaptive, and remain fairly stable under environmental changes. This gives much of animal sexual behavior a *stereotypical,* "automatic" quality whereby one fish or rat behaves pretty much like another. For example, the mating behavior of the stickleback progresses quite predictably: The male fish builds a nest and defends his territory against other males; he allows the egg-laden female to enter it (distinguishing her from males because the male has a red belly while the female does not); the female follows the male to the nest where the pair go through additional moves and countermoves; the sequence culminates with the male fertilizing the eggs laid by the female (Figure 1.2).

In these instances animals exhibit specific, stereotyped behaviors believed to be mediated through neurophysiological mechanisms called *innate releasing mechanisms* (IRM). The external cues that stimulate these mechanisms are specific *cue stimuli* or *social releasers* that trigger the behavior. The applicability of such principles to human sexual behavior is by no means self-evident. Nonetheless, the assumption that there are innate patterns underlying human sexual behavior is an important element in all biologically oriented theories of sexuality.

The term *innate* ("present at birth") is often used in contrast with *acquired:* the former is said to represent *nature,* or heredity; the latter, *nurture,* or the influence of the environment through learning. While there is a certain face validity to this distinction, it readily leads to the false dichotomy of whether one or another behavior is genetically or environmentally shaped: in effect all behavior, like all bodily structures (the *phenotype*), is the product of genetic endowment (*genotype*) and environ-

ment. The continuing search for biological factors in sexual behavior is not meant to exclude the significance of environmental influences but to put the interactional patterns between both sets of variables in a biosocial perspective.

Psychoanalytic Theory of Sexual Behavior

The concept of a *sexual instinct* has been central in psychoanalytic theory. Freud viewed sex as a psychophysiological process (like hunger) with both physical and mental manifestations. By *libido* (Latin for *lust*) he meant the psychological aspects, the erotic longing induced by the sexual instinct. His libido theory evolved into a broad conceptual scheme that purported to explain the nature and manifestations of sexuality throughout psychosexual development. Freud initially formulated a *single-instinct* theory, whereby the sexual interest served to sustain life and reproduction. He then shifted to a *dual-instinct* concept where *eros* represented the libidinal or "life instinct" and *thanatos* stood for aggression or the "death instinct."[7]

Freud did not divest sex of its commonly understood sense of pleasure through orgasm. Rather he extended the term to other pleasurable experiences ordinarily considered to be nonsexual. This was comparable to his broadening of the concept of "mental" from its ordinary sense of conscious experience to include the entire range of unconscious mental activity as well. Sexuality, in this broader sense, is the central theme of psychoanalytic theory, and infantile sexuality is the cornerstone of psychological development and mental health.

---

[7]Freud, *Beyond the Pleasure Principle,* in vol. 18 of *Collected Works.*

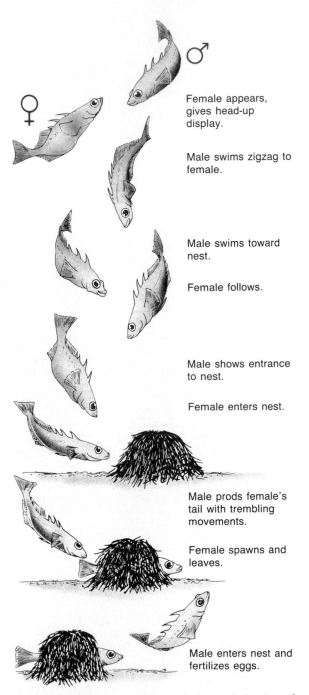

Female appears, gives head-up display.

Male swims zigzag to female.

Male swims toward nest.

Female follows.

Male shows entrance to nest.

Female enters nest.

Male prods female's tail with trembling movements.

Female spawns and leaves.

Male enters nest and fertilizes eggs.

Figure 1.2   The courtship and mating behavior of the three-spined stickleback.

Drive Theory

While biologists pursued the study of innately determined behaviors and psychoanalysts elaborated on sexual instincts, psychologists, by the 1920s, came to favor the concept of drives. A *drive* is a state of psychological arousal elicited by some physiological need (such as for food and water) or the avoidance of pain that motivates behavior to satisfy the need.

The concept of drives is no less hypothetical than that of instinct; its existence can only be inferred from certain behaviors. Yet by being more circumscribed, the concept of drives has proved more useful and compatible with the broader physiological principle of *homeostasis* ("equal state"), which refers to the tendency of the body to maintain a constant internal environment with respect to the chemical composition of its internal fluids like blood, its temperature, and so on.

A drive can be thought of as the psychological representation of a physiological need created by disturbances of the homeostatic balance, such as hunger or thirst. According to Clark L. Hull's *drive reduction theory*, formulated some 40 years ago, the basic motives for behavior emanate from the need to reduce such internal bodily tensions. In this homeostatic model, sexual arousal generates tension that disturbs the state of inner calm. The impetus to achieve orgasm is then due to the need to reduce sexual tension and regain internal calm (Hull, 1943).

Drive theory as applied to sexual behavior has several shortcomings. Unlike hunger or thirst, there is no evidence that sexual interest is generated by a physiological lack or imbalance; nor is sex essential to sustain individual life. Yet because sexual behavior has a "driven" quality to it and is linked with specific physiological processes, it remains convenient to refer to the *sex drive* as a motivating force.

Another problem is the reliance of drive theory on tension reduction in both physiological and psychological terms. But tension reduction is clearly not a universal goal of human behavior; on the contrary there are many occasions where we seek tension through excitement; particularly in sexual terms, we go to great lengths to arouse ourselves. Furthermore, organisms can be "pushed" into behavior by internal drives, while behavior can also be elicited by the "pull" of environmental incentives. For instance, one eats when hungry as well as when enticed by the offer of something one likes to eat; the same is true for sex.

In view of these considerations, *incentive theory* is now seen as a more satisfactory explanation of sexual behavior. The *positive incentive* in sex is the pleasure one anticipates in the experience; one does not necessarily have to be sexually hungry to be aroused (although the element of deprivation can clearly enhance sexual desire). A *negative incentive* on the other hand has an inhibiting effect because of the threat of pain or anxiety. If a person is taught that masturbation has ill effects, he or she will tend to shun it. The interplay of positive and negative incentives thus motivates the organism to behave in certain ways under given conditions.

Theories of Learning

Psychologists define *learning* as a relatively permanent change in behavior that occurs as a result of practice. But not all behavioral changes are due to learning. They may be due to *maturation*. For instance, infants begin to walk at about 15 months because at that time their neuromuscular system reaches the necessary level of maturity. Behavioral change may also come about through temporary conditions such as fatigue or the influence of drugs. The limits of what we learn and how we learn it are determined by the nature of our brain. But the substance of what we learn and the methods of learning are culturally determined. For example, our mental capacity to use language allows us to speak, but in order to speak a particular language, we need to learn it. Presumably a similar situation prevails with respect to sexual learning.

Human learning is so complex that psychologists have had to study the mechanisms of learning in simpler situations using experimental animals like rats and pigeons. Therefore, our knowledge of how sexuality is learned is based on inferences from general *theories of learning* or *behavior theory*. These theories focus on the acquisition of behavior patterns through various forms of learning rather than through the influence of biological determinants. And unlike psychoanalytic theory, they focus on conscious, cognitive processes, rather than unconscious motives.

Of the various theories of learning, *social learning* models are most readily applicable to subtle and complex behaviors (Bandura, 1986). For instance, observational learning involves learning by imitation from the behavior of others who act as models. Social learning is the primary process of *socialization* through which sexual behavior is molded throughout development. In this process, experience plays a key role in learning many of our sexual responses (Williams, 1987).

Some behavioral scientists have used the metaphor of "scripts" for the shaping and expression of sexual behavior. Through childhood and adolescence every individual develops a *sexual script* that acts as a record of past sexual activities, a standard for present behavior, and a plan for the future. Laws and Schwartz (1977) define sexual scripts as a "repertoire of acts and statuses that are recognized by a social group, together with the rules, expectations, and sanctions governing these acts and statuses" (p. 2). Like a blueprint your script regulates five key variables: with whom you have sex; what you do sexually; when sex is appropriate (both time of life and specific timing); where is the proper setting for sex; and why you have sex (Libby, 1976); Gagnon, 1977).

Scripting of sexual behavior occurs at several levels (Simon and Gagnon, 1986). The cultural scripts consist of the general social expectations for a given sexual relationship, for instance, that a married couple should engage in sex or that a couple should engage in sex play before having sexual intercourse. These social expectations can be quite specific and closely adhered to. This is why the sexual behavior of many people is quite predictable, as with actors on stage enacting their scripted roles.

In order for the cultural script to be played out by the individual, it should be made part of his or her own personal script. It is in this process that variations and improvisations take place. While scripting theory recognizes this, its main focus is on the predictability rather than spontaneity of sexual behavior.

The theory of sexual scripts challenges other schemes of psychosexual development on several grounds (Gagnon and Simon, 1973; Gagnon, 1977). First, it denies a biologically based sexual drive. Sexual behavior is the expression of a social script, not of some primordial urge. We are born with the capacity to behave sexually, or to do so in a particular way. It is sexual learning and social contingencies that shape our erotic life.

Second, unlike the stage theories, which trace development along a continuum, it is argued that sexual experiences are far more discontinuous. The infant fondling his penis is not masturbating in an adult sense, but merely engaging in an act that is diffusely pleasurable. Childhood and adolescent sexual behaviors, therefore, do not necessarily have a developmental linkage. We impose continuity because we want to bring the past into harmony with our present identities, roles, and needs.

Finally, the scripting theory rejects sharp distinctions between what is "sexual" and what is "nonsexual." Beyond reproductive functions, what is defined as sexual is culturally determined and socially scripted. In short, sex is what society says it is. Indeed, cultures define a wide diversity of signals, symbols, and behaviors as "sexual." Sexual activity need not be associated with erotic arousal, nor arousal with behavior you call sexual.

Learning theory is the dominant intellectual orientation in psychology, but its adherents come from other disciplines as well. For instance, Kinsey, despite being a biologist, viewed most aspects of human sexual behavior as the product of learning, whereby exposure to various sexual attitudes and experiences were assumed to shape an individual's sexual preferences and behavior. Masters and Johnson also base their therapeutic approaches on behavior modification techniques derived from learning theory (Chapter 8).

All of these theoretical approaches offer useful insights into one or another aspect of sexual behavior. But there is as yet no generally acceptable theory that in a comprehensive and internally consistent manner explains why we behave sexually the way we do.

THE STUDY OF HUMAN SEXUALITY

Sexuality is such a pervasive element in human life that there is virtually no field of study that could not be related to it in one manner or another. This very scope of the topic has resulted in its diffusion: by being everybody's business, the study of sex has ended up being nobody's business.

If one defines the field of human sexuality in terms of the store of sexual information and misinformation gathered over the centuries, then the field is vast and its origins ancient. But this sexual information, which has been mainly produced as an incidental by-product of other fields of study, lacks the coherence of a primary field of inquiry. Geneticists studying sex chromosomes, anthropologists charting lineages, and art historians analyzing erotic symbolism have very little to do with each other's work, even though each of them may be said to be dealing with the field of sexuality.

The study of sexuality as a specialized field is by comparison modest in scope and has a much shorter history. It is like a tiny statue standing on the massive pedestal formed by the broader approaches. But what it lacks in size, it makes up in a clearer identity.

The information in this text comes from both general and specialized sources. The basic data are from the general combinations of various fields, which are then augmented by the findings of sex research. We first briefly review all the disciplinary approaches that have contributed to this area. Then we dwell in greater detail on the field of sex research itself.

Biological Perspectives

The more obvious contributions from the *biomedical sciences* to the study of sexuality have been with respect to the structure, functions, and diseases of the sexual organs. But there are also important linkages to the study of sexual behavior through medical specialties like psychiatry, and through the comparative studies of animal sexual behavior by biologists and primatologists.

Medical Approaches to Sexuality   Medicine is as old as civilization, but we trace the origins of our basic medical concepts to the Greeks. *Aristotle* (384–322 B.C.) was the great codifier of ancient science who laid the foundations of biological study; words like *species* or *genus* are Latin translations of terms Aristotle employed. *Hippocrates* (c. 460–377 B.C.) (the "Father of Medicine") established medicine as an empirical science. The links between behavior and bodily humors that he elaborated on are the precursors of our modern concepts of the influence of hormones on sexual behavior. The discovery of Greek medicine during the Renaissance through retranslations of Arabic texts laid the foundation of modern medical sciences. The works of great anatomists like *Andreas Vesalius* (1514–1564) and artists like *Leonardo da Vinci* (1452–1518) transformed the representation of the body, including the genital organs, into an accurate science (Figure 1.3).

With the rise of medical specialities during the 18th and 19th centuries, a number of fields took on a more active interest in sexual functions. Currently, sexuality remains part of the concerns of every aspect of medical practice, but some specialties have a more direct involvement. *Anatomists* study genital structures; *embryologists* intrauterine development; *physiologists*, sexual functions; *geneticists*, the hereditary mechanisms underlying sexual development and behavior. These fields represent the *basic sciences* in medicine. Investigators working in these areas are often not physicians but specialists in these particular sciences.

More strictly medical are the applied or *clinical fields. Endocrinology* is the study of hormones and their disturbances; *urology* is the study of diseases of the reproductive tract; *obstetrics-gynecology*, of female reproductive functions and disorders; *dermatology*, of sexually transmitted diseases (or VD) as part of broader concerns with skin diseases. (This association is due to the fact that sexually transmitted diseases often manifest themselves through skin lesions.)

Specialists in *epidemiology* and *public health* are concerned with the prevention of illness. They have a particular interest in the sexually transmitted diseases because of the large numbers of individuals afflicted by them and because of the availability of effective means of preventing them.

The medical study of sexual behavior and the treatment of disturbed sexual functions and behavior are the domain of *psychiatry*. Particularly influential have been the contributions of *psychoanalysis* because of the centrality that sexuality plays in its theories.

Biological Approaches to Sexuality   Biologists have been primarily concerned with the study of plants and animals, not human beings. Yet the study of animals is of enormous significance to our understanding of human sexual functions and behavior. First, the study of animals leads to the development of methods that can be adapted to humans; most medical experiments and treatment trials are first worked out with animals. Since animal patterns of behavior are simpler, they can be more readily described and analyzed; the insights gained can then be used for the study of human social behavior.

Second, animals can be used in experiments where ethical considerations exclude the use of humans. For instance, the effects of separation of monkey infants from their mothers have led to important findings about sexual development. Similarly, the administration of sex hormones to pregnant animals has led to the elucidation of how the reproductive system develops and the effects of hormones on gender identity; such experiments could not con-

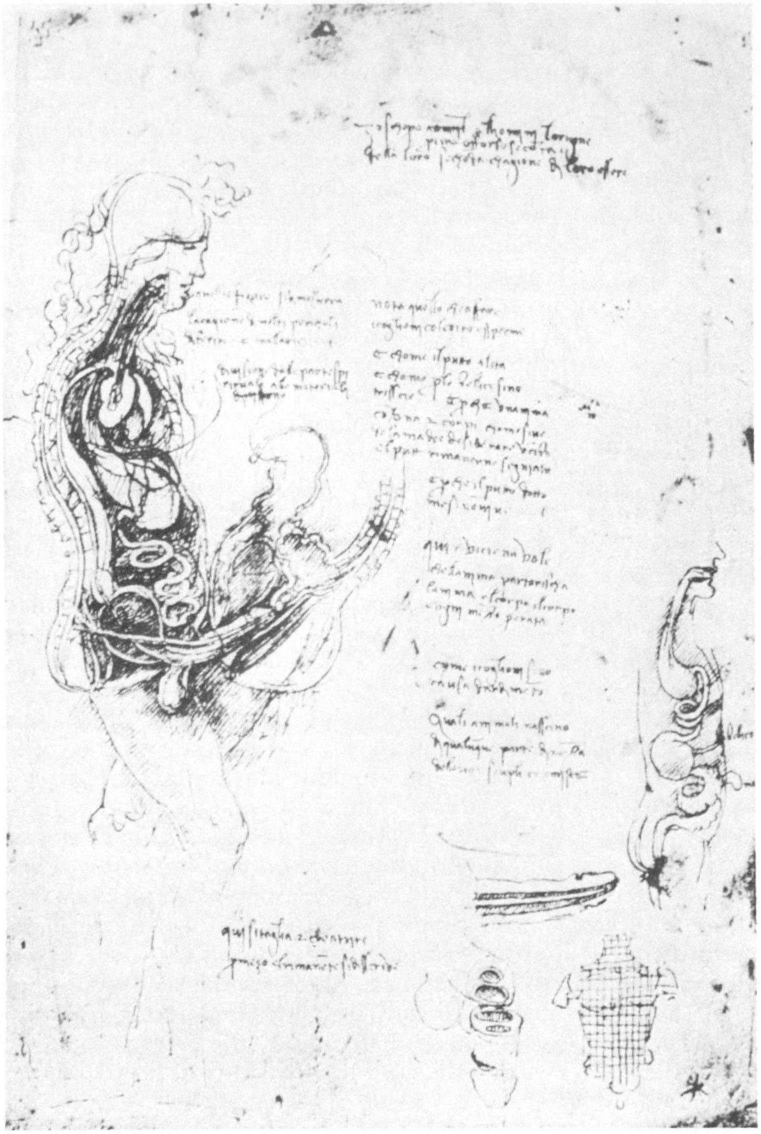

Figure 1.3   Leonardo da Vinci: Figures in coitus and anatomical sketches. The Royal Library, Windsor Castle.

ceivably be carried out with human mothers and infants.

Third, the study of animal behavior can generate principles and rules of behavior whose applicability to humans can be tested. For example, the discovery that infant monkeys obtain "contact comfort" by clinging to their mothers has led to a better appreciation of the importance of early nurturance for hu-

man babies. The study of social associations between male and female primates has likewise suggested ways of understanding certain aspects of human sexual relationships (Symons, 1979).

There are, however, important limitations in applicability of animal studies to human sexuality or any other form of behavior. Even though we are genetically very similar to our close primate cousins, like the chimpanzees, there is a chasm that separates humans from animals: our language and culture set us apart from the rest of the living world. This distinction is further enhanced by the religious belief that human beings have a soul that inhabits their animal-like bodies.

There is nothing we learn from the study of animals that can be automatically extended to people: what applies to humans must be demonstrated among humans. Furthermore, animals are so diverse in their behaviors that evidence can be found to bolster the "naturalness" of any sexual pattern, such as monogamy or promiscuity. Superficial behavioral comparisons and use of animal behavior as a metaphor are likely to lead to false conclusions. The seemingly same sexual behavior among various species may serve different purposes; likewise, the same basic sexual purpose may be attained by different behaviors.

The study of animal sexuality is not the exclusive domain of biologists. Experimental psychologists also do much of their research with animals. Similarly, many primatologists were trained in anthropology. And some sociologists now follow a biosocial orientation. The biological perspective on sexuality is therefore not a narrowly defined enterprise but the product of a multidisciplinary approach to the study of sexuality across the entire spectrum of living organisms.

Psychosocial Perspectives

Various disciplinary approaches to the study of sexuality must be seen as complementary rather than competitive. However, because of historical reasons, conceptual differences, methodological diversity, and disciplinary self-interest, this ideal has yet to be achieved. Thus, even though there is wide recognition of the interaction and interdependence of the biological and psychosocial determinants of sexual behavior, there has been much tension with respect to the relative importance of such variables (Box 1.1).

At its worst, the biological approach portrays individuals as soulless machines, whereas the psychosexual approach depicts them as disembodied souls. Yet, for most behavioral scientists, the role of biology in sexual behavior is similar to its role in the acquisition of language. We have a system of vocal cords that allows the articulation of human sounds, and our brain has the capacity to learn languages. But whether or not as individuals we learn to speak a given language is a function of whether or not we are taught that language. Whether we learn English or Swahili has nothing to do with biological factors but simply reflects the culture into which we are born. By the same reasoning, one would say that we are born with sex organs and the capacity to behave sexually, but our sexual behaviors and orientation are largely acquired through the social circumstances of our lives.

During the more recent past, there has been a tremendous upsurge of interest in sexuality among psychologists. Both with respect to sexual behavior, as well as to related topics like gender, identity, and sex roles, there has been a great deal of research done by psychologists. Of all the specialties represented in the fields of sex research and sex education, psychologists probably now constitute the largest group. New methods of sex therapy that have been developed over the past two decades by Masters and Johnson and others largely depend on behavior modification techniques developed by psychologists.

Sociologists have been greatly preoccupied with studying institutions like the family, yet they have traditionally done little with respect to sexual behavior as such. The most famous sociological study of sex was conceived and carried out by Alfred C. Kinsey, who was not a sociologist but a biologist, even though his methods and interpretations were distinctively sociological in nature. This neglect has been subsequently rectified; currently there are many sociologists working in the field of human sexuality.

Sex does not have to be discovered by societies; its existence is one of the undeniable realities of life. But as a biological given, sex in human society is but a behavioral potential; it is only through socialization that it assumes form and meaning (Davenport, 1977, p. 161). All cultures shape sexual behavior, but no two cultures do it exactly the same way. Since no single society can be regarded as representative of the human race as a whole, no serious understanding of sexuality is possible in an ethnocentric context. The only way to know the human family is to know something bout its many members (Ford and Beach, 1951, p. 250).

Cultural anthropologists have traditionally studied societies markedly different from our own. Referred to by various labels ("primitive," "tribal," "preliterate," and so on), these societies have usually been relatively small, homogeneous, technologically less advanced, and changing at a slower pace; thus, easier to study.

Of all the social sciences, anthropology ("science of man") has shown the most long-standing interest in sexuality; but even at that, sexual behavior has received less attention in the field than, for instance, kinship. Though this has changed over the past several decades, the cross-cultural study of sexual behavior is still in its formative stage.

The field of sexuality, as commonly defined, is dominated by the biomedical and the psychosexual perspectives. Yet there is a long and rich tradition of literary, artistic, and historical exploration of sexual themes that constitutes a source of deep insights into human sexuality. The behavioral sciences themselves owe their origins to the intellectual movement known as *Humanism* that flourished during the Renaissance (though the term was not coined until the 19th century).

The humanities currently subsume the study of literature, art, music, philosophy, religion, and history. The contributions of these fields may be thought of as the earliest and broadest attempts to make sense of sexual and all other human experience.

The Field of Human Sexuality

The *field of sexuality* (or *sexology*), as a specialized area of study, constitutes a more discrete entity than the multidisciplinary studies of sexuality we have discussed so far. Nevertheless, the is-

sue remains unresolved as to what direction the study of sexuality should take in the future: should sex be part of the concern of every relevant field, from biology to theology; or should there be a specialized discipline that concerns itself with various aspects of sexuality? These questions and issues currently engender considerable tension within the field of sexuality.

The modern study of sexuality can be traced to the broader concerns of Enlightenment figures, like Jean Jacques Rousseau (1712–1778), who addressed themselves to problems of sexual relationships and their proper place in society. Yet approached on such a general plane, the study of sexuality was merely part of larger intellectual inquiries into human nature and behavior.

More focused attention to sexual behavior developed in central Europe in the 19th century. This was primarily an effort on the part of physicians, mostly psychiatrists, to extend the benefits of scientific study to the area of sexuality. Some of the basic concepts of sexual pathology (like the notion of "degeneracy") had been expounded in France. But it was in Germany, in the second half of the 19th century, that the foundations of sexology were laid and then reached fruition at the turn of the 20th century.

In the United States, by the start of World War II, a number of behavioral investigations had been carried out along with numerous clinical studies.[8] But the full resurgence of these efforts, which marked the beginning of the modern field of sex research, did not start until Kinsey's work in the 1940s.

Alfred C. Kinsey (1894–1956) was a zoologist at Indiana University (Figure 1.4). After twenty-five years of work with the gall wasp, he turned to the systematic statistical study of human sexual behavior. The study was prompted by the need to find answers to the questions his students put to him in a course on marriage in which Kinsey had been persuaded to participate. From these early beginnings, Kinsey and his collaborators (Wardell B.

[8]For a review of this early work, see Kinsey et al. (1948).

Figure 1.4   Alfred C. Kinsey (1894–1956).

Pomeroy, Clyde E. Martin, and Paul H. Gebhard) collected over 16,000 sex histories from people in all walks of life across the United States—a feat unprecedented and unequaled. Kinsey alone collected 7,000 such histories—an average of two a day for 10 years. He died, however, long before he could fulfill his goal of interviewing 100,000 individuals.[9]

The Kinsey studies on the sexual behavior of the male and the female, despite the passage of almost four decades, remain the most comprehensive and systematic source of information on human sexual behavior. A number of studies have been done since that glibly compare themselves to Kinsey's work but come nowhere near it in scope and thoroughness.

The *Institute for Sex Research* founded by Kinsey (renamed recently as "The Kinsey Institute for Research in Sex, Gender, and Reproduction") has pursued other extensive investigations of sexual behavior, including studies of sex offenders and homosexuality. The institute library performs an outstanding archival and educational function with its extensive collection of books, films, and pictorial materials and artifacts relevant to sexuality.

The next threshold crossed in sex research was through laboratory investigations of human sexual physiology. Kinsey had anticipated the need for direct observation of sexual activity, but the progress from interviewing to observing did not take place until the 1960s, in the studies of *William Masters* (a gynecologist) and his research associate, *Virginia Johnson* (Masters and Johnson, 1966).

Working with 694 volunteer men and women, aged 18 to 89, these investigators observed, monitored, and filmed the responses of the body during 10,000 orgasms attained through masturbation or coitus. Their findings established at least a preliminary basis in the physiology of sex, a matter long neglected by experts in physiology and sexology. The subsequent work by Masters and Johnson with the treatment of sexual dysfunction established the modern field of sex therapy.[10]

Beyond these narrower confines of sex research, there have been enormous advances over the past three decades in the study of reproduction, contraception, and the treatment of the sexually transmitted diseases. Contraceptive devices like the birth control pill and antibiotics like penicillin have created new realities unprecedented in human history. Discoveries in molecular biology and manipulations of the reproductive process that have produced "test-tube babies" and embryo transfers are events that the sexologists earlier in the century could have barely imagined.

The field of human sexuality today has three major components: research, education, and therapy.

[9]Kinsey left no autobiography. There are many brief accounts of his career: see Brecher (1969): detailed biographies have been published by Christenson (1971) and Pomeroy (1972). Also see Robinson (1976).

[10]See Brecher (1969) for a brief account of this work, and Robinson (1976) for a more critical analysis of the ideological aspects of the work of Masters and Johnson.

Sex Research   Most research in sexuality is undertaken in universities. Biomedical scientists continue to explore the biological aspects of sex, but the majority of investigators of sexual behavior are behavioral scientists, many of them psychologists.

In some ways the field of human sexuality remains marginal compared to other disciplines. The quality of sex research, though rapidly improving, is still largely substandard compared to better established fields like psychology or biology. Its practitioners include both highly trained professionals and people with meager credentials. Reputations are all too often made through best-selling books rather than solid research or scholarly accomplishments.

There are two primary journals devoted to sex research: *Archives of Sexual Behavior* and *Journal of Sex Research*. Over a dozen more specialized journals and newsletters also furnish useful information—for instance, the SIECUS *Report*. Furthermore, important research on sexuality is just as likely to be reported in the scholarly publications of other fields.

Sex Education   In various ways, educators have provided instruction in sexuality at the college level since the start of the century. Pioneers like Prince A. Morrow were primarily concerned with the prevention of venereal diseases. In the period following World War II, courses in marriage and the family became popular, but it was not until the 1960s that human sexuality courses like the one you are taking made their appearance.[11]

Human sexuality courses have proven highly popular with students; they are gener-

ally perceived by the faculty as academically marginal. Except for a few fledgling attempts to develop graduate-level programs, this area remains a neglected field in higher education. Most established academics who work in this area have their primary appointments in some other department.

Sex education for adolescents and children has been fraught with more controversy. Who should provide sexual instruction to youngsters, in what form, and for what purposes continue to be vexing problems. There have been important developments in sex education over the past several decades. Yet there is still reason to be concerned with the level of adequacy of programs, which often operate with limited funding, self-taught instructors who have to work under a cloud of potential social disapproval, and so on.

The threat of AIDS now represents an unprecedented challenge, as well as opportunity, for sex education at all levels. Never before has there been as much public willingness to expose children to explicit sexual topics in school. How society will eventually deal with this issue remains to be seen.

Sex Therapy   Most people with sexual dysfunctions like impotence and failure to reach orgasm have been treated by medical practitioners, clinical psychologists, and other counselors. Now a new field of sex therapy has emerged over the past two decades, with its own methods (Chapter 8). It is unclear whether this *new sex therapy* will evolve into a full-fledged specialty or be absorbed into the mainstream of established disciplines.

Despite its shortcomings, the field of sexuality is now poised to expand and occupy its rightful place as a subject worthy of study and instruction. After centuries of neglect and oppression, this vital topic deserves to be treated with the same honesty, rigor, and integrity that have been brought to bear on other aspects of human life.

---

[11]For an overview of sex education in the United States, see Kirkendall (1981). How human sexuality came to be taught in one major university is described in Katchadourian (1981). Also see McCary (1975), Anderson (1975), and Sarrel and Coplin (1971).

Sexual Anatomy

Praise be given to God, who has placed man's greatest pleasure in the natural parts of woman, and has destined the natural parts of man to afford the greatest enjoyment to woman.

NEFZAWI, *The Perfumed Garden* (15th century)

OUTLINE

In physical terms, sexual organs are no different from the organs of other bodily systems. They are all built to carry out a set of specialized functions. Vital organs like the heart and lungs keep us alive as individuals. Sexual organs let us survive as a species.

In this chapter we will focus on the anatomy of the reproductive system, with special attention to organs like the vagina and the penis, which play a key role in our sexual activities. Remember, though, that sexual behavior involves far more than the use of sex organs. Your whole body is part of your sexual being. So are your ideas and your feelings.

One day we may be able to think and talk about penises and vaginas as comfortably as we do about noses and mouths; but that day is not here yet. If you are approaching this chapter with mixed feelings, you are not alone.

Your attitudes toward your body were established in childhood, but you can change them as an adult. Traditionally, children in our society have been given little accurate information about their genitals, often not even a correct name. Currently, parents are more forthright, but explanations can still be confused and awkward.

We absorb a great many negative attitudes toward our sexual parts. We learn that they are somehow "dirty," maybe because of their closeness to the outlets of feces and urine. At the same time we learn how desirable it is to be "well-developed." Our society has a love-hate relationship toward sexual functions and behavior.

Since the 1960s sexual attitudes in Western societies have become more open, accepting, and informed. Nudity and genital exposure are now fairly commonplace in pictures, in films, and at some beaches. But some people worry that we are losing our modesty and blunting our sexual sensibility.

Traditionally, men have been fascinated and excited by female genitals, but simultaneously they have feared and deprecated them. Males often are ambivalent about their own genitals as well, both exaggerating their importance and worrying about their size, shape, and capacity for performance. Female attitudes toward the genitals, their own or those of males, have had much less public expression, but women too have combined pride and pleasure with shame and confusion. Such attitudes have changed considerably in recent times.

It is not necessary to know detailed anatomy to have an active sex life, but some knowledge of anatomy is useful for other reasons.

First, by learning what your sex organs consist of and how they work, you can stop wondering and worrying.

Second, sexual activity can cause pregnancy and disease. To control the first and to avoid the second you need to know the parts of the body that are involved.

Finally, a responsible use of your sexuality requires that you accept your body and your partner's body. Beyond the practical benefits of learning about the "plumbing," a knowledge of anatomy is your foundation for a healthy acceptance of sexuality.

THE REPRODUCTIVE SYSTEM

One of the basic functions of the sex organs is *reproduction,* the generation of offspring. Of course, not every single one of us can or should have children. Even if you want to become a parent someday, much of your sexual activity will have nothing to do with having children.

Nonetheless, the shape and structure of our sex organs can be best understood in terms of their reproductive functions. The reason is that the sex organs have evolved over millions of years to enhance their reproductive success. The type of organs that have been most successful in this respect are the ones that have survived.

Sexual Reproduction   Does all life depend on sex, then? No—sexual reproduction was not the original form nor is it the only method of generating offspring. *Asexual reproduction* is still the way that many simpler forms of life perpetuate themselves. An amoeba splits in two or the arm of a starfish breaks off and gives rise to a new organism without any sex. The offspring is identical to the parent, and there is no division into male and female forms (Campbell, 1987). Perhaps you have read about cloning. Scientists are trying to reproduce creatures from just one body cell. Cloning also would be asexual.

Sexual reproduction first evolved among marine organisms. It is the predominant form by which most species reproduce today. No one knows how and why the shift from asexual to sexual reproduction took place, but its evolutionary consequences have been enormous. Without it, complex organisms like ourselves could not have evolved (Raven and Johnson, 1986).

Sexual reproduction does not always mean "having sex." For example, female fish release a mass of mature *ova,* or eggs, into the water, and then male fish release *sperm* over the eggs. This *external fertilization* would not work for land animals, because eggs and sperm would dry up without seawater. Instead, land animals rely on *internal fertilization.* The male deposits sperm directly into the female's body to fertilize the egg. This act of *copulation,* or sexual intercourse, is the way reptiles, birds, and mammals (which include humans) reproduce.

The key feature of sexual reproduction is that two different cells—one from a male, the other from a female—combine. Because two different parents contribute genetic material, their offspring have tremendous diversity. Think of the differences among brothers and sisters you know. Such genetic diversity helps individuals to adapt and survive in many environments.

Becoming a human parent is a complex experience in psychological and social terms. In a plain biological sense, it is a process hundreds of millions of years old; we share it with a wide variety of other life forms.

The Basic Plan   Whether you are female or male, your reproductive system is built on the same basic plan to fulfill similar functions. The first is the production and transport of *germ cells* (*sperm* in the male; *ova* in the female). The second is the production of *sex hormones,* which are secreted into the bloodstream (Chapter 4).

The reproductive system of both sexes centers in the *bony pelvis.* The main parts of each system and their relationship to the bony pelvis are shown in Figure 2.1. The male pelvis has a heavier bone structure; the female pelvis has a broader outlet to allow the passage of the baby during birth.

The bones of the pelvis consist of the *sacrum* (the triangular end of your backbone) and a pair of *hip bones* that are attached to the sacrum behind and to each other in front, at the *symphysis pubis.* The sexual organs are held in place by muscles that stretch across the opening of the pelvis and ligaments and sheets of connective tissues that attach them to the surrounding tissues and the pelvic bone. Although part of the same reproductive system, the organs that are located outside the body pelvis are referred to as the *external sex organs;* those inside the abdomen are the *internal sex organs.* The external sex organs are also called the *genitals.* They are the primary objects of sexual arousal and stimulation. If you are a woman, a mirror can help you to see your genitals more easily.

FEMALE SEX ORGANS

External Sex Organs
A woman's genitals are collectively called the *vulva* ("covering"). They include the *mons pubis,* the *major lips* and *minor lips,* the *clitoris,* and the *vaginal introitus* or opening (Williams and Warwick, 1980; Hollingshead and Rosse, 1985).[1]

The Mons Pubis   The mons pubis (or *mons veneris,* "mound of Venus") is the soft, rounded elevation of fatty tissue over the pubic sym-

---

[1]The many colloquial names for the female genitals include: cunt, pussy, slit, box, quim, snatch, twat, beaver, and bearded clam. Ancient sex manuals like *The Perfumed Garden* have even more fanciful terms like: crusher, silent one, yearning one, glutton, bottomless restless, biter, sucker, wasp, hedgehog, starling, hot one, delicious one, and so on (Nefzawi, 1964 ed.). Such designations reveal a good deal about cultural attitudes toward female sexuality.

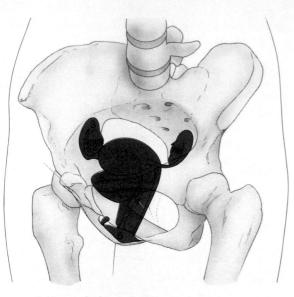

1. Uterus 2. Ovary 3. Fallopian tube
4. Vagina 5. Bladder 6. Labia

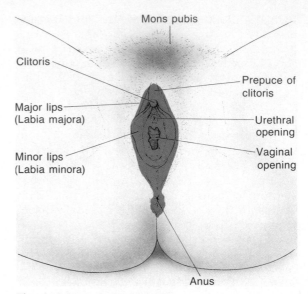

Figure 2.2   Female genitals.

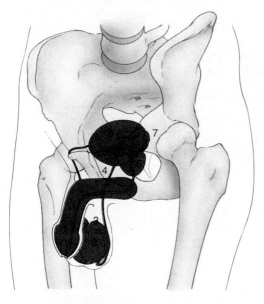

1. Penis 2. Testicle 3. Epididymis
4. Spermatic cord 5. Bladder 6. Prostate
7. Seminal vesicle

Figure 2.1   The reproductive system and the bony pelvis. (Top) Female organs: (1) uterus, (2) ovary, (3) fallopian tube, (4) vagina, (5) bladder, (6) labia majora and labia minora. (Bottom) Male organs: (1) penis, (2) testicle, (3) epididymis, (4) spermatic cord, (5) bladder, (6) prostate, (7) seminal vesicle.

physis. After it becomes covered with hair during puberty, the mons is the most visible part of the female genitals. It is quite responsive to sexual stimulation.

The Major Lips   The major lips or *labia majora* are two elongated folds of skin whose appearance varies a great deal: some are flat and hardly visible behind thick pubic hair; others bulge prominently. Ordinarily they are close together. The space between the major lips is the *pudendal cleft*; it becomes visible only when the lips are parted.

The outer surfaces of the major lips, covered with skin of a darker color, grow hair at puberty. The inner surfaces are smooth and hairless. Within these folds of skin are bundles of smooth muscle fibers, nerves, and blood vessels.

The Minor Lips   The minor lips or *labia minora* are two lighter-colored hairless folds of skin between the major lips. Into the space between them open the vagina and urethra, as well as the ducts of Bartholin's glands (discussed below). The upper portions of the minor lips form a single fold of skin over the clitoris,

Box 2.1

FEMALE CIRCUMCISION

The practice of female circumcision is far less known than its male counterpart, yet it has been widespread in some cultures and continues to be practiced, mainly on the African continent. Over 20 million African women are estimated to have undergone some version of this procedure (Remy, 1979).

As the counterpart of male circumcision (Box 2.4), the term "female circumcision" should be restricted to the removal of the prepuce of the clitoris, but it is usually extended to other procedures, such as the amputation of the clitoris (*clitoridectomy*) and *infibulation*, which involves cutting and sewing the edges of the major lips together, which blocks access to the vaginal area (except for a small opening to let out urine and menstrual blood). This closure makes coitus impossible; when the woman is deemed entitled to engage in intercourse, the orifice is enlarged by stretching it open.

Although the term "Pharaonic circumcision" is sometimes applied loosely to these procedures, there is no evidence that they were practiced by ancient Egyptians. It is not required in Judaism; only the Falashas, the black Jews of Ethiopia, observe it, presumably in imitation of neighboring groups (Gregersen, 1983).

These practices have been widely criticized. They interfere with female sexual responsiveness and health. The idea of sewing up the female genitals until the "rightful owner" can have access to them is a flagrant example of women being treated as property. Yet these procedures are defended by the societies that practice them on the grounds that cultures have the right to fashion their own rituals.*

The Western world has had its own version of female genital alterations. Early in the 19th century, "declitorization" was used both in Europe and the United States as a medical treatment for female masturbation, lesbianism, lack of sexual response ("frigidity"), and "excessive" sexual desire ("nymphomania").

*For more detailed accounts of these practices, see Gregersen (1983); Hayes (1975); Huelsman (1976); Paige (1978b); and Taba (1979)

---

which is called the *prepuce of the clitoris*, or the clitoral hood. The minor lips consist of spongy tissues, which become engorged (filled with blood) and swollen during sexual excitement. They are also highly sensitive to erotic stimulation.

From front to back the minor lips surround the clitoris, the external urethral opening, and the vaginal opening. The anus, which is completely separate from the external genitals, lies farther back.

The Clitoris   The *clitoris* ("enclosed") consists of two masses of erectile spongy tissue (*corpora cavernosa*).[2] Most of it is covered by the upper folds of the minor lips, but its free, rounded tip, the *glans*, projects beyond it.

The clitoris becomes engorged with blood during sexual excitement. Richly endowed with nerves, it is highly sensitive, an important focus of sexual stimulation, which is its sole function. Given its importance for female sexual arousal, the clitoris has recently become the focus of research (Lowry and Lowry, 1976; Lowry, 1978). Some cultures subject it to ritual alteration (see Box 2.1).

The Urethral Opening   The external urethral opening or *meatus* ("passage") is the small, median slit of the female *urethra*, which conveys urine and is totally independent of the reproductive system. Among some women, however, there may be a discharge of fluid through the urethra during orgasm, which some people consider to be a female ejaculate (Chapter 3).

[2]Colloquial terms for the clitoris include "clit," "button," and more fanciful terms like "little boy in the boat" (Rodgers, 1972).

The Vaginal Opening   The vaginal introitus, or opening, is visible only when the inner lips are parted. It is easy to tell from the urethral opening by its larger size and lower location. Its appearance depends to a large extent on the shape and condition of the *hymen*. This delicate membrane, which only exists in the human female, has no known physiological function, but its psychological and cultural significance as a "sign" of virginity has been enormous.

The hymen varies in shape and size and may surround the vaginal orifice, bridge it, or serve as a sievelike cover (Figure 2.3). Most girls' hymens will permit passage of a finger (or tampon), but cannot accommodate an erect penis without tearing. However, a flexible hymen will occasionally withstand intercourse. On the other hand, the hymen may be torn accidentally. These possibilities make the presence of the hymen unreliable evidence for virginity. In childbirth the hymen is torn further; only fragments remain attached to the vaginal opening. There is almost always some opening to the outside through the intact hymen. However, in rare instances the hymen is a tough fibrous tissue that has no opening (*imperforate hymen*). This condition is usually detected after a girl begins to menstruate and the products of successive menstrual periods accumulate, swelling the vagina and uterus. It is corrected by surgery, with no aftereffects.

Underneath the major and minor lips are several sets of muscles (especially the *pubococcygeus*) that are important to sexual function in women (Figure 2.4). They form a muscular ring around the lower end of the vagina. Such muscular rings that constrict bodily orifices are known as *sphincters*. Women can voluntarily flex these muscles or involuntarily tense them, narrowing the vaginal opening. The level of control and tension can be of prime importance (Chapter 8).

Underneath the more superficial bulbocavernosus muscles are two elongated masses of erectile tissue called the *vestibular bulbs* (shown on the right in Figure 2.4). These structures, connected at their upper ends with the clitoris, become congested with blood during sexual arousal, increasing sexual responsiveness. Together with the vaginal sphincter, they determine the size, tightness, and "feel" of the vagina (Box 2.3).

Internal Sex Organs

Let us turn to the internal sex organs of the female: the paired ovaries, the two uterine or fallopian tubes, the uterus, the vagina, and a pair of bulbourethral glands.

The Ovaries   The *ovaries* are the *gonads* or reproductive glands of the female. They produce *ova* ("eggs") and sex hormones (estrogens and progestins). They lie in the abdomen (Figures 2.5 and 2.6) on each side of the uterus, and are about an inch and a half long. The ovaries are held in place by folds and ligaments; these ligaments are solid cords, not to be confused with the fallopian tubes, which open into the uterine cavity.

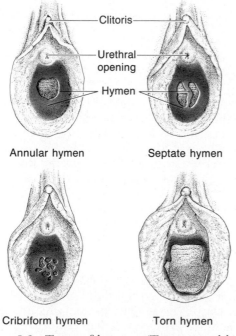

Figure 2.3   Types of hymens. (Top two and lower left) Intact hymens. (Lower right) The remnants of the hymen in a woman who has given birth.

The ovary has no tubes leading directly out of it. The ova leave the ovary by oozing out through its thin surface and becoming caught in the fringed end of the fallopian tube. Before puberty the ovary has a smooth, glistening surface; after the start of the ovarian cycle in puberty, its surface becomes increasingly scarred and pitted.

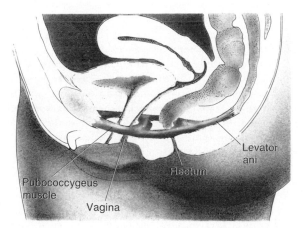

The Fallopian Tubes The two *fallopian tubes* or oviducts are about four inches long. The tubes are named after the 16th-century Italian anatomist Gabriello Fallopio, who mistakenly thought they were "ventilators" for the uterus. In fact, the fallopian tubes connect the ovaries and the uterus. The ovarian end of the tube, called the *infundibulum* ("funnel"), is cone-shaped and fringed by irregular projections, the *fimbriae,* which cling to the ovary but are not attached to it (Figure 2.6). Leaving the ovarian surface, the ovum finds its way into the opening of the fallopian tube—a remarkable feat, considering that the ovum is about the size of the tip of a needle and the opening of the uterine tube is a slit about the size of a printed hyphen.

The lining of the fallopian tube is covered with tiny hairlike structures (*cilia*). The ovum, unlike the sperm, cannot move on its own; the sweeping of these cilia and the contractions of the tube push it along. If the ovum were the size of an orange, the cilia would be as small as eyelashes.

The ovum is usually fertilized in the outer third of the fallopian tube. Although the fal-

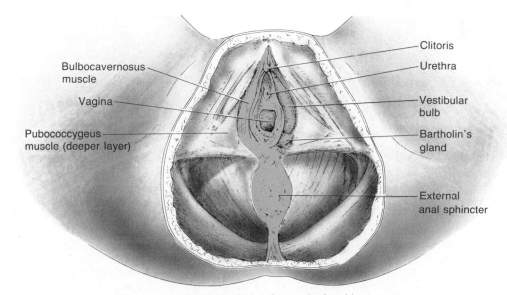

Figure 2.4 Muscles and structures in the vaginal area beneath the skin.

Box 2.2

DEFLORATION

The hymen is an exclusively human body part. No other animals have it. Why and how the hymen evolved is not clear, but most societies seem to have made the most of it. The old custom of parading the blood-stained sheets on the wedding night as proof of the bride's chastity has been practiced in both Western and Eastern cultures, and persists in some societies. Egyptian peasants test for the virginity of the bride before the wedding night by the bridegroom wrapping a piece of cloth around his index finger and inserting it into the bride's vagina (Gregersen, 1983).

Defloration ("stripping of flowers") is the tearing of the hymen through intercourse. Where it has been thought to pose a magical threat, special men or women have been assigned to carry it out. In various cultures, horns, stone phalluses, or other implements have been used in ritual deflorations. Among the seminomadic Yungar of Australia, girls were deflowered a week before marriage by two old women. If a girl's hymen was discovered at this time not to be intact, she could be starved, tortured, mutilated, or even killed.

Mosaic law took proof of virginity seriously. To refute trumped-up charges that the bride was not a virgin, the parents had to show her stained garment. If convicted of falsely accusing the bride, the groom had to pay 100 pieces of silver to her father and lost the right ever to divorce her ("because he has given a bad name to a virgin of Israel"). If the accusation could not be refuted, the woman was stoned to death at the door of her father's house (Deuteronomy 22:1–21).

Actually, the hymen is not a reliable badge of virginity (or "maidenhead"). It can be ruptured by vigorous physical activity and masturbation, but sexual intercourse sometimes leaves it intact.

Under optimal circumstances, first coitus is an untraumatic event: in the heat of sexual excitement the woman feels minimal pain, and bleeding is generally slight. What sometimes makes the experience painful is the muscular tension that an anxious, unprepared, or unwilling woman experiences in response to clumsy and forcible attempts at penetration. In anticipation of such difficulties, some women with no premarital sexual experience used to have their hymens stretched or cut surgically before their wedding nights with the knowledge and consent of their grooms. By contrast, some women who have lost their hymen have had it "restored" through plastic surgery, a procedure that is reportedly popular in Japan.

---

lopian tubes are surgically not as accessible as the vas deferens of the male, they are still the most convenient sites for female sterilization (Chapter 7).

The Uterus   The *uterus* or womb is a hollow, muscular, pear-shaped organ in which the embryo (called a fetus after the eighth week) is sheltered and nourished until birth. It stretches as the fetus grows and shrinks again after childbirth.

The Greek word for uterus is *hystera,* a term that supplies the root for words like "hysterectomy" (surgical removal of the uterus) and "hysteria," a psychological condition in which the ancient Greeks supposed the uterus wandered through the body in search of a child (Veith, 1965).

The uterus is usually tilted forward or anteverted (Figure 2.5). Attempts at self-abortion or abortion by unqualified individuals often end in disaster because of this tilt. When a probe or long needle is pushed blindly into the vagina, the instrument pierces the roof of the vagina, and instead of entering the uterus, penetrates the abdominal cavity, causing infection.

The uterus has several parts (Figure 2.6): the *fundus* ("bottom") is the rounded portion that lies above the openings of the uterine tubes; the *body* is the main part; and the *cervix* ("neck") is the lower portion, which projects

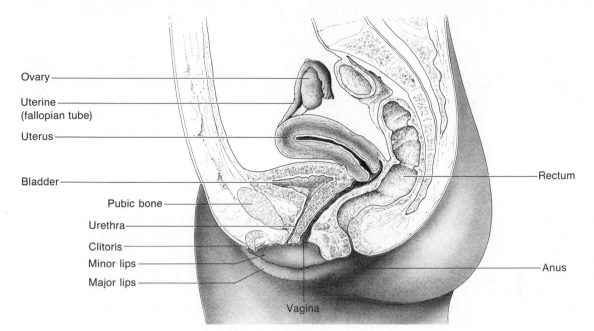

Ovary

Uterine
(fallopian tube)

Uterus

Bladder

Pubic bone

Urethra

Clitoris

Minor lips

Major lips

Vagina

Rectum

Anus

Figure 2.5   The female reproductive system (side view).

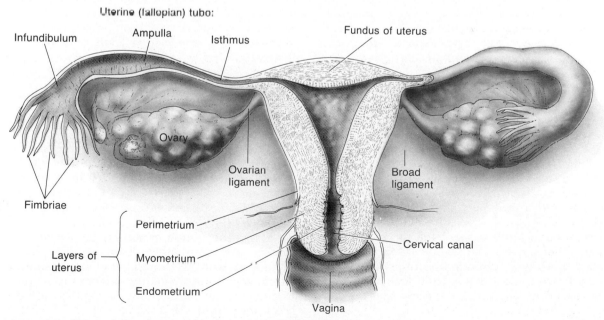

Uterine (fallopian) tube:

Infundibulum

Ampulla

Isthmus

Fundus of uterus

Ovary

Ovarian
ligament

Broad
ligament

Fimbriae

Perimetrium

Layers of
uterus

Myometrium

Endometrium

Cervical canal

Vagina

Figure 2.6   Internal female reproductive organs (front view).

Box 2.3

SIZE OF THE VAGINA

The size of the vagina, like that of the penis, has been the object of much interest and speculation. Taoist sex manuals categorize vaginas into eight types, depending on their depth; they range from the Black Pearl (4 inches) to the Deep Chamber (6 inches), the Inner Door (7 inches), and the North Pole (8 inches) (Chang, 1977, p. 52). Popular notions differentiate between tight and relaxed vaginas, those that can actively "grasp" the penis and others that are passive. However, there has not been any formal research done to substantiate such notions.

Functionally it makes more sense to consider the vaginal entrance separately from the rest of its body. The vagina beyond it is a soft and stretchy organ; although it looks like a flat tube, it actually functions more like a balloon. Normally there is no such thing as a vagina that is permanently "too tight" or "too small." Properly stimulated, any normal adult vagina can accommodate the largest penis.

The claim that some vaginas are too large is more tenable. A vagina may not return to normal size after childbirth, and tears produced during the process can weaken the vaginal walls. Even in these instances, however, the vagina expands only to the extent that the penis requires. In short, most of the time there is no problem of "fit" between penis and vagina.

By contrast, the introitus is highly sensitive. The degree of congestion of the erectile tissues of the bulb of the vestibule and the level of tension of the vaginal sphincter make a great deal of difference in how relaxed or tight the vagina will feel to the woman and her partner. If these muscles tense up, they cause coital discomfort; if they are too lax, orgasm may not occur. To enhance her sexual experience, a woman can learn to relax or tighten her vaginal muscles and strengthen them with special exercises (Chapter 8).

There is a long-standing controversy as to whether the penis can be "trapped" inside the vagina (*penis captivus*). The theme is not limited to the Western world. Natives of the Marshall Islands in the Pacific believe that incestuous relations lead to vaginal spasm, which traps the penis. Other cultures have similar fears that penile entrapment will lead to the discovery of illicit relationships (Gregersen, 1983). Probably this notion is a misconception, arising from the observation of dogs. (The penis of the dog expands into a "knot" inside the vagina and cannot be withdrawn until loss of erection.) Occasional reports continue to refer to it in humans (Melody, 1977).

---

into the vagina. The opening of the *cervical canal* into the vagina is about the size of the lead in a pencil (Figure 2.5), but it stretches for childbirth.

The uterus has three layers. The inner mucosa or *endometrium* consists of numerous glands and a rich network of blood vessels. This is where the embryo develops. Its structure varies with the phases of the menstrual cycle (Chapter 4). The second layer, the *myometrium*, consists of smooth muscles. These muscles contract during childbirth, pushing the baby out. The third, the *perimetrium*, is the external cover.

The Vagina   The *vagina* ("sheath") is the female organ of copulation. Through it pass the menstrual discharge and the baby during birth. The vagina in its unstimulated state is a collapsed muscular tube, a potential, rather than permanent, space (Box 2.3). In anatomical illustrations it appears as a narrow cavity in the side views (Figure 2.5).

The inner lining of the vagina, called the *vaginal mucosa*, resembles the inside skin of the mouth. In contrast to the uterine endometrium, it contains no glands, but during sexual excitement the vaginal mucosa exudes a clear lubricating fluid (Chapter 3).

In adult women the vaginal walls have soft ridges and furrows (*rugae*). After women cease to menstruate during the menopause (Chapter 4), these become thinner and smoother. The vaginal walls are poorly supplied with nerves, so they are relatively insensitive. In gynecological examinations conducted for Kinsey, 98 percent of the women could feel a touch on the clitoris; in contrast, fewer than 14 percent could detect a touch in the vagina (Kinsey et al., 1953). However, the area surrounding the vaginal opening may be highly excitable. There also have been recent claims that the anterior wall of the vagina has an erotically sensitive zone (the *Grafenberg spot*). (See Box 3.3 in Chapter 3.)

Underneath the vaginal mucosa is the *muscular layer* of the vaginal wall, which can stretch considerably during coitus and especially in childbirth.

Bulbourethral Glands   The *bulbourethral glands,* or *Bartholin's glands,* are two small structures located below the vestibular bulbs (Figure 2.4). Their ducts open on each side in the ridges between the edge of the hymen and the minor lips. These glands were formerly assumed to lubricate the vagina; now they are considered to play at most a minor role in this process.

Breasts

The *breasts* are not sex organs, but they have important reproductive and erotic significance. Characteristic of mammals, which suckle their young, they contain milk-producing organs called *mammary glands.* Among female primates, only women have large breasts even when not suckling. Within the breast, loosely packed fibrous and fatty tissues surround the mammary glands (Figure 2.7).

Although we generally associate breasts with females, males also have breasts, alike in structure but usually less developed. If a male is given female hormones, he will develop female-looking breasts.

The *nipple* is the prominent tip of the breast, into which the milk ducts open. It has smooth muscle fibers, which make it erect in

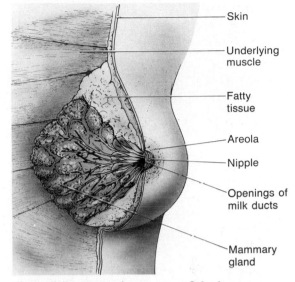

Figure 2.7   Internal structure of the breast.

response to stimulation. The *areola* is the circular area around the nipple. The nipples, richly endowed with nerve fibers, are highly sensitive; they can play an important part in sexual arousal. Many (but not all) women, and only some men, find the stimulation of their nipples sexually arousing.

Nipples vary in size and may be pushed inward or *inverted.* These are usually harmless anatomical variations which do not usually interfere with breastfeeding.

The size and shape of the breasts vary widely among women and even in the same woman, with age, weight, and other factors (Figure 2.8). Small and large breasts go in and out of "fashion." Size and shape have no bearing on their capacity to produce milk or on their erotic responsiveness. In addition to personal preference, the sensitivity of the breast depends on the hormonal levels, which fluctuate with the menstrual cycle and in pregnancy.

The female breasts develop during puberty (Chapter 4, Figure 4.4). Sometimes one grows faster than the other, but eventually the two sides become approximately equal in size.

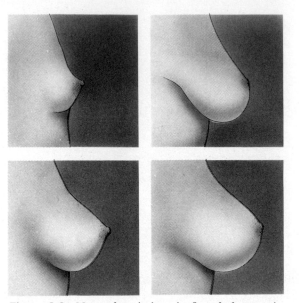

Figure 2.8   Normal variations in female breast size and shape.

With age, the breasts undergo other natural changes. As their supporting ligaments stretch, they tend to sag; following the menopause, they become smaller and less firm. Such changes, though physiologically normal, bother some women.

Exercises, creams, and similar methods that are claimed to augment breast size do not work. If a woman is truly unhappy, she can turn to plastic surgery. It can make breasts larger or smaller, and correct asymmetries (that occur naturally) and deformities (that may follow breast surgery). Breast enlargement with liquid silicone injections has led to numerous complications. A much safer approach now utilizes soft silicone implants; the materials introduced into the breast are in an inert sac and do not come into direct contact with breast tissue.

Can cosmetic surgery on healthy body parts be justified? It is a matter of personal choice. Some women feel happier following surgery; other women who have lost both breasts to cancer still consider themselves completely feminine, just as they are.

MALE SEX ORGANS

External Sex Organs
The external sex organs of the male are the *penis* and *scrotum*. You have probably heard many slang terms for them.[3]

The Penis   The *penis* ("tail") is the male organ for copulation and urination. It consists of three parallel cylinders of spongy tissue, through one of which runs the *urethra,* conveying both urine and semen (Figure 2.9).

The three cylinders of the penis are structurally similar. Two of them are called the cavernous bodies (*corpora cavernosa*), and the third, which contains the urethra, the spongy body (*corpus spongiosum*). Each cylinder is wrapped in a fibrous coat, but the cavernous bodies have an additional common covering that makes them appear to be a single structure for most of their length. In erection, the spongy body stands out as a distinct ridge on the underside of the penis.

As the terms "cavernous" and "spongy" suggest, the penis is a cluster of irregular spaces, like a dense sponge. These tissues are connected to a rich network of blood vessels. During sexual arousal they become engorged with blood. Pressing against their tough fibrous coat, they cause the penis to become erect and stiff.

The smooth, rounded head of the penis is known as the *glans* ("acorn"). The glans is formed entirely by the free end of the spongy body, which expands to shelter the tips of the cavernous bodies (Figure 2.10). Like the clitoris, the glans penis has particular erotic importance. It is richly endowed with nerves and

[3]Modern colloquial terms for the penis include prick, poker, pecker, rod, tool, cock, dick, dong, joy stick, boner, and weenie (Haeberle, 1978, p. 491). Fifteenth-century descriptions were more exotic: housebreaker, ransacker, rummager, pigeon, shamefaced one, the indomitable, and swimmer (Nefzawi, *The Perfumed Garden,* 1964 ed., pp. 156–157). A dictionary of gay slang lists several pages of synonyms, many of them terms for food (salami); male names (Peter, Mickey, Mr. Wong); and weapons or tearing instruments (dagger, hammer, spear, gun) (Rodgers, 1972).

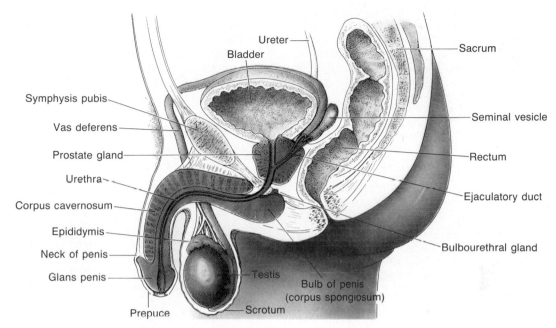

Ureter
Bladder
Sacrum
Symphysis pubis
Vas deferens
Prostate gland
Urethra
Corpus cavernosum
Epididymis
Neck of penis
Glans penis
Prepuce
Testis
Scrotum
Bulb of penis
(corpus spongiosum)
Seminal vesicle
Rectum
Ejaculatory duct
Bulbourethral gland

Figure 2.9   The male reproductive system (side view).

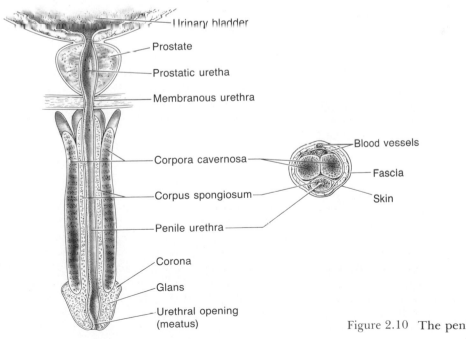

Urinary bladder
Prostate
Prostatic uretha
Membranous urethra
Corpora cavernosa
Corpus spongiosum
Penile urethra
Corona
Glans
Urethral opening
(meatus)
Blood vessels
Fascla
Skin

Figure 2.10   The penis (longitudinal section).

highly sensitive. At its rim (*corona*), the glans slightly overhangs the neck of the penis, which forms the boundary between the body of the penis and the glans. At the tip of the glans is the longitudinal slit for the *urethral meatus* or opening.

The skin of the penis is hairless and unusually loose, which permits expansion during erection. Although the skin is fixed to the penis at its neck, some of it folds over and covers part of the glans (like the sleeve of an academic gown), forming the *prepuce*, or *foreskin*. Ordi-

Box 2.4

MALE CIRCUMCISION

Circumcision, cutting off the penile prepuce or foreskin, has been practiced around the world as a religious ritual or a medical measure. About half of the males in the world are circumcised. In the circumcised male the glans and the neck of the penis are completely exposed (see figure).

Circumcision was performed in ancient Egypt as early as 2000 B.C. It long antedates the well-known practice among Jews and Moslems.* The Greek historian Herodotus, who traveled to Egypt in the 5th century B.C., wrote that Egyptians "circumcise for reasons of cleanliness," (quoted in Manniche, 1987, p. 8).

Circumcision is just one form of widespread attempts in many cultures to alter the shape of the genitals by cutting, piercing, slicing, or inserting objects. For example, the Burmese would insert tiny bronze bells under the penis; the Dayaks of Borneo would pierce the glans with a rod ("ampallang") with balls or brushes fixed to its end, to stimulate the vagina during coitus. Among Pacific islanders the foreskin is slit lengthwise (*superincision*). Other societies use *subincision* to slice the underside of the penis all the way to the urethra; as a result men have to squat like women when urinating (Gregersen, 1983).

Circumcision for medical purposes in the United States dates back to the 19th century. Its original justification was to help combat masturbation. After this rationale was discredited, its advocates endorsed the practice as hygienic, because circumcision prevents the accumulation of smegma under the prepuce. Further support for the practice came from reports that cancer of the penis ap-

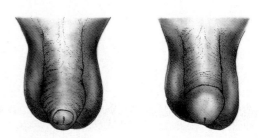

The penis before and after circumcision.

peared to be less frequent among circumcised men, and cancer of the cervix to be less common among their spouses.

The validity of these associations has been called into question. Physicians are currently divided as to the necessity of circumcision in infancy. Although circumcision is not recommended by the American Academy of Pediatrics, nearly 1 million newborn boys (or 60 percent) undergo the procedure a year in the United States; by comparison, only 20 percent of Canadian infants are circumcised. The operation is performed without anesthesia.

Circumcision remains a medical necessity when the foreskin is so tight that it cannot be easily retracted over the glans (*phimosis*). This condition is rare and impossible to predict in infancy. It takes several years for the foreskin to become retractable among the majority of boys (Paige, 1978b).

It is often assumed that because of his fully exposed glans penis, the circumcised male is more rapidly aroused during coitus and more likely to ejaculate prematurely. Current research has failed to support this belief: there seems to be no difference between the excitability of the circumcised and uncircumcised penis (Masters and Johnson, 1966).

*The basis for the practice among Jews is set forth in Genesis 17:9–15. "You shall circumcise the flesh of your foreskin, and it shall be the sign of the covenant between us."

Box 2.5

SIZE OF THE PENIS

The average penis is three to four inches when flaccid and about twice as long when erect. Its diameter in the relaxed state is about 1¼ inches, with an increase of another ¼ inch in erection. Penises can, however, be considerably smaller or larger (Dickinson, 1949).

Variation in size and shape is the rule for all parts of the human body. Nevertheless, the size and shape of the penis are often the cause of special curiosity and concern. Representations of enormous penises can be found in numerous cultures, including some from remote antiquity (as in this Greek statuette of a satyr). These anatomical exaggerations may be caricatures or monuments to male vanity or symbols of fertility and power. Symbolic representations of the penis have often been used for religious and magical functions.

The size and shape of the penis, contrary to popular belief, are not related to a man's body build, race, virility, or ability to give and receive sexual satisfaction. Furthermore, variations of size tend to be less in the erect state: the smaller the flaccid penis, the proportionately larger it tends to become when erect (Masters and Johnson, 1966). The penis does not grow larger through frequent use, pills, creams, or "exercise."

Many women do not care about the size of their sexual partner's penis, although this theme is popular in pornography (usually created by men). The size of the penis appears more important for enhancing sexual attractiveness for some gay men.

---

narily the prepuce pulls back easily to expose the glans. *Circumcision* is the cutting away of the prepuce. In the circumcised penis the glans is always totally exposed (Box 2.4).

Small glands in the corona and the neck produce a soft yellowish substance called *smegma,* which has a distinctive smell. This local secretion accumulates under the prepuce. It has no known function and must not be confused with semen, which comes out through the urethra.

Several popular ideas about the penis are myths. Size is not all-important (Box 2.5), as

men often worry. The human penis (unlike that of a dog) has no bone, nor does it have voluntary muscles within it. The muscles that surround the base of the penis externally squirt out urine and semen, but they play no significant role in erection.

The Scrotum   The *scrotum* is a multilayered pouch. Its skin is darker than the rest of the body, has many sweat glands, and at puberty becomes sparsely covered with hair. Underneath it there is a layer of loosely organized muscle fibers (*cremasteric muscle*) and fibrous tissue. These muscle fibers are not under voluntary control, but they contract in response to cold, sexual excitement, and other stimuli, making the scrotum appear compact and heavily wrinkled. When the inner side of the thigh is stimulated, the muscle contracts reflexively (the *cremasteric reflex*). Otherwise the scrotum hangs loose, and its surface is smooth.

The scrotal sac contains two separate compartments, each holding a single *testicle* and its *spermatic cord*. The spermatic cord contains the vas deferens, blood vessels, nerves, and muscle fibers. When these muscles contract, the spermatic cord shortens and pulls the testicle upward within the scrotal sac, an important feature in sexual arousal (Chapter 3).

Internal Sex Organs

The internal sex organs of the male consist of a pair of testes or testicles, with their duct systems for the storage and transport of sperm: the paired epididymis, vas deferens, and ejaculatory ducts and the single urethra. The male also has paired seminal vesicles and bulbourethral glands and the single prostate gland.

The Testes   The *testes* are the gonads, or reproductive glands, of the male. *Testis* is derived from the root for "witness." In ancient times a man would place his hand on his genitals when taking an oath—hence our word "testify."

The two testicles are about the same size (about 2 inches long), although the left one usually hangs somewhat lower than the right (a fact noted by classical sculptors). They produce sperm and sex hormones (androgens). Each testicle is enclosed in a tight, whitish fibrous sheath that penetrates inside the testicle, subdividing it into conical lobes (Figure 2.11). Each lobe is packed with convoluted *seminiferous* ("sperm-bearing") *tubules*. These threadlike structures are the sites at which sperm are produced. Each seminiferous tubule is one to three feet long, and the combined length of the tubules of both testes extends over a quarter of a mile. This elaborate system of tubules allows for the production and storage of billions of sperm.

Sperm are produced within the seminiferous tubules; male sex hormone is produced in between them by *interstitial cells* (or *Leydig's cells*). The cells responsible for the two primary functions of the testes are entirely separate. Yet the hormones produced by Leydig's cells are essential for sperm to develop.

The testes contain a third type of cell called *Sertoli cells,* which are interspersed

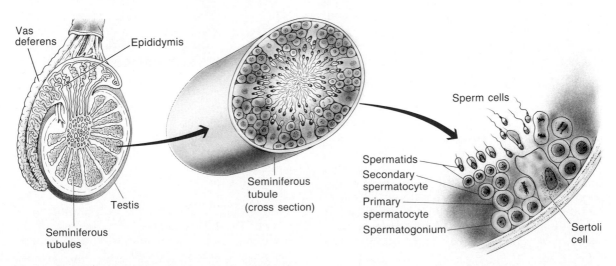

Figure 2.11   Testis and sperm developing within it.

among the developing sperm cells within the seminiferous tubules. Their function in sperm development will be discussed shortly. Sertoli cells also produce a hormone called *inhibin,* which we shall deal with in Chapter 4.

The Epididymis   The seminiferous tubules converge into an intricate maze of ducts which culminate in a single tube, the *epididymis.* The epididymis is a remarkably long tube (about 20 feet), which is so tortuous and convoluted that it appears as a C-shaped structure not much longer than the testis to whose surface it adheres (Figure 2.9). Its structure allows for the storage of large numbers of sperm.

The slow contraction of the seminiferous tubules moves sperm into the epididymis, where they mature and become able to move on their own with whiplash movements of their tails.

The Vas Deferens   The *vas deferens* is the much shorter continuation of the epididymis. It travels upward in the scrotal sac before entering the abdominal cavity; its portion in the scrotal sac can be felt as a firm cord. The fact that the vas is surgically so accessible makes it the most convenient site for sterilizing men (Chapter 7).

The vas deferens joins the duct of the seminal vesicle to form the *ejaculatory duct* (Figure 2.9). This duct is a very short (less than one inch), straight tube, which runs its entire course within the prostate gland and opens into the urethra.

The Urethra   The *urethra* starts at the base of the urinary bladder, crosses the prostate gland, and then runs through the penis to its external opening (Figures 2.9 and 2.10). The two ejaculatory ducts and the multiple ducts of the prostate gland open into it (Figure 2.10). Here the various components of semen coming from the testes, seminal vesicles, and the prostate gland mix before ejaculation.

Semen and urine use the same urethral passage, but they are never mixed. Two urethral sphincters keep them apart: an *internal sphincter* (where the urethra enters the prostate), and an *external sphincter* (right below

where the urethra exits from the prostate gland) (Figure 2.9). During urination, the internal sphincter and the external sphincter (which is under voluntary control after infancy) relax, and the contraction of the bladder wall pushes urine out. During ejaculation, the internal sphincter remains closed, the external one open, allowing only semen to flow out of the penis.

Accessory Organs   Three other organs contribute to the production of semen. They are the prostate gland, two seminal vesicles, and two bulbourethral glands.

The *prostate* is about the size and shape of a chestnut. Its base is against the bottom of the bladder (Figure 2.10). It contains smooth muscle fibers and glandular tissue whose secretions contribute to the seminal fluid and its characteristic odor. *Prostaglandins,* hormones produced by the prostate (and many other tissues), have far-ranging effects on the functions of the body (Chapter 4).

The *seminal vesicles* are two sacs (Figure 2.9), each of which ends in a straight, narrow duct, which joins the tip of the vas deferens to form the ejaculatory duct. The seminal vesicles contribute much of the fluid in semen. Their secretions are rich in carbohydrates—especially fructose, which provides energy to sperm and enhances their mobility.

The *bulbourethral glands* (*Cowper's glands*) are two pea-sized structures attached to the urethra in the penis through their tiny ducts (Figure 2.9). During sexual arousal these glands secrete a clear, sticky fluid that appears in droplets at the tip of the penis. Medieval theologians called it the "distillate of love." A Latin poet wrote the following epigram (quoted in Ellis, 1942):

> You see this organ . . . is humid
> This moisture is not dew nor drops of rain,
> It is the outcome of sweet memory
> Recalling thoughts of a complaisant maid.

There usually is not enough of this secretion to serve as a coital lubricant. However, as it is alkaline, it may help to neutralize the acidity of the urethra (which is harmful to sperm) before semen passes.

Although this secretion must not be confused with semen, it can contain stray sperm. For this reason if a condom is used for contraception, a man should put it on before he enters the vagina, no matter when he intends to ejaculate.

HOW GERM CELLS DEVELOP

Now that you know the basics of male and female anatomies, we can return to their common purpose—producing the cells that make new life. How do these germ cells or *gametes*— the sperm and ova—develop? Their development (*gametogenesis*) follows the same basic principles in both sexes, although the sperm and egg are dissimilar cells, each with a specialized function (Moore, 1982; Sadler, 1985).

In the nucleus of every living cell are *chromosomes*. They carry *genes,* which transmit all hereditary characteristics. Gametogenesis relies on *mitosis,* or ordinary cell multiplication. In mitosis a cell divides into two cells with the same number of chromosomes. In addition, germ cells undergo a special kind of reduction division called *meiosis,* through which their number of chromosomes is halved.

All human cells (other than ova and sperm) have 46 chromosomes. Twenty-two pairs of autosomes are similar in both sexes; one pair of *sex chromosomes* is not. The cells of the female body have two X sex chromosomes; those of males one X and one Y sex chromosome. The genetic configuration (*genotype*) of female body cells is thus 44 + XX; that of male body cells, 44 + XY.[4]

The germ cells have half this number of chromosomes: ova have 22 + X; sperm have either 22 + X or 22 + Y chromosomes. Therefore, when sperm and egg merge during fertilization, the normal number of chromosomes is recreated instead of doubled (Chapter 6).

Development of Sperm

Sperm production (*spermatogenesis*) takes place within the seminiferous tubules, starting at puberty. Before then the tubules are solid cords with dormant germ cells. A cross-section of an adult's tubule shows germ cells in various stages of development (Figure 2.11). Sperm that are fully formed are released into the center of the tubule and transported to the epididymis, where they mature further. Spermatogenesis is in progress simultaneously in all the tubules, so generations of sperm reach maturity in successive waves. The process takes approximately 64 days, and cycles follow each other uninterruptedly. The development of spermatozoa is guided by the Sertoli cells, which "nurse" them by providing physical support and nutrition (Kessel and Kardon, 1979).

Spermatogenesis has three phases (Figure 2.12). In the first phase, the earliest cell or *spermatogonium,* multiplies through mitosis. It is transformed into the *primary spermatocyte,* with the full 46 chromosomes. In the second phase, each primary spermatocyte undergoes meiosis (through two specialized divisions called the first and second meiotic divisions), giving rise to two *secondary spermatocytes.* In turn each splits into two *spermatids.* These four cells now have half the normal complement of chromosomes: two are 22 + X; two are 22 + Y. The third phase entails no further division. An extensive process of differentiation transforms spermatids into sperm.

Mature sperm have a head, middle piece, and a tail (Figure 2.13). The head contains the chromosomes; it is the only part kept in fertilization. The middle piece contains the part of the cell that produces energy. The tail moves the sperm with whiplash movements. Sperm are smaller than one-tenth of a millimeter, so they are not visible to the naked eye.

---

[4]One of the two X chromosomes in female cells is inactive. It appears as a small dark area inside the nucleus, and is visible under a microscope. Called a *Barr body,* it provides a convenient way to verify that the person is a genetic female. It is sometimes used for that purpose with athletes (Arms and Camp, 1987).

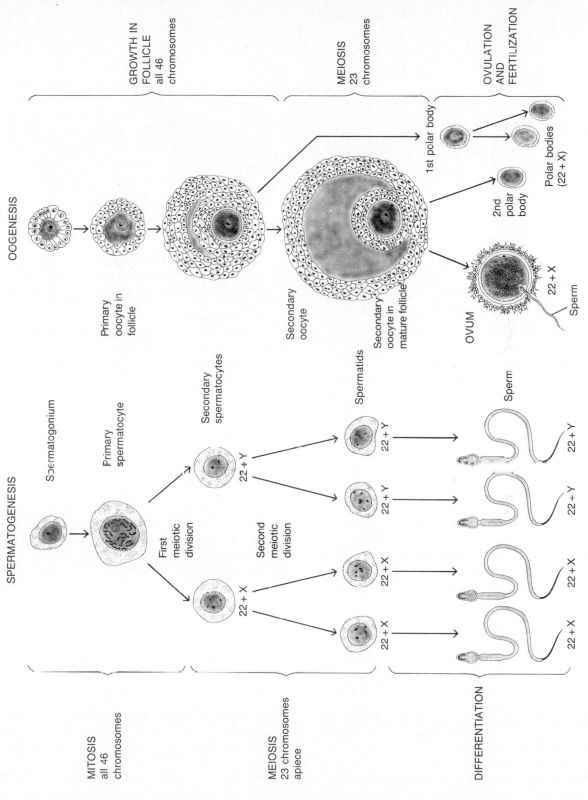

Figure 2.12 How sperm and egg develop.

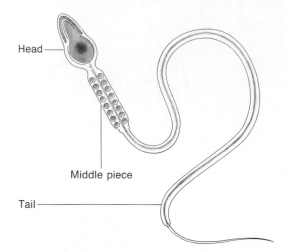

Figure 2.13   Human sperm. Component parts.

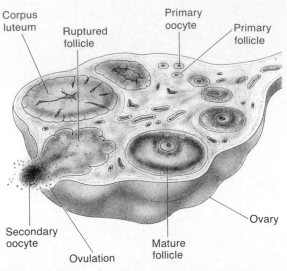

Figure 2.14   "Time-lapse" view inside of ovary. Each month one ovum matures. The follicle ruptures, becomes the corpus luteum, and then turns into scar tissue.

Development of the Ovum

Unlike men, who produce billions of sperm during their lifetime, women are born with all the immature ova they will ever produce. These cells number some 2 million, of which about 40,000 survive to puberty; about 400 reach maturity during a woman's reproductive lifetime, and only a few ever get fertilized.

Figure 2.12 shows the maturation of the ovum (*oogenesis*). The development of the ovum begins before birth. The female infant is born with *primary oocytes*. The primary oocyte with its surrounding cells make up the *primary follicle*. Primary oocytes begin the first meiotic division before birth, but the process is suspended until puberty. After puberty, each month a cluster of follicles begins to mature. Usually one of them gets ahead and becomes progressively larger (while the others regress), until it is a mature follicle or *Graafian follicle* (Figure 2.14). This liquid-filled vesicle contains the ovum and *granulosa cells*, which surround it and line the follicle wall. (It is these granulosa cells that produce the female hormone estrogen). During ovulation (Chapter 4), the wall of the follicle breaks, freeing the ovum to enter the fallopian tube.

Shortly before ovulation, the primary oocyte completes the first meiotic division re-

sulting in one *secondary oocyte* (with 22 + X chromosomes) and the *first polar body* (a small nonfunctional cell). At ovulation, the secondary oocyte begins the second meiotic division, but it does not complete it unless it is fertilized by a sperm. In that case cell division is completed and results in a *mature oocyte* (or ovum) and the *second polar body*. The first polar body meanwhile divides in two. Thus, the end result of oogenesis is one mature oocyte and three polar bodies which degenerate. This is unlike spermatogenesis, where four mature sperm develop from a simple primary spermatocyte.

To complete the story, let us now return to the follicle. After ovulation, the rest of the follicle is transformed into the *corpus luteum* ("yellow body"), a gland that takes over hormone production. At the end of the ovarian cycle, the corpus luteum turns into scar tissue.

The ovum is one of the largest cells in the body. Next to it the sperm looks minuscule (Figure 2.15). Nonetheless, the ovum is still scarcely visible to the naked eye. All the ova needed to repopulate the world would fill two gallon jugs; all the sperm needed for the same

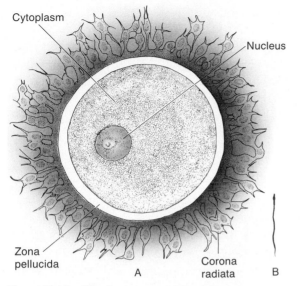

Cytoplasm

Nucleus

Zona pellucida

Corona radiata

A   B

Figure 2.15   The human egg (A). A human sperm is in the lower right corner (B), for comparison.

purpose would fit into an aspirin tablet. The volume of DNA needed to produce the entire next generation of the world is less than one-tenth of an aspirin tablet (Stern, 1973).

The *nucleus* of the ovum, which contains the genetic material, is surrounded by a large amount of *cytoplasm*. This material is necessary for sustaining life after fertilization. (The sperm has a minimal amount of cytoplasm; to cover its long journey it must travel light.) A clear area, the *zona pellucida*, forms a protective layer surrounded by follicular cells that remain attached to it after ovulation to form the *corona radiata* ("radiating crown").

HOW THE REPRODUCTIVE SYSTEM DEVELOPS

We have seen how the male and female gametes develop. How does an embryo develop its male or female anatomy?[5]

[5]Discussion in this section is based on Sadler (1985); Gordon and Ruddle (1981); and Moore (1982). The modern understanding of the differentiations of the reproductive system was first envisioned by Alfred Jost (1953).

The genital system of both sexes makes its appearance during the fifth to sixth week of intrauterine life, when the embryo is 5 to 12 millimeters long. At this stage any embryo—male or female—has a pair of undifferentiated *gonads,* two sets of *genital ducts,* and a *urogenital sinus*—a common opening to the outside for the genital ducts and the urinary tract (Figure 2.16). It also has the rudiments of external genitals (Figure 2.17).

At this time we cannot reliably tell the sex of the embryo even under a microscope; the gonads have not yet become either testes or ovaries, nor have the other structures differentiated. Of course, the sex of the child is already decided; genetic sex is determined at the moment of fertilization by the chromosomes of the sperm. If the fertilizing sperm carries a Y chromosome, the child will be male; otherwise it will be female (Chapter 6).

Differentiation of the Gonads

How does the Y chromosome initiate the process of testicular development? It had been assumed that a testicular organizing substance, called the *X-Y antigen*, causes the undifferentiated gonad to develop into testes. Now it has been established that the factor that initiates male development is a single gene on the Y chromosome, called the *testes determining factor (TDF)* (Page et al., 1987). The testes determining gene acts as a biological "master switch," deciding whether or not other genes related to sexual development are turned on. "Maleness" is therefore determined by a minute piece of the Y chromosome.

It is in response to this initial "push" by the testes determining factor that gonadal cells organize into distinct strands (*testis cords*), the forerunners of the seminiferous tubules. By about the seventh week, if the organ is not recognizable as a developing testis, we presume that it will develop into an ovary. More definitive evidence of ovarian structure comes at about the tenth week, when the forerunners of the follicles become visible.

If the undifferentiated gonad is going to develop into a testis, further development oc-

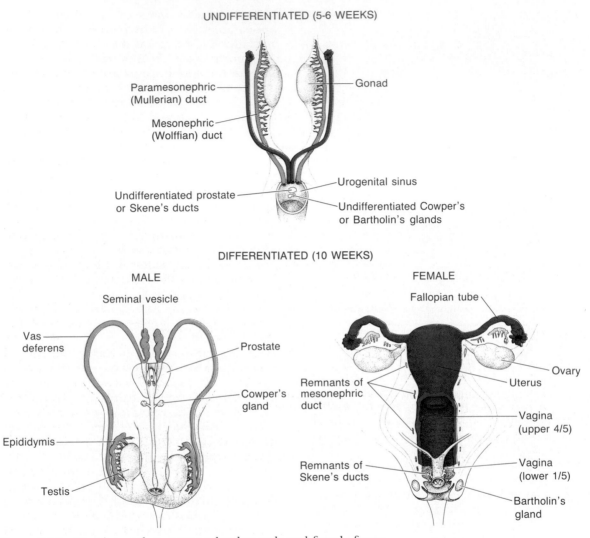

UNDIFFERENTIATED (5-6 WEEKS)

Paramesonephric (Mullerian) duct

Gonad

Mesonephric (Wolffian) duct

Undifferentiated prostate or Skene's ducts

Urogenital sinus

Undifferentiated Cowper's or Bartholin's glands

DIFFERENTIATED (10 WEEKS)

MALE

Seminal vesicle

Vas deferens

Prostate

Cowper's gland

Epididymis

Testis

FEMALE

Fallopian tube

Ovary

Uterus

Remnants of mesonephric duct

Vagina (upper 4/5)

Remnants of Skene's ducts

Vagina (lower 1/5)

Bartholin's gland

Figure 2.16   How internal sex organs develop male and female forms.

curs mainly in the inner or medullary portion of the gonad; if it is going to develop into an ovary, it is the peripheral or cortical portion of the gonad that develops, giving rise to the primitive germ cells which eventually become the ovarian follicles. Testosterone produced by the embryonal testis is necessary to promote the maturation of seminiferous tubules. The embryonal ovary also produces estrogenic hormones, but their role in the further development of ovaries is unclear.

Differentiation of the Genital Ducts

In the undifferentiated stage, the embryo has *two* sets of ducts: the *paramesonephric* or *Mullerian* (the potential female) *ducts* and the *mesonephric,* or *Wolffian* (the potential male) *ducts* (Figure 2.16).

Just as the Y chromosome directs the undifferentiated gonad to become a testis, the embryonal testes determine the future of the genital ducts. They supply the male hormone *testosterone,* produced by Leydig cells, which

UNDIFFERENTIATED (5-6 WEEKS)

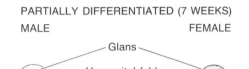

Genital tubercle
Urogenital fold
Labioscrotal swelling

PARTIALLY DIFFERENTIATED (7 WEEKS)

MALE   FEMALE

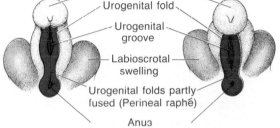

Glans

Urogenital fold

Urogenital groove

Labioscrotal swelling

Urogenital folds partly fused (Perineal raphé)

Anus

FULLY DIFFERENTIATED (12 WEEKS)

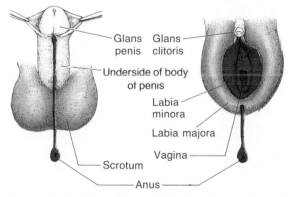

Glans penis   Glans clitoris

Underside of body of penis

Labia minora

Labia majora

Vagina

Scrotum

Anus

Figure 2.17 How the genitals develop male and female forms.

promotes the further differentiation of the Wolffian system; and *Mullerian regression hormone*, produced by Sertoli cells, which inhibits the further development of the Mullerian ducts. As a result the Wolffian duct on each side eventually becomes the epididymus, vas deferens, and seminal vesicle, while the Mullerian ducts degenerate.

In the absence of these testicular hormones the Wolffian ducts degenerate, and the Mullerian ducts form the fallopian tubes, uterus, and the upper two-thirds of the vagina. (The lower third of the vagina, the bulbourethral glands, the urethra in both sexes, and the prostate gland are derived from the embryonal urinary system.)

The differentiation of the reproductive tract therefore depends first on the Y chromosome and then on testicular hormones. Without them both, the undifferentiated system will develop the female pattern. Regardless of genetic makeup, if the gonads are removed in animal experiments early in life, the reproductive tract will develop the female pattern.[6]

Both the testes and the ovaries develop in the abdominal cavity. There the ovaries remain until birth; subsequently they move down until they reach their adult positions in the pelvis.

In the male this early internal migration is followed by the further descent of the testes into the scrotal sac. In about 2 percent of male births, one or both of the testes fail to descend into the scrotum before birth. This condition is called *cryptorchidism*. In most of these boys the testes do descend by puberty. If not, hormonal or surgical intervention becomes necessary to move the undescended testicle into the scrotum, because the higher temperature of the abdominal cavity interferes with spermatogenesis, causing sterility. Undescended testes are also more likely to develop cancer.

The canal through which the testes move out of the abdominal cavity and into the scrotal sac normally closes during early infancy. Should this fail to occur, loops of intestine may find their way into it resulting in *congenital inguinal hernia*. This is different from the type of hernia ("rupture") which usually occurs in adult men as a result of weakening of abdominal muscles and severe exertion (such as when lifting a weight). Both types of hernias are easily treated by surgery.

[6]Bernstein (1981). For further discussion of how chromosomes affect sex differentiation, see Gordon and Ruddle (1981) and Haseltine and Ohno (1981).

Differentiation of the External Genitals

Male and female genitals are also, at first, the same (Figure 2.17). Even after the gonads begin to be distinguishable by the second month of life, several more weeks are necessary for the more distinctive development of the external genitals. Not until four months are the genitals of the fetus unmistakably male or female.

The undifferentiated genitals have three main components: the *genital tubercle,* the *urogenital folds,* and the *labioscrotal swellings.* In the male, the genital tubercle grows into the glans penis; the urogenital folds elongate and fuse to form the body of the penis and urethra; and the fusion of labioscrotal swellings forms the scrotal sac. The genitals in the female undergo less marked changes in appearance: the genital tubercle becomes the clitoris; the urogenital folds, the labia minora; and the labioscrotal swellings, the labia majora.

The genitals, like the internal sex organs, differentiate under the influence of androgen. Its presence leads to the male pattern; its absence to the female pattern. However, whereas testosterone masculinizes the internal organs, one of its derivative hormones, called *dihydrotestosterone,* must transform the external ones.

Because the reproductive systems of male and female develop from the same embryonal origins, each part has its developmental counterpart, or *homologue,* in the other sex. With the aid of Figures 2.16 and 2.17, try to match the homologous pair of organs in male and female. In principle it should be possible to identify the homologue to every part, even though the degenerated remnants of the Wolffian ducts in the female and the Mullerian ducts in the male are inconsequential structures. You can find a complete listing of all homologous pairs in Moore (1982, p. 216).

Seven homologous pairs of organs and parts are clearly functional in both sexes:

testis—ovary
Bartholin's gland—Cowper's gland
glans penis—glans of clitoris
corpora cavernosa of penis—corpora cavernosa of clitoris
corpus spongiosum of penis—bulb of the vestibule
underside of penis—labia minora
scrotum—labia majora

The homologue of the prostate, the urethral and paraurethral glands, "Skene's glands," in the female, may be functional in some women, allowing them to "ejaculate" during orgasm, as we discuss in the next chapter.

The basic differences between the bodies of males and females are in the reproductive system. Yet the similarities underneath these differences suggest that men and women are after all more alike than our culture may lead us to believe.

REVIEW QUESTIONS

1. List the organs of the female and male reproductive systems and identify their functions.

2. What are the major similarities and differences between the female and male sex organs?

3. Which organs in the reproductive systems of males and females produce fluids? What is the function of each?

4. List the successive cell stages in spermatogenesis and oogenesis.

5. Match the homologous organs and parts of the male and female reproductive systems.

THOUGHT QUESTIONS

1. Should a woman with healthy breasts have surgery for cosmetic purposes?
2. Should female circumcision be prohibited by international law?
3. How do you explain the gigantic penises in the art of various cultures?
4. Why do you think the hymen evolved among humans?
5. What would you tell a young man who is concerned about the size of his penis?

SUGGESTED FURTHER READINGS

Dickinson, R. L. (1949). *Atlas of human sex anatomy* (2nd ed.). Baltimore: Williams and Wilkins. A unique collection of sketches and measurements of the sex organs.

Kessel, R. G., and Kardon, R. H. (1979). *Tissues and organs: A text-atlas of scanning electron microscopy*. San Francisco: W. H. Freeman. Photographs in Chapters 15 and 16 show genital tissues in extraordinary detail.

Moore, K. L. (1982) *The developing human* (3rd ed.). Philadelphia: Saunders. Sadler, T. W. (1985) *Medical embryology* (5th ed.). Baltimore: Williams and Wilkins. The embryonal development and differentiation of the genital system are succinctly described and well illustrated in both books.

Netter, F. H. (1965). *Reproductive system*. The Ciba Collection of Medical Illustrations, Vol. 2. Summit, N.J.: Ciba. Brief description of reproductive organs. Profusely illustrated with excellent color plates.

Nilsson, L. (1973) *Behold man*. Boston: Little, Brown. Superb color photographs of the reproductive organs.

Sexual Physiology

Whatever the poetry and romance of sex and whatever the social significance of human sexual behavior, sexual responses involve real and material changes in the physiological functioning of an animal.

ALFRED C. KINSEY

A machine "turns on" at the flip of a switch. Does your sexual self "turn on" like a machine? The image of the body as a machine is a common metaphor, but it hardly does justice to the complexity of human sexuality. Many physiological processes underlie sexual activity. When discussing anatomy, we looked at *structure*. Physiology deals with *function*—the way your sexual parts and their control mechanisms operate.

Too often people make sharp distinctions between the *physiological* or bodily aspects of sex and its *psychological* or mental aspects. They speak as if mind and body were completely separate. Of course, that is not the case:

> Physiology and psychology relate to different levels of organization and not to different kinds of causal agents. At the physiological level we study the organization and interrelations of organs and organ systems; at the psychological level we concentrate upon functions of the total individual (Beach, 1947, p. 15).

As we explore the physical basis of sexuality, remember that we still need to deal with the psychological side. But psychology by itself cannot do justice to sexual experience. The two approaches must complement each other, not compete.

SEXUAL AROUSAL

At the core of every sexual experience is the phenomenon of sexual arousal. Arousal is hard to define because it entails a variety of physiological and psychological states. Bancroft (1983) has proposed that sexual arousal has four basic components:

1. *Sexual drive*—the motivation to act sexually and the level of our sexual responsiveness or arousability.
2. *Central arousal*—the state of alertness in the brain when our attention is focused on the sensations of sexual stimulation (being "turned on").
3. *Genital responses*—the reactions of the sexual organs to erotic stimulation, such as erec-

tion of the penis and lubrication of the vagina.
4. *Peripheral arousal*—other bodily responses, such as increases in the heart rate and blood pressure.

The concept of sexual drive is intriguing but not easy to test. We have already discussed its connection with theories of sexual behavior (Chapter 1) and will consider its possible linkage to sex hormones in Chapter 4. This chapter will deal with the process of sexual stimulation, the responses of the genital organs and the rest of the body to sexual arousal, and the control mechanisms that regulate these activities.

A useful approach to the study of sexual behavior is to view it as an interaction between a *stimulus* and its behavioral *response*. This model helps us to distinguish cause and effect more clearly.

Much of our sensory stimulation comes from the environment. Various sights, smells, and sounds strike us as sexy. In addition, there are internal triggers for sexual arousal. They include erotic thoughts and feelings in sexual fantasies.

No matter what the erotic stimulus, the sexual response pattern of the body follows an orderly and predictable set of physiological changes: sexual arousal leads toward orgasm, followed by its aftermath. Although most sexual experiences stop short of orgasm, we shall discuss the whole sexual response cycle as our general model.

Physical Stimulation

There are five basic sources or forms of sensation—sight, hearing, taste, smell, and touch—and all of them transmit messages that may be interpreted by the brain as erotic. There are no specialized sensory nerves for receiving sexual sensations.

Though vision and hearing are especially important in communicating verbal and nonverbal erotic messages, touch remains the most direct physical mode of erotic stimulation. It figures in almost all sexual arousal leading to orgasm. It is, in fact, the only type of stimulation to which the body can respond by *reflex*, independent of higher brain centers. Even if a man is unconscious or has a spinal cord injury that prevents any genital sensations from reaching the brain (but leaves sexual coordinating centers in the lower spinal cord intact), he may still have an erection when his genitals or inner thighs are caressed; similarly, a woman with a spinal cord injury may respond with vaginal lubrication.

Touch   The erotic component in being touched is part of the broader, more fundamental need for bodily contact that goes back to our infancy and to our primate heritage. A crucial component in infant care is touching, caressing, fondling, and cuddling by the adult caretakers. The problems that result from the deprivations of such contact have been well documented in studies of human infants in institutions (Spitz and Wolf, 1947) and other primates (Harlow, 1958). Indeed, touch plays such a major part in primate life that the practice of grooming has been called the "social cement of primates" (Jolly, 1972, p. 153). Thus, the basis of sexual arousal may well be the more fundamental need for security and affection.

Touch sensations are received through special nerve endings in the skin and deeper tissues called *end organs*. They are distributed unevenly, so some parts (like the fingertips) are more sensitive than others (like the skin of the back).

The more richly endowed with nerves a part of the body is, the greater is its potential for stimulation. Some of the more sensitive areas of the body are especially susceptible to sexual arousal. These *erogenous zones* include the clitoris; the labia minora; the vaginal introitus; possibly parts of the front wall of the vagina; the glans penis, particularly the corona and the underside of the glans; the shaft of the penis; the area between the anus and the genitals; the anus itself; the buttocks; the inner surfaces of the thighs; the mouth, especially the lips; the ears, especially the lobes; and the

breasts, especially the nipples.

Although it is true that these areas are more highly responsive to sexual stimulation, they are not the only sensitive zones: the neck, the palms and fingertips, the soles and toes, the abdomen, the groin, the center of the lower back, or any other part of the body may well be erotically sensitive for you (Goldstein, 1976). In unusual cases, some women have been reported to reach orgasm when their eyebrows are stroked or pressure is applied to their teeth (Kinsey et al., 1953, p. 50).

The concept of erogenous zones is very old. Explicit or implicit references to them are plentiful in ancient love manuals such as the *Kama Sutra* (Vatsyayana, 1966) and the *Ananga Ranga* (Malla, 1964). Some knowledge of erogenous zones has been rightly assumed to enhance effectiveness as a lover. However, people are profoundly affected by previous experience and the mood of the moment. No one can approach a sexual partner solely guided by an "erotic map" of the body and expect to elicit sexual arousal. For instance, even though the female nipple is generally highly responsive to erotic stimulation, not all women enjoy such stimulation at all times. The same holds true for the sensitive parts of the male, such as the glans penis.

Because individuals vary in what they find sexually arousing in general, or on a specific occasion, it is necessary that partners communicate their needs, likes, and dislikes to each other.

Sights, Sounds, and Smells   What we see, hear, smell, and taste are also important sources of erotic stimulation. It is generally believed by most behavioral scientists that these means of sexual arousal, in contrast to touch, do not operate reflexively: we learn to experience certain sights, sounds, and smells as erotic and others as sexually neutral or offensive.

An alternative viewpoint suggests that although learning helps shape our arousal responses, nonetheless some responses to particular sexual cues are inborn. Just as animals react to certain "sexual triggers," humans presumably respond likewise to certain sexual cues, such as nudity, more or less universally (Morris, 1977). Cross-cultural and cross-species comparisons are offered in support of these views (Gregersen, 1983). Scientists who favor the view that arousal patterns are learned point to cultural diversity; those who favor biological models dwell on the similarities across cultures and between humans and animals (Symons, 1979).

The *sight* of the nude body in general and of the genital organs in particular is a nearly universal source of erotic excitement. Though there are great cross-cultural, individual, and gender differences in what sights are considered erotic, the importance of visual stimuli in sexual arousal can hardly be over-estimated. Our preoccupation with physical attractiveness, cosmetics, and dress testifies to that. What most people consider "sexy" is largely, but by no means exclusively, a matter of physical appearance.

The effect of *sound* is less evident but quite significant. Tone and softness of voice help determine the erotic effect of what is said. The sighs, groans, and moans uttered during sex can in themselves be highly arousing to the participants (and to others within earshot). Similarly, certain types of music with pulsating rhythms or romantic qualities often serve as erotic stimuli, or set the mood for sex.

The importance of the sense of *smell* has declined in humans relative to other species. Among animals chemical substances called *pheromones* act as powerful erotic stimulants through the sense of smell. An intriguing possibility exists that humans too secrete pheromones (Box 4.4), but even without them, the use of scents and preoccupation with body odors in most cultures attest to the erotic importance of the sense of smell (Hopson, 1979).

We have dwelt so far on the positive effect of sensory stimulation in generating erotic arousal. Sights, sounds, and smells can exert an equally powerful inhibiting influence. Some people are more sensitive than others in this respect; for them even a minor unpleasant intrusion may act as a turn-off. For example, a person may look quite sexy but lose erotic ap-

peal because of a slight bad odor. Effective sexual arousal thus is equally dependent on what signals we send and avoid sending.

Psychological Stimulation

Despite all the power of physical stimulation, the key to human sexual arousal is psychological. Sexual arousal is, after all, an emotional state greatly influenced by other emotional states. Stimulation through the senses will normally result in sexual arousal if, and only if, it is accompanied by the appropriate emotional conditions. Affection and trust enhance, and anxiety and fear inhibit erotic response in most cases.

Given our highly developed central nervous systems, we can also react sexually to purely mental images—a dream, a wish, an idea. Sexual fantasy is the most common erotic stimulant. Our responsiveness is not based solely on the external situation, but also on the store of memories from the past and thoughts projected into the future.

As human beings we share common developmental influences as well as a common biology. For example, we are all cared for as infants by adults who become the first and most significant influences in our lives. What was said earlier about erogenous zones is therefore applicable to the psychological realm as well. Just as most of us are likely to respond to gentle caressing on the inside of our thighs, we are also likely to respond positively to an expression of affectionate sexual interest. Still, as each of us is unique in our developmental histories, our sexual response will vary depending on the persons we deal with and the social circumstances under which we interact.

The issue of sexual stimulation is closely tied to sexual attractiveness. As we said above, "sexiness" is usually perceived in physical terms. But there are equally important behavioral factors (for instance, how we move or talk) as well as psychological considerations (what kind of person we are) that determine how sexually attractive a person appears to others.

SEXUAL RESPONSE

How would you go about studying sexual response? We can examine changes in the sex organs, other bodily responses, or subjective reports of "how it feels."

The study of the physiology of sexual function was largely neglected until recently. In the fourth century B.C., Aristotle observed that the testes are lifted up within the scrotal sac during sexual excitement and that contractions of the anus accompany orgasm; more than 23 centuries passed before such observations were confirmed under laboratory conditions.

It is common knowledge that sexual arousal leads to orgasm and is followed by sexual satiety. The pioneer sexologist Havelock Ellis characterized this process as a two-phase sequence, based on a male model: *tumescence* ("swelling") entailed the enlargement of the penis in erection and *detumescence* the return to its unstimulated state (Ellis, 1942). Nevertheless there were few systematic investigations of the physiology of orgasm until the research conducted by Masters and Johnson in the 1960s (Box 3.1). Since then, other studies have supplemented and sometimes amended their work (for instance, Bohlen et al., 1980; Bohlen, 1981) and new technologies have opened up opportunities for physiological research (Box 3.2).

We shall be dealing throughout this chapter with typical patterns of human sexual response. Do not take them as standards of normality. Many variations are perfectly normal. No one else reacts exactly like you.

Response Patterns

Sexual arousal and orgasm "feel" highly pleasurable, but it is difficult for us to be aware of their effects fully and objectively. Hence, these effects must be observed under laboratory conditions. Figures 3.1 and 3.2 summarize such observations by Masters and Johnson (1966). The sexual response pattern for males (Figure 3.1) and the three patterns for females (Figure 3.2) include the same four phases: *excitement*, *plateau*, *orgasm*, and *resolution*. These patterns show up no matter what type of stimulation produces them. The basic physiology of orgasm is the same, regardless of whether it is brought about through masturbation, coitus, or some other activity.

The sexual responses of men and women are basically similar; yet there are also a number of differences. The first major difference is the greater variability of female response patterns. You probably noticed that although a single sequence characterizes the basic male pattern (Figure 3.1), three alternatives are shown for females (Figure 3.2). The second difference involves a *refractory period* in the male, but not the female, cycle. This obligatory rest period immediately follows orgasm and extends into the resolution phase. During this period, regardless of how intense the sexual stimulation, the male cannot achieve full erection and another orgasm; only after the refractory period has passed can he do so. The length of the refractory period seems to vary greatly among males and with the same person on different occasions. It may last anywhere from a few minutes to several hours. The interval usually gets longer with age and with successive orgasms during a sexual episode (Kolodny et al., 1979).

Arousability is a function not only of the refractory period but also of complex interactional factors. In many mammalian species, sexual response is strongest with new sexual partners, and the introduction of a new partner will revive sexual interest (Michael and Zumpe, 1978). This phenomenon (known as the "Coolidge effect") is usually stronger among males than females.[1] Whether or not a

[1]The term comes from a probably made-up story recounted by G. Bermant as follows: One day the President and Mrs. Coolidge were visiting a government farm. Soon after their arrival they were taken off on separate tours. When Mrs. Coolidge passed the chicken pens she paused to ask the man in charge if the rooster copulates more than once a day. "Dozens of times," was the reply. "Please tell that to the President," Mrs. Coolidge requested. When the President passed the pens and was told about the rooster, he asked "Same hen every time?" "Oh no, Mr. President, a different one each time." The President nodded slowly, then said, "Tell that to Mrs. Coolidge" (Symons, 1979, p. 211). The Coolidge effect is discussed in detail in Wilson (1982).

Box 3.1

THE MASTERS AND JOHNSON STUDY OF SEXUAL RESPONSE

William Masters, a gynecologist and research director of the Reproductive Research Foundation in St. Louis, Missouri, published *The Human Sexual Response* with Virginia Johnson, his research associate, in 1966. Like the Kinsey reports, the book created a sensation among both professional circles and the general public. For the first time, it presented a detailed account of the reaction of the body during the sexual response cycle.

Masters and Johnson were primarily interested in investigating the physiology of orgasm in a laboratory setting. Their subjects were 694 normally functioning volunteers of both sexes between the ages of 18 and 89 years. The group included 176 married couples and 106 women and 36 men who were not married at the beginning of their participation in the project (though 98 in this group had been married before)—a total of 382 women and 312 men. Many were from a university community in St. Louis, Missouri, and the group was predominantly white.

Applicants were screened by detailed interviews and physical examinations, and all those considered to have physical abnormalities or to be emotionally unstable were eliminated. The subjects thus did not constitute a random sample of the general population, and in this sense they were not "average people." However, they were not specifically selected for their sexual attributes: the only requirement was that they be sexually responsive under laboratory conditions. In socioeconomic terms the group was, on the whole, better educated and more affluent than the general population, though there was some representation across social classes.

The research procedure was to observe, monitor, and sometimes film the responses of the body as a whole and the sex organs in particular to sexual stimulation and orgasm. Both masturbation and sexual intercourse were included in the experiment. In order to observe vaginal responses, a special camera was used; it was made of clear plastic, which permitted direct observation and filming of the inside of the vagina. All research subjects were told in advance about the exact nature of the procedures in which they would participate, and the unmarried subjects were assigned mainly to studies that did not involve coitus.

The laboratory in which the research took place was a plain, windowless room containing a bed and monitoring and recording equipment. The subjects were first left alone to engage in sex, and only when they felt comfortable in this setting were they asked to perform in the presence of the investigators and technicians monitoring the equipment (recording heart rates, blood pressures, brain waves, and so on). It was the type of setting in which hundreds of experiments of all kinds were conducted in medical centers all over the world. The only unique element was the specific physiological function under study.

During almost a decade (beginning in 1954) at least 10,000 orgasms were investigated. Because more of the subjects were women and because females were sexually more responsive than males under the circumstances, about three-quarters of these orgasms were experienced by women.

Although no one has yet attempted to replicate Masters and Johnson's study fully, the general validity of their physiological findings has been widely accepted.

In 1970 Masters and Johnson published a second volume, *Human Sexual Inadequacy*, which dealt with the treatment of sexual dysfunction (Chapter 8). Their work provided the basis of our modern methods of sex therapy.

---

similar mechanism operates among humans has not been formally established, but a yearning for new or multiple sexual partners may be a manifestation of it.

Females do not have refractory periods (Figure 3.2). Even in the pattern closest to that of the male (pattern A), soon after the first orgasm is over, the level of excitement can lead again to another climax. Women can therefore have "multiple orgasms," that is, consecutive

Box 3.2

NEW TECHNOLOGY IN SEX RESEARCH

To monitor and measure the physiological changes during the sexual response cycle, a number of new instruments have been developed. The objective data obtained from their use supplements, in important ways, the subjective reports of men and women undergoing sexual arousal and orgasm. These instruments measure the two basic physiological processes that underlie the sexual response cycle—vasocongestion and increased muscular tension.

Genital vasocongestion in the male is assessed by measuring the degree of erection. A *penile strain gauge* consists of a flexible band that fits around the base of the penis and expands with the swelling penis. To study sexual arousal outside of the laboratory, the *Rigiscan monitor* collects data on penile rigidity and tumescence for up to three 10-hour sessions. The subject wears the battery-operated monitor strapped to his thigh with two loops attached to the tip and base of the penis (see the figure). This device easily records erections during sleep.

In the female, the *vaginal photoplethysmograph* accomplishes the same purpose by recording the color changes that result from vasocongestion. A transparent acrylic cylinder is placed within the vaginal opening (see the figure). The light within the cylinder illuminates the vaginal wall; the changes in color produced by the increased number of red blood cells are detected by the photoelectric cell. A more sophisticated device is a *bioimpedance analyzer*, which measures blood flow by changes in the electrical conductivity of a part of the body (Bradford, 1986).

Increases in muscular tension and the contractions of orgasm can be measured by a *perineometer*. Placed in the vagina, this instrument will register changes in pressure. *Electronic perineometers* do this by detecting electrical activity when muscle fibers of the vagina and anus are activated (Ladas et al., 1982).

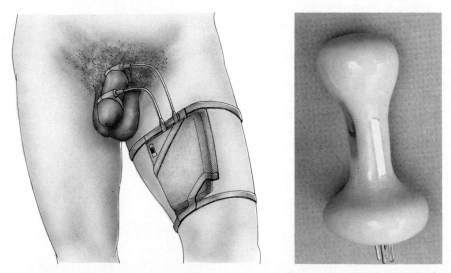

(Left) The battery-operated Rigiscan monitor collects data on penile erection. (Right) Vaginal probe monitors arousal two ways. A photoplethysmograph (note the photoelectric cell in the knob) measures vasocongestion, while a myograph (note the metal strip in the shaft) measures myotonia.

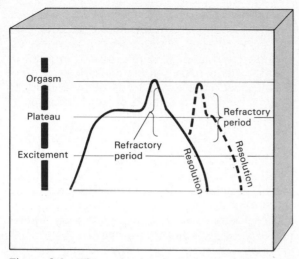

Figure 3.1   The male sexual response cycle.

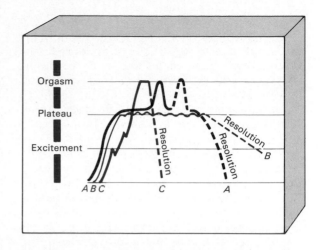

Figure 3.2   The female sexual response cycle.

orgasms in quick succession. Of course, not all women, all of the time, can or want to be multi-orgasmic.

Male capacity for multiple orgasm is very limited because of the refractory period. Among the subjects studied by Masters and Johnson (1966), only a few men seemed capable of repeated orgasm with ejaculation within minutes. Other evidence suggests that multiple orgasm is not rare if a man experiences orgasm without ejaculation (Robbins and Jensen, 1976). We shall return to this issue shortly.

Apart from these differences, the basic physiological response patterns in the two sexes are the same. In males (Figure 3.1) and females (Figure 3.2, pattern A) excitement mounts in response to effective stimulation. If erotic stimulation is sustained, the level of excitement becomes stabilized at a high point, which is the plateau phase, until orgasm follows. This abrupt release is followed by a gradual dissipation of pent-up excitement during the resolution phase.

The lengths of these phases can vary greatly. In general the excitement and resolution phases are the longest. The plateau phase is usually relatively short, and orgasm usually is measured in a matter of seconds. The overall time for one complete response

cycle may range from a few minutes to hours. Of course, not all episodes of arousal reach the point of orgasm. Among the Kinsey subjects, men reported reaching orgasm during sexual intercourse generally within four minutes, whereas it took women 10 to 20 minutes to do so. However, when women relied on self-stimulation they could reach orgasm as fast as the men (Kinsey et al., 1948, 1953).

In alternative female patterns, the woman attains a high level of arousal during the plateau phase, but instead of an orgasmic peak, there is a period of protracted orgasmic release (pattern B in Figure 3.2). The term *status orgasmus* refers to this intensive orgasmic experience. A single prolonged orgasmic episode is superimposed on the plateau phase or a series of orgasms follow each other without discernible plateau-phase intervals (Masters and Johnson, 1966). Finally, pattern C shows a more abrupt orgasmic response, which bypasses the plateau phase and is followed by a quicker resolution of sexual tension.

What really goes on during the phases of the sexual response cycle? Numerous physiological changes occur in the genitals and in other parts of the body. We shall describe separately the changes for the male and female genital organs as well as other parts of the body common to both sexes.

Physiological Mechanisms

Two physiologic mechanisms explain how the body and its various organs respond to sexual stimulation: vasocongestion and myotonia (Masters and Johnson, 1966; deGroat and Booth, 1980).

Vasocongestion is the filling of blood vessels and tissues with blood. When the flow of blood into a region exceeds the capacity of the veins to drain the blood away, it becomes *engorged*. Sexual excitement is accompanied by widespread vasocongestion in surface and deep tissues. Its most obvious manifestation is the erection of the penis. The physiology of erection has been the subject of considerable research, yet its precise mechanisms remain unclear (Benson et al., 1981; Newman and Northup, 1981; Krane and Siroky, 1981).

The primary cause of erection is increased arterial blood inflow and decreased venous outflow. When the penis is flaccid, the blood coming in through the arteries is drained through the deep veins within the penis, bypassing the spongy tissues of the penis. During sexual arousal, the increased blood inflow is shunted into the spongy tissue, causing the corpora cavernosa to expand (just as a garden hose becomes stiff when filled with water under pressure).

Women experience vasocongestion too: the clitoris enlarges, the labia swell, and the vagina moistens. We assume that similar physiological processes are at work, but new research may reveal different mechanisms (Levin, 1980). So far we know less about female than male sexual physiology (apart from reproduction). One reason is that much of the experimental work in this area has been done with animals. It is easy to see a male animal's erection, but hard to see a female's response. This same consideration applies to the exploration of nervous mechanisms, discussed in the next section.

Myotonia means increased muscular tension. Even when you are completely relaxed, your muscles maintain a certain degree of muscle tone. From this baseline, muscular tension increases during voluntary actions or involuntary contractions. During sexual activity increased myotonia is widespread. It affects both smooth (involuntary) and skeletal (voluntary) muscles. Myotonia culminates in orgasmic contractions, which involve muscles of both varieties. Although myotonia is present from the start of sexual excitement, it tends to lag behind vasocongestion and disappears shortly after orgasm.

Excitement and Plateau Phases

Although described as separate phases by Masters and Johnson, excitement and plateau are a continuous process. The plateau phase is a time of sustained, intense excitement. Therefore, we will discuss them together.

In response to effective sexual stimulation, a sensation of heightened arousal develops. Thoughts and attention turn to the sexual activity at hand, and the person becomes progressively oblivious to other stimuli and events. Most people attempt to exert some control over the intensity and tempo of mounting sexual tensions. They may try to suppress it by diverting attention to other matters, or to enhance it by dwelling on its pleasurable aspects. If circumstances are favorable to fuller expression, these erotic stirrings are difficult to ignore. On the other hand, anxiety or distraction may easily dissipate sexual arousal during the early stages.

Although excitement sometimes intensifies rapidly and relentlessly, it usually mounts more unevenly. In younger adults the progression is steeper, whereas in older people it tends to be more gradual. As the level of tension reaches the plateau phase, external distractions become less effective, and orgasm is more likely to occur. The prelude to orgasm is pleasurable in itself, and following a period of sustained excitement one may voluntarily forego the climax. Pelvic congestion unrelieved by orgasm can be a source of minor and transient discomfort for both sexes. Men experience localized heaviness and tension in the testes ("blue balls"), whereas women have a more diffuse sense of pelvic fullness with feelings of restlessness and irritability.

The behavioral and subjective manifestations of sexual excitement vary so widely that no one description can possibly encompass them all. With mild sexual excitement, few reactions may be visible to the casual observer; on the other hand, in intense excitement behavioral changes are dramatic. The person in the grip of intense sexual arousal appears tense from head to toe. The skin becomes flushed, salivation increases, the nostrils flare, the heart pounds, breathing grows heavy, and the person feels, looks, and acts quite differently than usual.

Sexual arousal may also have more personal effects: some stutterers speak more freely when sexually aroused; the gagging reflex may become less sensitive (which explains the ability of some people to take the penis deep into their mouths); persons afflicted with spastic paralysis coordinate their movements better; those suffering from hay fever may obtain temporary relief; bleeding from cuts decreases. the perception of pain is also blunted during sexual arousal (which may help people to withstand sadomasochistic practices).

Male Sex Organs   The penis undergoes striking changes during sexual excitement. Erection is the most obvious (Figure 3.3).

Erection is not an all-or-nothing phenomenon; there are many gradations between the totally flaccid penis and the maximally congested organ immediately before orgasm. As a penis becomes erect, it first increases to its full length and circumference, and then attains its maximum rigidity or hardness (Wein et al., 1981).

A certain degree of rigidity is necessary to enter the vagina. However, a less than fully erect penis can go in ("soft entry"). Especially among older men, it is not unusual for penetration to begin with only a partially erect penis; full rigidity follows during intercourse.

Erection can occur with remarkable rapidity: in younger males, less than ten seconds may be all the time required. Older men generally respond more slowly. During the plateau phase there is further engorgement, primarily in the corona of the glans, and its color deepens. Erection is now more stable, and the man

may temporarily turn his attention away from sexual activity without losing his erection.

The scrotum also contracts and thickens during sexual arousal. During the plateau phase there are no further scrotal changes. The changes undergone by the testes, though not as visible, are marked. First, during the excitement phase both testes are lifted up within the scrotum, mainly as a result of the shortening of the spermatic cords and the contraction of the scrotal sac. During the plateau phase this elevation progresses farther until the testes are actually pressed against the abdomen. For reasons that are unclear, full testicular elevation is a precondition for orgasm. The second major change is a marked increase in size (about 50 percent) because of vasocongestion.

The Cowper's glands show no evidence of activity during the excitement phase. During the plateau phase, drops of clear fluid produced by these glands appear at the tip of the penis; some men produce enough of it to wet the glans or even dribble freely; others produce very little fluid. The prostate gland enlarges during the plateau phase.

Female Sex Organs   The vagina responds like the penis to sexual stimulation. During the excitement phase the vagina shows lubrication, expansion of its inner end, and color change (Figure 3.4). Moistening of the vaginal walls is the first sign of sexual response in women; it may be present within 10 to 30 seconds. The amount of vaginal lubrication alone does not reflect the degree of sexual arousal; some women, though highly aroused, produce little fluid. The production of vaginal fluid gets less after the menopause. The resulting vaginal dryness may require lubricants, but it need not interfere with sexual enjoyment.

Moistening by the clear, slippery, and mildly scented vaginal fluid is important for a woman's enjoyment of coitus. As the fluid is alkaline, it also helps neutralize the vaginal canal (which normally tends to be acidic) in preparation for the transit of semen.

The vaginal wall has no secretory glands, so it was assumed that the lubricant emanated

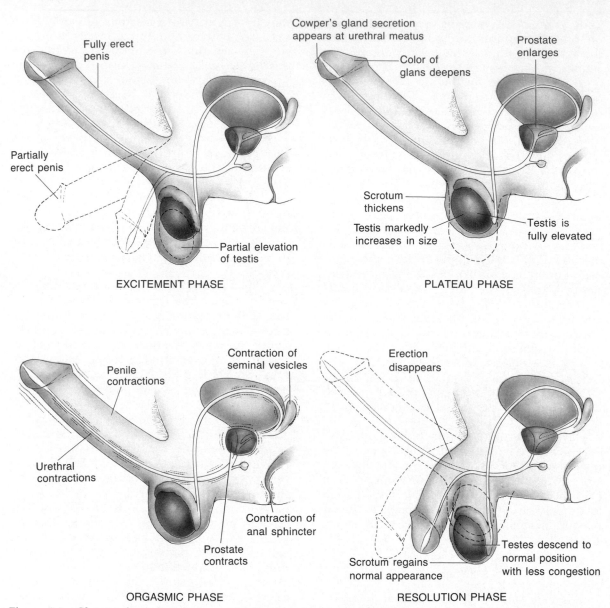

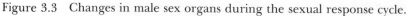

Figure 3.3   Changes in male sex organs during the sexual response cycle.

from either the cervix, the Bartholin's glands, or both, until it was discovered by Masters and Johnson (1966) that this fluid comes out mainly from the vaginal walls. Although the vaginal lubricatory reaction is often compared to "sweating" of the skin, the analogy is not accurate. Sweat or perspiration is produced by sweat glands in the skin; the vaginal lubricant is not produced by glands. It seeps out of the congested blood vessels (capillaries) near the surface of the vaginal wall, and is known as a *transudate*.

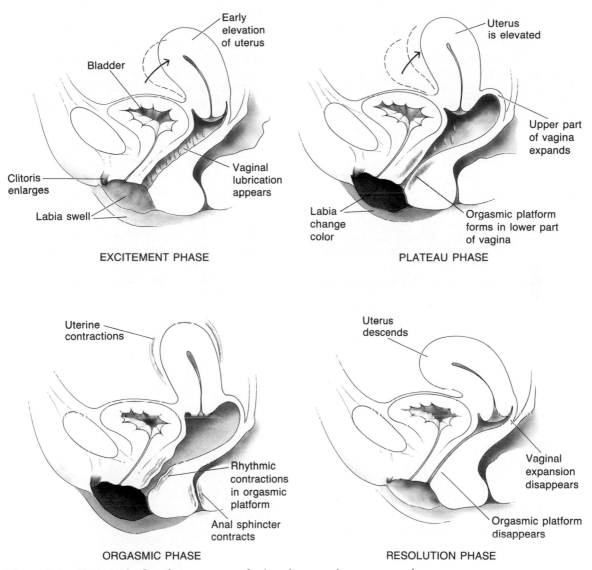

Figure 3.4   Changes in female sex organs during the sexual response cycle.

The second major vaginal change during the excitement phase is the lengthening and expansion of the inner two-thirds of the vagina, known as "tenting," which creates the space where the ejaculate will be deposited. Finally, the vagina undergoes a color change. Its walls take on a darker hue reflecting the effects of progressive vasocongestion.

During the plateau phase the outer third of the vagina becomes swollen, forming the *orgasmic platform* and narrowing the vaginal opening. Meanwhile the tenting effect at the inner end attains full vaginal expansion. Vaginal lubrication tends to slow down, and if the plateau phase is unduly protracted, further production of vaginal fluid may cease altogether.

Recently, considerable interest has been

elicited by reports that there exists an erotically sensitive area deep in the anterior (front) wall of the vagina that becomes swollen during sexual arousal. Named the *Grafenberg spot*, the exact physiology of this dime-sized region and its connection with "female ejaculation" are yet to be fully determined (Box 3.3).

The clitoris too becomes markedly congested, late in the excitement phase. The clitoral glans may double its diameter. At the same time the minor labia swell. (At this point the penis has been erect for some time and the vagina is fully lubricated.)

During the plateau phase the clitoris is retracted under the clitoral hood, receding to half its unstimulated length (which may be misinterpreted by the sexual partner as indicating loss of sexual tension). When excitement abates, the clitoris reemerges from under the hood. During protracted plateau phases there may be several repetitions of this retraction—emergence sequence.

The labia of women who have not given birth (nulliparous labia) respond somewhat differently from those of women who have (parous labia). Nulliparous major lips become flattened, thinner, and more widely parted during excitement, revealing the congested moist tissue within. Parous major lips are larger and, instead of flattening, become markedly engorged during arousal.

As the excitement phase progresses to the plateau level, the minor lips become severely engorged and double or even triple in size in both parous and nulliparous women. They also turn pink, even bright red in light-complexioned women. In parous women the resulting color is a more intense red or a deeper wine color. This vivid coloration of the minor lips has been called the *sex skin*. Like full testicular elevation of the male, the presence of the sex skin in the sexually aroused woman heralds impending orgasm.

Bartholin's glands secrete a few drops of

Box 3.3

THE GRAFENBERG SPOT AND FEMALE EJACULATION

The question of whether or not women ejaculate has intrigued people for a long time. Until the 17th century, the vaginal lubricant fluid was assumed to be a female "ejaculate" analogous to male semen and therefore essential for conception. This misunderstanding was the basis of theological tolerance of female masturbation during coitus; if a woman could not reach orgasm during coitus it was permissible that she did so by manipulation; otherwise she would be unable to complement the male's semen with her own ejaculate to make conception possible. When the nature of human reproduction became better understood this notion was discarded, and women were declared incapable of ejaculation.

Still, the controversy lingered on. Reports kept appearing of women emitting fluid at orgasm, but there was much confusion about its composition. The suggested alternatives included urine, fluid

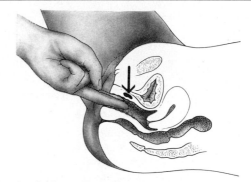

The Grafenberg Spot (G-spot).

from Bartholin's glands, vaginal lubricant expelled forcibly during orgasmic contractions, and fluid from the urethra.

In 1950, Ernest Grafenberg published an article in which he made two claims (Grafenberg, 1950). First, there exists an "erotic zone" on the front wall of the vagina along the course of the urethra. This area becomes enlarged during sexual excitement and protrudes into the vaginal canal; following orgasm, it reverts to its nonstimulated state. Second, "large quantities of a clear transparent fluid" gush out of the urethra during orgasm, at least in the case of some women.

Grafenberg explained both phenomena by the hypothesis that there persisted in the female the homologue of the erectile tissues surrounding the male urethra (the corpus spongiosum) and the homologue of the prostate gland (known as "Skene's glands"). The existence of the first would explain the swelling of the anterior (front) vaginal wall. Similarly, the Skene's glands would function as a "female prostate," providing the fluid in female ejaculation (just as the ejaculatory fluid in the male is provided by the prostate). Grafenberg's hypothesis was plausible on the grounds of embryonal development, but he offered no anatomical evidence and little objective clinical data to support it.

Attention was refocused on this issue by Sevely and Bennett (1978). Based on a review of the literature, they concluded that women do ejaculate and that the fluid they emit is similar though not identical to prostatic fluid in the male. Perry and Whipple (1981) pursued further this line of research, and they named the "erotic zone" in the anterior vaginal wall the *Grafenberg spot* (G-spot). According to them and other researchers, the G-spot is located about two inches from the vagina's entrance (see the figure). When properly stimulated, it swells and leads to orgasm (often a whole series of them) in many women. During orgasm, many women emit a fluid through the urethra. These women are embarrassed because they believe they are losing urinary control; hence, to avoid it, they try to suppress their orgasms (Ladas et al., 1982).

What is the fluid these women emit? Some researchers have reported it to contain higher levels of acid phosphatase (an enzyme secreted by the prostate gland) and glucose (sugar), and lower levels of urea and creatinine (end products of protein metabolism excreted in urine) than contained in urine (Belzer, 1981). Acid phosphatase levels, though higher than in urine, have never been found as high as in the male ejaculate. Hence, in this view, women emit urethral fluid with some similarities to prostatic fluid in men (Addiego et al., 1981).

Attempts to validate these claims by other researchers have not generally been successful. In one study, the female "ejaculate" turned out to be urine (Goldberg et al., 1983). Even if a sensitive area exists in the anterior vaginal wall, no anatomical evidence has linked it to Skene's glands (Alzate and Hoch, 1986). Instead of just a discrete "spot," the entire anterior vaginal wall appears to be highly sensitive, and its stimulation likely to lead to orgasm (Hoch, 1986).

If further studies validate the existence of the G-spot and female ejaculation, it will help us understand female sexual physiology and treat problems. Meanwhile, the controversy may be causing undue anxiety if a woman believes that she is lacking sexually if she does not "ejaculate" or does not appear to be blessed with a G-spot.

---

fluid rather late in the excitement phase. The function of this fluid is not clear. It contributes little to vaginal lubrication.

The uterus responds to sexual stimulation by elevation from its tilted position, which pulls the cervix up and contributes to the tenting effect in the vagina. Full uterine elevation is achieved during the plateau phase and is maintained until resolution.

Other Reactions Erection of the nipple is the first response of the female breast in the excitement phase. (A man's nipples too sometimes grow erect in the late excitement and plateau phases.) Engorgement of blood vessels is also responsible for the swelling of the woman's breasts. In the plateau phase the engorgement of the areolae is more marked. As a result, the erect nipples appear relatively smaller.

The breast swells further during this phase, particularly if it has never been suckled. It may increase by as much as a fourth of its unstimulated size.

Sexual arousal results in definite skin reactions. The *sex flush* is more common and pronounced in women. It starts in the excitement phase as a rash-like redness of the skin in the center of the lower chest, which spreads to the breasts, the rest of the chest wall, and the neck. This sexual flush reaches its peak in the late plateau phase and is an important component of the excited, straining facial expression characteristic of the person about to experience orgasm. The appearance of tension is also conveyed by the musculature of the body in the plateau phase. The muscles of the feet, in particular, tense up, with rigid extension of the toes (carpopedal spasm).

The heart rate increases in the excitement phase, and by the plateau phase it reaches 100 to 160 beats a minute. (The normal resting heart rate is 60 to 80 beats a minute.) The blood pressure similarly increases. Changes in respiratory rate lag somewhat behind those in heart rate. Faster and deeper breathing becomes apparent only in the plateau phase.

These changes in cardiovascular function are comparable to levels reached in moderate physical effort, such as taking a brisk walk or going up a flight of stairs. The strain on the cardiovascular system is easily handled by most individuals. Nonetheless, a substantial proportion of men lose sexual interest, or develop problems with potency, following heart attacks. In most of these cases the problem is psychological; there is usually no medical reason why men and women with heart disease should not maintain an active sexual life, although it makes sense to check with the doctor (Kolodny et al., 1979).

Orgasm

In physiological terms *orgasm* is the discharge of accumulated neuromuscular tension which results from sexual arousal. Subjectively, it is a high pitch of erotic tension, one of the most intense and profoundly satisfying human sensations; it lasts a fraction of a minute, yet it feels like an eternity. The term "orgasm" is derived from the Greek *orgasmos* ("to swell," "to be lustful"). Colloquial terms include "to climax," "to come," and "to spend."

The intensity of orgasmic experience varies somewhat with age, physical condition, and context, but it is usually felt by both men and women as an intensely pleasurable experience. Davidson (1980) characterizes it as a form of "altered state of consciousness."

In adult males the sensations of orgasm are linked to ejaculation, which occurs in two stages. First, there is a sense that ejaculation is imminent (or "coming") and unstoppable. Second, there is a distinct awareness of throbbing contractions at the base of the penis, followed by the sensation of fluid moving out under pressure.

Orgasm in the female starts with a feeling of momentary suspension or "stoppage." Sensations of tingling and tension in the clitoris then reach a peak and spread to the vagina and pelvis. This stage varies in intensity. It may also involve sensations of "falling," "opening up," or emitting fluid. (Some women compare it to mild labor contractions.) It is followed by a suffusion of warmth spreading from the pelvis through the rest of the body, culminating in throbbing sensations in the pelvis. One woman describes it as follows:

> There are a few faint sparks, coming up to orgasm, and then I suddenly realize that it is going to catch fire, and then I concentrate all my energies, both physical and mental, to quickly bring on the climax—which turns out to be a moment suspended in time, a hot rush, a sudden breathtaking dousing of all the nerves of my body in pleasure (Hite, 1976).

Due to the tell-tale evidence of ejaculation, a man can hardly miss identifying an orgasm, yet some women feel uncertain whether they have had one. Are all orgasms the same? No two sexual experiences of any kind are ever the same, but the male orgasmic experience seems to follow a more standard pattern than the female. Possibly women have different kinds of orgasms, based on different physiological mechanisms (Box 3.4). Other evidence suggests that the orgasmic experience in the two sexes is not that dissimilar. When brief descriptions of orgasm by men and women were submitted to a panel of psychologists, they were unable to identify the sex of the authors from the descriptions (Vance and Wagner, 1976).

Male Sex Organs   The characteristic rhythmic muscular contractions of orgasm begin in the prostate, seminal vesicles, and vas deferens, then extend to the penis (Figure 3.3), involving the entire length of the urethra, as well as the muscles covering the root of the penis. At first the contractions occur fairly regularly at intervals of approximately 0.8 second, but after the first several strong throbs they become weaker, irregular, and less frequent.

Studies subsequent to the work of Masters and Johnson have studied male orgasm as manifested in contractions of the anal sphincter. Masters and Johnson reported that two to four anal contractions occurred during this period. Later Bohlen and his associates found an average of 17 anal contractions (Figure 3.5) (Bohlen et al., 1980). They occur over a period of 26 seconds, a close estimate of the length of male orgasm, although subject to considerable individual variation.

Ejaculation ("throwing out") entails the ejection of semen. It is the most obvious manifestation of orgasm in males following puberty, when the prostate and other glands produce seminal fluid. Before puberty, boys do not ejaculate, even though they can experience orgasm.

Ejaculation has two distinct phases. During the first phase, called *seminal emission* (or *first-stage orgasm*), the smooth muscles of the prostate, seminal vesicles, and vas deferens contract, pouring their contents into the dilated urethral bulb. At this point the man feels ejaculatory pressure building up. In the second, or *expulsion phase (second-stage orgasm)*, the semen is expelled by the vigorous contractions of the muscles surrounding the root of the penis, other muscles of the pelvic region, and

the genital ducts. The inner sphincter of the urinary bladder simultaneously closes, preventing semen from flowing into the bladder.

The amount of fluid and the force with which it is ejaculated are popularly associated with strength of desire, potency, and so on, but these beliefs are difficult to substantiate. There does seem, however, to be a valid association between the emission phase and the onset of the refractory period. Some men are able to inhibit the emission of semen while they experience the orgasmic contractions—in other words, they have *nonejaculatory orgasms*. Such orgasms do not seem to be followed by a refractory period, so these men can have consecutive or multiple orgasms, like women. Only if emission of semen occurs with ejaculation does a refractory period follow. Ancient Chinese sex manuals have long extolled the virtues of nonejaculatory coitus (Box 3.5). The Indian practice of *Karezza* or *coitus reservatus* is a similar protracted coitus without ejaculation.

However, the matter is even more complicated. The fact that no ejaculate comes out of the penis does not necessarily mean that emission did not take place. In cases of *retrograde ejaculation,* the flow of semen is reversed, so that instead of flowing out of the urethra, it is emptied into the urinary bladder. The sensation of orgasm in this condition is unchanged, and a refractory period presumably follows normally. This condition occurs in some illnesses and with the use of certain common tranquilizers and blood pressure drugs (Chapter 8).

Female Sex Organs   During female orgasm, the most visible effects occur in the orgasmic platform (Figure 3.4). This area contracts rhythmically and with decreasing intensity (initially at approximately 0.8-second intervals). The more frequent and intense the contractions of the orgasmic platform, the more intense is the subjective experience of orgasm. At particularly high levels of excitement these rhythmic contractions are preceded by nonrhythmic (spastic) contractions of the orgasmic platform that last several seconds. Whether or not women ejaculate during orgasm is discussed in Box 3.3.

Orgasmic contractions in the uterus start at the top (the fundus) and spread downward; although these contractions occur simultaneously with those of the orgasmic platform, they are less distinct and more irregular. It has often been assumed that the contractions of the uterus during coitus cause the semen to be sucked into its cavity. Such an effect was reported by Beck as early as 1874 (Levin, 1980). Masters and Johnson found no evidence to

Box 3.4

VARIETIES OF FEMALE ORGASM

Probably more has been written on the female orgasm than any other aspect of human sexuality (Levin, 1981). Until fairly recently, our ideas about the female orgasm were strongly influenced by psychoanalytic theory. According to Freud women experience two types of orgasm: clitoral and vaginal. *Clitoral orgasm* is attained exclusively through direct clitoral stimulation, usually by masturbation; *vaginal orgasm* results from vaginal stimulation, usually through coitus. This dual orgasm theory claims that in young girls the clitoris is the primary site of sexual excitement. With psychosexual maturity the sexual focus is said to shift from the clitoris to the vagina, so after puberty the vagina emerges as the dominant orgasmic zone. That makes vaginal orgasm more "mature" than clitoral orgasm. With some variations, this model was reiterated by Freud's followers. Modern psychoanalysts, however, do not all adhere to this view (Salzman, 1968).

Kinsey and his associates (1953) doubted the validity of the Freudian concept of dual orgasm. Masters and Johnson (1966) reaffirmed these doubts and established that physiologically there is only one type of orgasm: orgasmic response to clitoral, vaginal, or any other form of stimulation is all the same (Masters and Johnson, 1966). Others further endorsed the single-type orgasm model. For instance, Sherfey (1973) favored it in her theory of the evolution of female sexuality. Furthermore, given the greater female capability for multiple orgasms, she hypothesized that females are endowed with an insatiable sexual drive, which has been socially suppressed so as not to interfere with maternal functions.

Investigators have nevertheless continued to distinguish different types of orgasm. Fisher (1973), based on his appraisal of psychological and physiological studies, has restated the distinction between clitoral and vaginal orgasm, while rejecting its psychoanalytic implications. Singer and Singer (1972), based on their physiological studies, have proposed three types of female orgasm. First is the *vulval orgasm*, characterized by involuntary rhythmic contractions of the vaginal entrance, which they consider to be the same as orgasms described by Masters and Johnson. Second, there is the *uterine orgasm*, which results from the repetitive displacement of the uterus and is dependent on coitus or a close substitute. The third type is the *blended orgasm*, combining elements of the other two.

As an extension of their work with the G-spot (Box 3.3), Ladas, Whipple, and Perry (1982) have suggested that a *continuum of orgasmic response* would integrate earlier orgasmic models with their own findings. At one end is the *clitoral orgasm* (the Singers' "vulval orgasm"); triggered by clitoral stimulation, it involves contractions of the circumvaginal muscles and is felt primarily at the orgasmic platform. At the other end is the *vaginal orgasm* (Singers' "uterine orgasm"); triggered by stimulation of the G-spot, it involves contractions of the uterus and is hence experienced in the region of the pelvic organs. The vaginal orgasm is terminative—one is enough to attain sexual satiety. The clitorial orgasm is not and calls for repetition (Fisher, 1973; Bentler and Peeler, 1979).

This issue has also found its way into literature. In *The Golden Notebook*, the novelist Doris Lessing writes:

> A vaginal orgasm is emotion and nothing else, felt as emotion and expressed in sensations that are indistinguishable from emotion. The vaginal orgasm is a dissolving in a vague, dark, generalised sensation like being swirled in a warm whirlpool. There are several different sorts of clitoral orgasms, and they are more

powerful (that is a male word) than the vaginal orgasm. There can be a thousand thrills, sensations, etc., but there is only one real female orgasm and that is when a man, from the whole of his need and desire, takes a woman and wants all her response. Everything else is a substitute and a fake, and the most inexperienced woman feels this instinctively (Lessing, 1962, p. 186).

Whatever the experiential differences, it is best to avoid the notion that one form of female orgasm is "better" than other. Women were needlessly burdened in the past by the idea that clitoral orgasm was less "mature." Now they are made to feel they must have multiple orgasm and ejaculatory orgasm. Every woman should discover her own possibilities. No woman should have to "measure up" to anyone else's standards.

---

support this idea, but others claim to have done so (Fox and Fox, 1969).

Through new laboratory studies, we now have more precise measures of orgasm. One such approach is to simultaneously record anal and vaginal pressures (which reflect the activity of contracting muscles) during orgasm. As shown in Figure 3.5, pressures in both the anus and vagina erupt into regular and synchronized contractions with the onset of orgasm; in some but not all women, they are followed by irregular contractions to the end of the orgasm. The lengths of recorded orgasms range from 13 to 51 seconds; the period of signaled orgasms (that is, the interval between when the women say they started and stopped orgasm) ranges from about 7 to 107 seconds (Bohlen et al., 1982b). These investigators, unlike Masters and Johnson, found no correlation between the number of orgasmic contractions and level of satisfaction, perceived intensity, or sexual gratification.

Other Reactions   The breasts show no further changes during orgasm; but if a woman is nursing, milk may be ejected from her nipples. Temperature changes in the skin lead to feelings of pervasive warmth following orgasm; and there are popular references to sexual excitement as a "glow," "fever," or "fire." Superficial vasocongestion is the likely explanation of this sensation.

During orgasm, the heart rate reaches its peak, rising to 100 to 180 beats per minute. The respiratory rate may go up to 40 a minute (the normal rate is about 15 a minute, inhalation and exhalation counting as one). Breathing, however, becomes irregular; a person may momentarily hold the breath and then breathe rapidly (Fox and Fox, 1969). Some of the panting and grunts uttered during orgasm result from involuntary contractions of the muscles that force air through the respiratory passages. When the genital muscles contract, so do skeletal muscles in other parts of the body.

A Common Model   To integrate the patterns of male and female orgasm into a coherent scheme, Davidson (1980) has proposed a *bipolar hypothesis* of orgasm. He postulates that when sexual excitement passes a critical threshold, an orgasmic control area in the central nervous system triggers the orgasm. It sends neural impulses simultaneously in two directions: "upward" to higher brain areas in the cortex, producing an intense subjective experience, and "downward" to the genital-pelvic region, producing physiological reactions.

This model would apply to both sexes. It would link the psychological experience of orgasm with its physical aspects. It would explain why contractions without ejaculation in men and "vulval orgasm" in women would not induce a refractory period. On the other hand, a refractory period would follow ejaculation in the male and "uterine orgasm" in the female. Especially if the latter turns out to be accompanied by female ejaculation, then male and female orgasm would emerge as essentially identical.

Box 3.5

EJACULATORY CONTROL

The aspect of male orgasm that has attracted the most attention is ejaculation. It is essential for fertilization and represents the peak of male orgasmic pleasure, but ejaculation can also be fraught with anxiety. For many a man, getting an erection is a bit like becoming airborne in a glider; he strives to remain aloft as long as he can, but he does not have full control. Once orgasm is triggered, he cannot stop it. If it occurs before he and his partner are satisfied, he feels he has failed to "perform." Ironically, his "performance anxiety" itself makes it more likely that he will fail to get an erection or he will ejaculate prematurely (Chapter 8).

An even deeper anxiety has plagued men since ancient times over the loss of semen. The ancient Greeks were mindful of its value: Hippocrates commented on its "precious" character; Pythagoras called it a "flower of the blood"; and Galen held that an ounce of semen was worth 40 ounces of blood, a notion that persisted into the Middle Ages. In the 12th century, the Talmudic scholar and physician Maimonides wrote:

> Effusion of semen represents the strength of the body and its life, and the light of the eyes. Whenever it [semen] is emitted to excess, the body becomes consumed, its strength terminates, and its life perishes. . . . He who immerses himself in sexual intercourse will be assailed by [premature] aging. His strength will wane, his eyes will weaken, and a bad odor will emit from his mouth and his armpits. . . . His teeth will fall out and many maladies other than these will afflict him. The wise physicians have stated that one in a thousand dies from other illnesses and the remaining [999 in the thousand] from excessive sexual intercourse.

The horror of wasting sperm dominated thinking through the 19th century. Fantastic notions about "masturbatory insanity" and the ravages of "spermatorrhea" (nocturnal emission) were directly based on such theories of wasting from semen loss.

The notion of semen power is even more ancient in Asia (Gregersen, 1983). Hindu ascetics practiced continence to enhance their physical and psychic powers. To the Chinese, semen embodied the essence of *yang*, the basic male element in nature. As a vital essence, semen was not to be squandered. Rather than damning sexual intercourse, the Chinese devised a technique of engaging in protracted coitus without ejaculation, except at carefully controlled intervals. The point was not to avoid ejaculation, but to allow it at optimal intervals, depending on the age of the man, his state of health, the season of the year, and other factors.

This doctrine meant that a man extended his period of coital pleasure and provided his sexual partner with ample opportunity for her enjoyment. Because a woman supposedly did not ejaculate, there was no problem with her experiencing orgasm as often as she wished. Moreover, her vaginal secretions were a rich source of *yin*, the female element in nature. By extended exposure to vaginal secretions through prolonged coitus, a man absorbed the vital female elements; and by refraining from ejaculation, he conserved the male elements. As a result, he attained the ideal harmony of yin and yang, which led to good health and a long life. (Some modern applications of these ideas are discussed in Chang, 1977.)

In Taoist texts, these erotic wisdoms are often conveyed through dialogues between an emperor and his Tao of Loving advisors (usually women). In the following excerpts from *The Secrets of the Jade Chamber*, three advisors offer their views:

TSAI NU: It is generally supposed that a man derives great pleasure from ejaculation. But when he learns the Tao he will emit less and less; will not his pleasure also diminish?

P'ENG TSU: Far from it. After ejaculation a man is tired, his ears are buzzing, his eyes heavy and he longs for sleep. He is thirsty and his limbs inert and stiff. In ejaculation he experiences a brief second of sensation but long hours of weariness as a result. And that is certainly not a true pleasure. On the other hand, if a man reduces and regulates his ejaculation to an absolute minimum, his body will be strengthened, his mind at ease, and his vision and hearing improved. Although the man seems to have denied himself an ejaculatory sensation at times, his love for his woman will greatly increase. It is as if he could never have enough of her. And this is the true lasting pleasure, is it not? (Chang, 1977, p. 21).

SU NU: When a man loves once without losing his semen, he will strengthen his body. If he loves twice without losing it, his hearing and his vision will be more acute. If thrice, all diseases may dis-

appear. If four times, he will have peace of his soul. If five times his heart and blood circulation will be revitalized. If six times, his loins will become strong. If seven times, his skin may become smooth. If nine times, he will reach longevity. If ten times, he will be like an immortal (Chang, 1977, p. 44).

Resolution Phase and Aftereffects

Whereas the onset of orgasm is fairly distinct, its termination is less so. As the genital rhythmic throbs become progressively less intense and less frequent, neuromuscular tensions give way to a profound state of relaxation.

The manifestations of the postorgasmic phase are the opposite of those of the preorgasmic period. The entire musculature relaxes. The pounding heart and accelerated breathing revert to normal. Congested and swollen tissues and organs resume their usual color and size. As the body rests, the mind reverts to its ordinary state of consciousness.

The descent from the peak of orgasm may occur in one fell swoop or more gradually. Particularly at night, when the profound postcoital relaxation compounds natural weariness, some people tend to fall asleep. Others feel alert or even exhilarated.

It is not unusual to feel thirsty or hungry following orgasm. Smokers may crave a cigarette. There may be a need to urinate. Some people develop a headache, which is due to vasocongestion in the brain and may be treated with the appropriate drugs (Johns, 1986).

Regardless of the immediate postorgasmic response, a healthy person recovers fully from the aftereffects of orgasm in a short time. Protracted fatigue is often the result of activities that preceded or accompanied sex (drinking, drugs, lack of sleep), rather than of orgasm itself. When a person is in ill health, however, the experience itself may be quite taxing.

Reactions of Sex Organs   In the resolution phase the changes of the preceding stages are reversed (Figures 3.3 and 3.4). A man loses his

erection in two stages: a rapid loss of tumescence, which reduces the organ to a semi-erect state, is followed by a more gradual decongestion, in which the penis returns to its unstimulated size.

After ejaculation, if the penis remains in the vagina it stays erect longer. If the man withdraws, is distracted, or attempts to urinate, detumescence is more rapid. (A man cannot urinate with a fully erect penis, because the internal urinary sphincter closes reflexively to prevent intermingling urine and semen.)

In the woman, the orgasmic platform subsides rapidly. The inner vaginal walls return much more slowly to their usual form. Lubrication may in rare instances continue into this phase, and such continuation indicates lingering or rekindled sexual tension. With sufficient stimulation, another orgasm may follow if the woman so desires.

Other Reactions   Changes in the rest of the body also are reversed following orgasm. In the resolution phase the areolae of the breasts appear less swollen, and the nipples regain their fully erect appearance ("false erection"). Gradually, breasts and nipples return to normal size. The sexual flush disappears. The heart and respiratory rates gradually return to normal. The musculature relaxes.

In 30 to 50 percent of individuals, there is a sweating reaction. Among men this response is less consistent and may involve only the feet and the palms.

CONTROL OF SEXUAL FUNCTIONS

Underlying the patterns of sexual response are complicated mechanisms of control. Some of

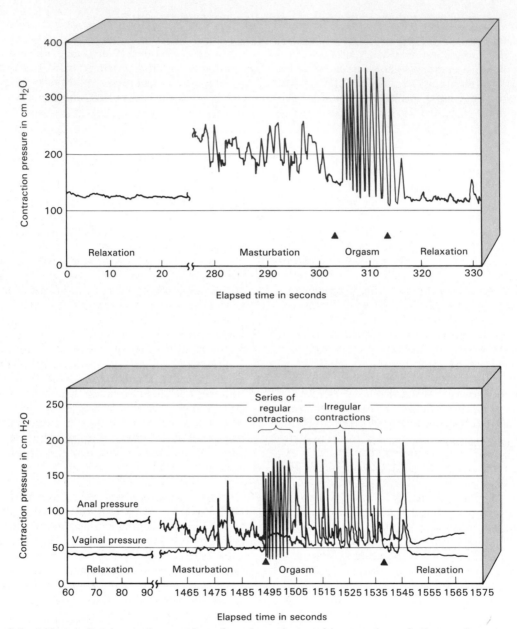

Figure 3.5   (Above) Computer-drawn plot of anal tension and contractions during male masturbation. (Below) Anal and vaginal pressures during female masturbation.

them are hormonal (Chapter 4). Others are neurophysiological and coordinated at two levels: in the spinal cord and in the brain (De-Groat, 1986).

Spinal Control

Sexual functions are controlled at their most elemental level by *spinal reflexes*. An example of a reflex action is the way you kick when the tendon below your knee is tapped. Reflexes are involuntary in the sense that they are automatic. They do not require an "order" to act by the brain; yet, if you try, you can modify or inhibit them to some extent.

The knee-jerk reflex involves the *voluntary nervous system*, muscles that we can move at will. Other reflexes, including those involved in sexual functions, belong to the *autonomic nervous system*. This system operates involuntarily to control many of the internal functions of the body. Its activities are normally carried out without our being aware of them (Guyton, 1986).

The autonomic nervous system has two main subdivisions, both involved in sexual function: the *parasympathetic* and the *sympathetic* system. Stimulation of one or inhibition of the other has the same overall effect. One of the basic functions of autonomic nerves is to control the flow of blood by constricting or dilating arteries. In the case of genital blood vessels, parasympathetic stimulation causes arteries to dilate. Sympathetic stimulation makes them constrict.

Much of the research on the spinal control of sexual functions has been done with animals, and inferences have been drawn to humans. Male animals are easier to study (because of the ease with which erections and ejaculations can be monitored), so we know more about these male processes, which is what we shall describe first.

Erection and Ejaculation   The reflex control centers of erection and ejaculations are located in the spinal cord. *Afferent* nerves bring in sensory impulses to these centers (such as sensations that arise by touching the genitals). These centers are activated by these impulses and in turn send out "orders" through *efferent* nerves that bring about changes in the genital organs (such as causing vasocongestion of the penis). Although the spinal reflexes can function independently, they are in close communication with the brain center through messages that go up and down the spinal cord.

There are two *erection centers* in the spinal cord (Figure 3.6). The primary, or *parasym-*

pathetic, center is located in the lowest, or sacral, portion of the spinal cord (between the segments S2 and S4). Into this center come the sensory impulses or sensations from the stimulation of the genitals or nearby erogenous zones. These nerve impulses travel to the spinal cord through the pudendal nerve. In response, the S2-S4 reflex center sends out impulses through efferent parasympathetic nerves which cause vasocongestion and erection of the penis.

A second erection center is located higher up, at the junction of the Thoracic and Lumbar segments of the spinal cord (T11–L2). This center is part of the *sympathetic* system and receives impulses from the brain that have been generated by psychogenic stimulation, such as through erotic sights or sounds and erotic fantasies. Sympathetic nerves then carry efferent impulses to the genital organs which stimulate erection. The exact mechanism by which this occurs is not clear.

Through messages that go up and down the spinal cord, the brain is in constant communication with the spinal centers. Impulses reaching the brain from the spinal cord are interpreted as the subjective sensation of sexual arousal. Thus, the brain communicates its own state of arousal to the spinal cord, as well as being aroused by sensations from the spinal cord.

The same two locations in the spinal cord contain the *ejaculatory centers* (Figure 3.7). The *sympathetic* ejaculatory center is located in the T11–L2 segments. Sympathetic efferent nerves go out from this center to the smooth muscles of the prostate gland, seminal vesicle, and vas deferens. The contraction of these muscles results in seminal emission, or the first phase of ejaculation. The second phase, or expulsion of semen, is triggered by the *voluntary* or *motor* (rather than the parasympathetic) center in the sacral segment of the spinal cord. These impulses go out through somatic efferent nerves to the muscles surrounding the base of the penis. Contractions of these muscles complete the ejaculation of semen (Hart and Leedy, 1985).

Emission and ejaculation are thus closely

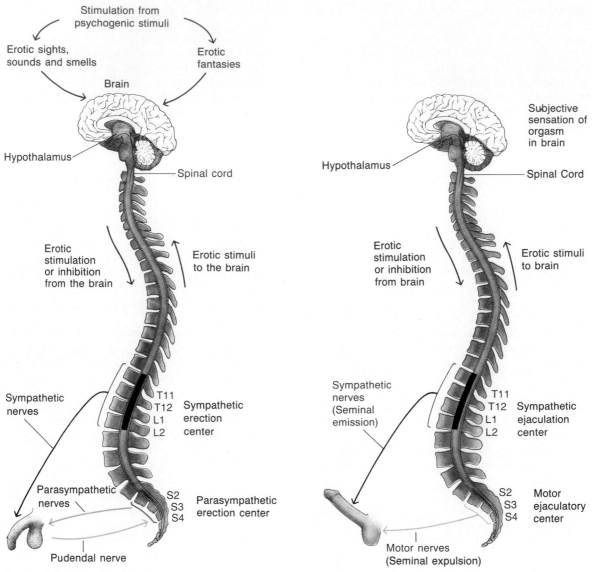

Figure 3.6   Spinal reflex for erection.

Figure 3.7   Spinal reflex for ejaculation.

integrated yet separate physiological processes. Ordinarily they occur together, but under some conditions one or the other may be experienced separately. Emission can take place without ejaculatory contractions; then semen simply seeps out of the urethra. Likewise, ejaculatory contractions can occur without semen

being emitted (Tarabulcy, 1972; Bancroft, 1983).

The processes described so far are controlled by reflexes. They are independent of the brain, in the sense that they can occur without its assistance. For instance, a man whose spinal cord has been cut in an injury above the level of the reflex erection center may still be capable of erection. He will not feel the stimulation of his penis; for that matter, he may even be totally unconscious. His penis will respond "automatically."

Obviously, these reflex centers still can be influenced by the brain. Intricate networks link the brain to the reflex centers in the spinal cord. Daydreaming may trigger erection without physical stimulation; worries may inhibit erection despite persistent physical stimulation. Usually mind and body work together. Spurred by erotic thoughts, the man touches his genitals, or excited by a touch, he has erotic thoughts.

The instances in which erection seems to be nonsexual in origin involve tension of the pelvic muscles (as when lifting a heavy weight or straining during defecation). Irritation of the glans or a full bladder may have the same effect. Erections in infancy are explained on a reflex basis also. An additional gruesome example is erection experienced by men during execution by hanging.

Reflexive Mechanisms in Women   It is generally assumed that there are spinal centers in women that correspond to the erection and ejaculatory centers in men. The reflexive centers in the female spinal cord are less well identified in part because the manifestations of orgasm are difficult to ascertain among female experimental animals.

The effects of spinal cord injury on sexual function have been less well studied among women. Clinical experience shows women often lose the ability to reach orgasm because of diminished vasocongestion and vaginal lubrication (Kolodny et al., 1979). Nonetheless, the ability of women to engage in coitus in spite of spinal injuries is likely to be less limited than that of men. Women can engage in coitus without adequate vasocongestion (using artificial lubrication), whereas men cannot do so without an erection. Apart from coitus, both women and men can engage in other satisfying sexual interactions (Chapter 8).

The Role of Neurotransmitters   Some of the most exciting recent discoveries in neuroscience have involved the chemicals that transmit impulses in the brain and by peripheral nerves. Such *neurotransmitters* may help explain both how the genitals function and the brain processes underlying sexual motivation (Thompson, 1985).

Some neurotransmitters, such as *acetylcholine*, have long been known to transmit impulses from autonomic nerves to muscle cells. When muscles in the walls of arteries relax, the walls dilate and more blood flows; this mechanism underlies vasocongestion and erection.

More recently, other neurotransmitters have been discovered that may have an even more direct bearing on sexual functions. One unlikely-sounding substance is *vasoactive intestinal polypeptide (VIP)*. First isolated in the gastrointestinal tract, VIP is also found in nerves of the genitourinary system, as well as in the vas deferens, prostate, and the cavernous bodies of the penis. It is hypothesized that VIP causes relaxation of cavernous smooth muscles, allowing increased blood flow; it may be important in starting and maintaining erection (Arsdalen et al., 1983; Goldstein, 1986). Other neurotransmitters may shed further light on many other aspects of sexual physiology. Our understanding of their role is just developing (Money, 1987).

Brain Control
All sensory input must be interpreted in your brain before you feel any sensation. No thought, however trivial, and no emotion, however fleeting, can exist in an empty skull. Therefore all bodily experiences must be finally understood at the level of brain activity. Such understanding will not substitute for other ways of conceptualizing human experience, as in psychological or ethical terms, but neither can we ever dispense with understanding the physical basis.

There are many ways to investigate mechanisms in the brain. One is electrical stimulation. Erection, ejaculation, and copulatory behavior are elicited by stimulating certain areas in the brain. Related studies use an electroencephalograph to monitor the natural electrical activity of the brain during orgasm (Cohen et al., 1976).

The second method of study is destruction of brain centers. For instance, scientists have been able to eliminate male copulatory behavior in a variety of mammalian species by destroying the medial preoptic region of the brain. Similarly, they have eliminated sexual receptivity in female mammals by causing hypothalamic lesions. If stimulation of a brain area results in a sexual reaction such as erection, and destruction of the region eliminates it, then we can infer that the area is at least one link in the chain of brain mechanisms controlling that function. (Animal studies like these may be cruel, but they are often important for understanding physiological functions.)

A third and complementary approach is to identify brain areas that inhibit a given function. How we act is the outcome of the "push" to behave in a certain way and the "pull" restraining us. An activity appears or increases if scientists remove or destroy the parts of the brain that inhibit it. For example, removal of the temporal lobes in a monkey results in a striking increase in autoerotic, heterosexual, and homosexual behavior. These monkeys often show penile erections even when sitting quietly (the *Kluver-Bucy syndrome*). Similar reactions occur in humans whose temporal lobes on both sides have been removed for treatment purposes (Terzian and Dale-Ore, 1955).

The central core of the brain, the first part to evolve, controls the most basic life-sustaining activities. Of particular interest to sexuality in this region is the *hypothalamus,* which exerts a crucial influence on nerves and hormones (Chapter 4), and the *limbic system.*

The limbic system surrounds the upper portion of the central brain core, deep within the cerebral hemispheres (Figure 3.8). This portion of the brain probably provides the central nervous control of our sexual behavior. Because we know little of its workings, it has been called the "black box" of our sexuality (Bancroft, 1983). Included in this system are the hippocampus, the fornix, the amygdala, the mammillary bodies, and the cingulate gyrus. Closely interconnected with the hypothalamus, the limbic system is involved in many

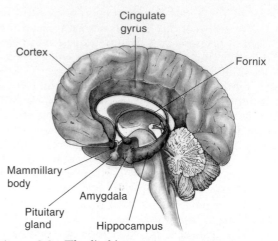

Figure 3.8   The limbic system.

"instinctive" activities like feeding, attacking, fleeing from danger, and sex. Electrical stimulation of parts of the limbic system elicits penile erection, mounting, and grooming in male animals (MacLean, 1976). Certain regions of the limbic system are also involved in the experience of emotions and may be linked to feelings of sexual pleasure.

In the hypothalamus and related regions of the rat brain, areas have been identified whose electrical stimulation appears to be highly rewarding; they have been called *pleasure centers* (Olds, 1956). Similar centers may exist in the limbic system of the human brain (Heath, 1972).

At the moment, we cannot adequately explain the way sexual sensations feel by neurophysiological processes. No one yet has precisely localized sexual sensations in the human brain. Even if researchers found "sex centers," we still would need to understand how all the systems related to sexuality work together. As Beach (1976) explains,

> Instead of depending on one or more centers, sexual responsiveness and performance are served by a net of neural subsystems, including components from the cerebral cortex down to the sacral cord. Different subsystems act in concert, but tend to mediate different units or elements in the normally integrated patterns (p. 216).

We are still a long way from understanding the physiological bases of the complex and subtle human sexual experience, but the rewards of such understanding will be enormous.

REVIEW QUESTIONS

1. Which sensory systems are involved in sexual arousal?

2. Which two basic physiological processes underlie most signs of sexual arousal?

3. What are the stages and main genital manifestations of the sexual response cycle?

4. What are the other reactions of the body during the sexual response cycle?

5. What are the spinal reflexes involved in erection and orgasm?

THOUGHT QUESTIONS

1. A perfume arouses you. Is your response due to psychological or physiological mechanisms?

2. How does ideology affect discussions of the varieties of female orgasm?

3. What studies would you conduct to settle the controversy about the G-spot?

4. What would be the psychological and social consequences if it were to be established that discrete brain centers control sexual behavior?

5. What are the ethical arguments for and against doing laboratory studies of sexual physiology with humans?

SUGGESTED READINGS

Masters, W. H., and Johnson, V. E. (1966). *Human sexual response*. Boston: Little, Brown. The original and still the standard source on sexual physiology.

Brecher, R., and Brecher, E. (1966). *An analysis of human sexual response*. A nontechnical and clear summary of the Masters and Johnson research on the sexual response cycle.

Davidson, J. M. (1980). "The psychobiology of sexual experience." In Davidson, J. M., and Davidson, R. (Eds.), *The psychobiology of consciousness*. New York: Plenum Press. A comprehensive and clear overview of the role of neuroendocrine factors in sexual experience.

Thompson, R. F. (1985). *The brain*. An excellent introduction to the neurosciences, including the neurophysiological mechanisms regulating sexual functions.

Sex Hormones

Hormones are chemical substances that are secreted directly into the bloodstream by ductless or *endocrine glands* and by specialized *neurosecretory cells*. Close to fifty hormones produced by some ten major endocrine sources are crucial to develop and sustain a vast range of vital physiological functions (Guyton, 1986). Sex and reproduction in particular simply could not exist without hormones.

The endocrine and the nervous systems, which are closely integrated, form a vast communication network within your body. The nervous system transmits electrical impulses through nerves; the endocrine system secretes hormones into the bloodstream to reach *target organs* and tissues whose development, sustenance, and functions it controls.

The concept of chemical control of bodily functions and temperaments can be traced to the ancient Greeks. They counted four *humors*—blood, mucous, yellow bile, and black bile—and ascribed a person's temperament to their balance in the body. Our modern science of endocrinology is young, though; the term *hormone* ("to excite") only goes back to the turn of this century (Crapo, 1985), and virtually everything we know about hormones, particularly their effects on sexual development and behavior, has been learned in the 20th century (Money, 1987).

HORMONAL SYSTEMS AND SEXUAL FUNCTIONS

Many of the body's hormones have a bearing on sex and reproduction. We will deal here only with the hormones most specifically and directly linked to sexual functions. These hormones are produced by the *gonads* (testes and ovaries), the *adrenal cortex*, the *pituitary gland*, and the *hypothalamus* (Figure 4.1).[1]

Gonadal Hormones
The hormones produced by the testes and ovaries are commonly referred to as *sex hormones*

[1]The general discussion of hormonal functions in this chapter is based on standard physiology and endocrinology texts, including West (1985), Greenspan and Forsham (1986), Guyton (1986), and Gilman et al. (1985).

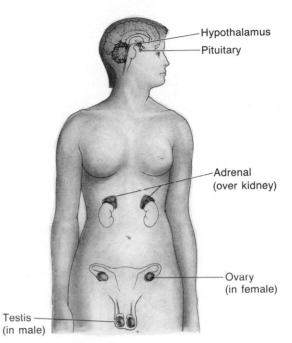

Figure 4.1 Endocrine glands that regulate sexual functions.

because of their gonadal origins and the crucial roles they play in sexual physiology. It is also common to refer to the hormones produced by the testis and by the ovary, respectively, as "male" and "female" sex hormones. Such designations can be misleading: sex hormones are also involved in functions other than sex; male and female hormones exist in both sexes (though in different concentrations). These hormones, as well as the related hormones produced by the cortex of the *adrenal glands* (located over the kidneys), belong to a family of chemical compounds called *steroids*, which have a common basic molecular structure (Gilman et al., 1985).

Androgens *Androgen* is the generic name for a class of compounds of which *testosterone* is the most important (hence the two terms are often used synonymously). In the male, androgens are produced mainly by the testes, secreted by the interstitial Leydig's cells located between the seminiferous tubules (Chapter 2). Androgens are also produced in both sexes by the adrenal cortex, which is the main source of androgens in women, and in smaller quantities by the ovary (Murad and Haynes, 1985a).

Androgens are responsible for the sexual differentiation of the male reproductive system before birth (Chapter 2), the sexual maturation of males at puberty, and some of the secondary sexual characteristics of the female. They are also general *anabolic agents* that promote the building up of tissues. Testosterone is associated with the male sexual drive and possibly with aggressive behavior.

The effects of *castration* (removal of testes) in humans and animals have been known since antiquity. The ancient Egyptians and the Chinese castrated boys to produce *eunuchs*. Bulls are turned into tamer steers and stallions into geldings by castration.

Some of the earliest experiments in endocrinology were done in this area: in 1771 Hunter masculinized hens by transplanting testes from roosters. In 1849 Berthold showed that transplanting male gonads into castrated roosters prevented the typical effects of castration—the first formal experimental evidence for the existence of an endocrine gland. In the 1930s the testicular substance responsible for these effects was isolated and called testosterone (Gilman et al., 1985).

Estrogens and Progestins   The two principal classes of female sex hormones are *estrogens* and *progestins*. Though commonly referred to in the singular, there is no single hormone called "estrogen." Of the three main estrogens in humans, *estradiol* is the most potent. Similarly, *progesterone* is the most important of the progestins. However, in common usage, "estrogen" and "progesterone" have come to represent the two classes of female hormones.

Because the ovaries are not as accessible as the testes, their experimental removal (while keeping the animal alive) was not possible in former times. The fact that the ovaries control the female reproductive system with hormones was established only in 1900. Since the early 1920s an enormous amount of research has been undertaken with these substances, show-

ing their critical roles in reproduction and contraception (Murad and Haynes, 1985b).

Estrogens and progestins are secreted by the granulosa cells of the ovarian follicles (Chapter 2). Estrogens are produced while the follicles are maturing; progestins are produced after ovulation, when the follicle develops into the corpus luteum. During pregnancy the production of these hormones is taken over by the *placenta,* the organ through which the fetus is attached to the uterus and sustained until birth.

Estrogens do not seem to play a critical role in the differentiation of the female reproductive system before birth (Chapter 2), but they are responsible for most of the sexual maturational changes in girls at puberty. Estrogens and progestins regulate the menstrual cycle and are essential for reproduction. Estrogens are necessary for the implantation of the fertilized ovum and for sustaining the endometrium (the inner layer of the uterus) and the embryo. Progestins in turn provide optimal conditions for implantation and the initial growth of the fertilized ovum (Chapter 6). The relationship of these hormones to female sexual drive and behavior is unclear. Small amounts of estrogens are produced in the male by the testes and adrenal cortex but fulfill no known function.

In addition to the natural or *endogenous* forms of these hormones produced by the body, there are many *synthetic* products with the same properties. Estrogens and progestins can be taken orally or injected. They are the main constituents of birth control pills (Chapter 7). Estrogens are also readily absorbed through the skin and mucous membranes and are used, for instance, as creams in treating the vaginal dryness that affects some menopausal women (Box 4.1). Testosterone is much more effective when injected than when taken orally. Unlike the female hormones, its medical uses are limited. Gonadal hormones are inactivated in the liver and their metabolic by-products are excreted mainly in the urine.

Inhibin   The ovarian follicles and the Sertoli cells in the testes also produce a hormone

Box 4.1

THE MENOPAUSE

Our perception of midlife women has been dominated by the events of the *menopause* ("stopping of menses"). For unknown reasons, ovarian function declines markedly between the ages of 45 and 55 (average age, 51 years) leading to infertility and a sharp decrease in the production of estrogens and progestins. The alternative term *climacteric* (from Greek for "crisis") more broadly encompasses the many biological and psychological changes of this period.

In physiological terms, the source of change is the ovary and not the pituitary, which continues producing gonadotropins. The ovarian follicles stop responding to gonadotropin stimulation and no longer produce estrogen.

The most common physical symptoms of the menopause are *hot flashes* (or flushes). A sensation of warmth rises to the face from the upper body, with or without perspiration and chilliness. These flashes come at about hourly intervals over a period of a few months or several years. Other physical symptoms include dizziness, headaches, palpitations, and joint pains. The bones tend to lose calcium and become more porous (*osteoporosis*) following midlife, making postmenopausal women more liable to suffer fractures (London and Hammond, 1986). To keep her bones strong a woman should remain physically active, eat an adequate diet supplemented with calcium, and avoid excessive use of tobacco, alcohol, and caffeine.

Among the psychological symptoms of the menopause, sadness is the most prominent (experienced by about 40 percent). It may range from mild moodiness to severe depression. Other symptoms are tiredness, irritability, and forgetfulness.

For sexual function, the most important menopausal changes include loss of elasticity of the vaginal wall and thinning of the vaginal lining, with marked decrease of the lubricatory response during sexual excitement. The vaginal changes may lead to painful coitus, but the problem of the "dry" vagina is easily remedied by the use of lubricants or estrogen replacement therapy.

Older women may also develop a tendency toward burning during urination following intercourse, because the penis irritates the bladder and urethra through the thinner vaginal wall. Women who remain sexually active seem less likely to manifest some of these changes.

Following the menopause, sterility does not set in abruptly. There is a period of several years of menstrual irregularity and relative infertility. Pregnancies are rare beyond age 47, though some have been medically documented as late as 61 years.

The menopause is not a disease. Many women experience no significant ill effects. Among those that do, only one out of ten is markedly inconvenienced. Nonetheless, some women do experience physical distress. Furthermore, cultural and psychological factors can worsen the experience.

An effective treatment for some of the symptoms of the menopause is *estrogen replacement therapy* (ERT). When such therapy became available several decades ago it was enthusiastically hailed; then, with the cancer scare in the 1970s, estrogen use fell into disrepute. Present opinion is that estrogen replacement is a legitimate form of therapy when needed, but it should be used in the lowest effective doses, for the shortest possible period of time, and under medical supervision (Mosher and Whelan, 1981; Judd, 1987).

---

called *inhibin*, which regulates the secretion of the pituitary hormone FSH, to be discussed shortly (Steinberger and Steinberger, 1976; DeJong and Sharpe, 1976). There is increasing evidence that the gonads produce other complex chemicals (polypeptides), which also affect gonadal action (Ying et al., 1986).

Pituitary Hormones

The *pituitary gland* (or the *hypophysis*) is a small

structure at the base of the brain (Figure 4.1). Its multiple hormonal functions have earned it titles like "the master gland" and "conductor of the endocrine orchestra"—quite a feat for a pea-sized organ that weighs in at less than one gram.

The pituitary actually consists of two main parts: the *anterior pituitary* (or adenohypophysis) and the *posterior pituitary* (or neurohypophysis). We are concerned here mainly with the anterior pituitary, which produces six hormones, two of which, known as *gonadotropins,* control gonadal functions. The two gonadotropins are the *follicle-stimulating hormone* (FSH) and the *luteinizing hormone* (LH). Though named after ovarian structures (because they were first discovered in females), these two hormones are identical in males and females. The anterior pituitary also produces *prolactin,* which stimulates milk production in lactating women (and may suppress sexual function in the male at abnormally high levels) (Bancroft, 1983). A posterior pituitary hormone, *oxytocin,* stimulates milk letdown and uterine contractions and may have a role in sexuality (Carmichael et al., 1987).

The placenta produces *human chorionic gonadotropin* (hCG) during pregnancy, which fulfills the same basic functions as the pituitary gonadotropins. (Its role in the detection of pregnancy is discussed in Chapter 6.)

Unlike the steroid hormones, which are relatively simple chemicals, the gonadotropins are complex polypeptides (or proteins). The names of the gonadotropins are descriptive of their functions in the female: FSH stimulates the maturation of ovarian follicles and LH promotes the formation of the corpus luteum. In the male, FSH is responsible for sustaining and regulating sperm production within the seminiferous tubules; LH stimulates the interstitial Leydig's cells to produce testosterone.

Hypothalamic Hormones

The *hypothalamus* is a part of the brain closely linked to the pituitary. It functions both as a nervous center mediating various sexual functions (Chapter 3) and as an endocrine gland, sending its hormones to the pituitary. The hypothalamus itself is subject to complex influences from other parts of the brain.

Hypothalamic hormones control the levels of pituitary hormones by influencing not their production but their release; hence they are known as *releasing factors* (RF) or *releasing hormones* (RH). A single hormone, *gonadotropin-releasing hormone* (GnRH), controls the output of both FSH and LH by the anterior pituitary. How GnRH controls their release at different levels is not known.

Like pituitary hormones, the hypothalamic releasers are made up of complex molecules; yet they are relatively less complex. For example, GnRH is a chain of ten amino acids; LH has closer to 100.

Control Systems

In one sense, the gonadal hormones do the footwork in influencing the target organs, the hypothalamus is in charge of the headquarters, and the pituitary hormones act as intermediaries. Hormones flow down from the hypothalamus to the anterior pituitary and on to the ovaries or testes. However, this flow is not one-way. The gonadal and pituitary hormones in turn travel upwards, influencing the output of the "higher" sources. Figure 4.2 diagrams this process for testosterone.

We call this kind of model a *cybernetic system* ("helmsman") whereby the components control each other (Hafez, 1980; Guyton, 1986). A basic mechanism in such models is *feedback.* A common example of feedback is a household thermostat: when the temperature in a room falls below the level at which the thermostat has been set, a sensing device turns on the furnace, which produces heat; when the room temperature goes above the set point, the furnace is turned off. In this way, room temperature and furnace mutually control each other so that the temperature remains at a predetermined level.

The concentration of hormones in the bloodstream is like the room temperature, and the endocrine gland is like the furnace. As the body uses up hormones, their concentration in

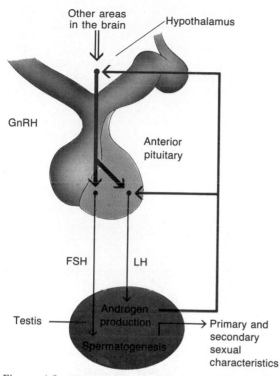

Figure 4.2   Hormonal control of gonadal function illustrated for the male.

the bloodstream falls below the physiological set point and more hormones are produced; as the hormone level goes up, its production is reduced. Later we shall look at the "thermostat" itself.

As Figure 4.2 shows, several feedback loops link the endocrine centers. *Long feedback* links up the gonads with the hypothalamus-pituitary complex. With testosterone as the example, we see that GnRH from the hypothalamus prompts the pituitary to release LH, which in turn increases the production of testosterone by the testes; higher levels of testosterone in the bloodstream then inhibit GnRH production through "negative feedback." When the level of testosterone falls, the cycle repeats. The suppression of FSH by the testes is actually due also to inhibin (Brobeck, 1979; Ramasharma and Sairam, 1982). *Short feedback* similarly links the hypothalamus with the pituitary.

We could draw almost the same diagram for the female. The same hormones are involved at the hypothalamic and pituitary level. The ovary produces estrogens in response to FSH and progestins in response to LH stimulation; both ovarian hormones in turn travel upward to inhibit the production of GnRH, FSH, and LH. However, this model of hormonal regulation cannot be applied as neatly to females, because hormonal levels vary markedly and cyclically during the monthly cycle.

The target organs and tissues are outside of these feedback loops. They are acted upon by gonadal hormones, which regulate their development and functions. Sexual functions, and presumably sexual behavior as well, are thus dependent on appropriate levels of these hormones.

Different hormones are constantly circulating in the bloodstream. How do various tissues tell one hormone from another? The ability of a given hormone to influence a particular tissue depends on *receptors* either within the cell or on the cell membrane. Hormones are like suitcases on luggage carousels in airports—they go round and round until their owner recognizes them and picks them up.

In Chapter 2, we discussed the effects of sex hormones on the growth and differentiation of the reproductive system during embryonal development. Now we can turn to the maturational changes that these hormones bring about during puberty and to their effects on sexual behavior.

PUBERTY

During the second decade of life your body was transformed into that of an adult. *Puberty* ("growth of hair") refers to the biological aspects and *adolescence* to the psychosocial side of this process. Maturation of the reproductive system and the development of secondary sexual characteristics are the core events in puberty, yet the changes during this phase are so pervasive that almost all tissues in the body are affected (Tanner, 1984).

The primary changes of puberty are: acceleration and then slowing down of the growth of the skeleton (the adolescent growth spurt); changes in body composition as a result of skeletal and muscular growth; changes in the quantity and distribution of body fat; developments in the circulatory system and musculature that lead to greater strength and endurance; development of the reproductive organs and secondary sex characteristics (Marshall and Tanner, 1974).

These changes have two major biological outcomes with profound psychosocial repercussions. First, the child attains the body of an adult, including the capacity to beget or bear children. Second, most of the physical sex differences between male and female, or *sexual dimorphism*, become established.

Somatic Changes

Puberty entails dramatic bodily, or somatic, changes. The first signs of puberty appear at about age 10 to 11 among girls and 11 to 12 among boys (Figure 4.3). There is, however, a wide range of normal timing. Boys and girls may normally enter puberty several years sooner or later than the average ages.

Some important changes of puberty are presented in Figures 4.4–4.7. Each of these developmental events has its own schedule and range of variability. Although the sequence of events shown in Figure 4.3 is generally consistent, it does not hold true in every case. For instance, the onset of menstruation is usually among the later events of puberty, but occasionally (and quite normally) it may be its first sign.

The pubescent *growth spurt* is among the most dramatic events in human development. Growth in height, or stature, is an ongoing process throughout childhood. By age ten boys have already attained 78 percent and girls 84 percent of their adult height. What makes the adolescent height spurt so striking is not the magnitude but the rate of growth. During the year of peak growth in height, a boy adds an average of three to five inches to his stature and a girl somewhat less (Tanner, 1984).

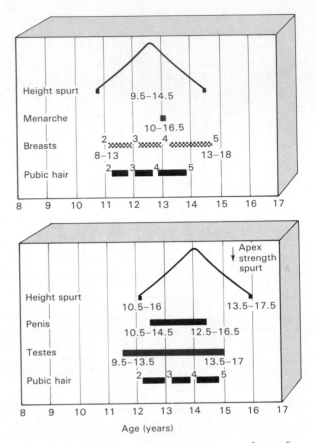

Figure 4.3 Timing of some events at puberty for an average Western girl (above) and boy (below). The normal range of ages for each event is given directly below it. Stages numbered 2–5 refer to the next three figures.

The weight gain during puberty follows a pattern similar to height gain, although it is less consistent. The factors that contribute to gain in weight are the increased size of the skeleton, muscles, and internal organs, and the greater amount of body fat. An important new sex difference is the amount and distribution of subcutaneous fat—the fat under the skin that contributes to shaping bodily contours. The average teenage girl enters adulthood with more subcutaneous fat than does the average teenage boy, especially in the region of the pelvis, the breasts, the upper back, and the backs of the upper arms. This fat accounts for the generally more rounded and softer contours of the female body.

There is a striking increase in the size and strength of the musculature at puberty in both sexes, but it is more pronounced for males. Among pubescent boys the increase in the number of muscle cells is fourteen-fold, among girls ten-fold. Similarly, among females maximum muscle cell size is reached by age 10 to 11, whereas in males, muscle cells continue to enlarge until the end of the third decade.

During puberty the body not only grows faster but undergoes marked changes in proportions. The various parts of the body grow at different rates. For example, legs accelerate in growth a year before the trunk, contributing to the stereotype of the gangling adolescent.

The face becomes distinctively adult as the profile becomes straighter, the nose and the jaw more prominent, and the lips fuller. These changes are more marked among males, whose facial appearance is further altered by the growth of facial hair and the recession of the hairline of the head.

Numerous internal changes accompany the external manifestations of puberty. The heart, like other muscles of the body, increases in size until its weight nearly doubles. Blood volume, hemoglobin, and the number of red blood cells are all increased. The same is true for lung size and respiratory capacity. All of these changes affect both sexes, but are more marked in the male.

The net effect of these and related physiological developments in puberty is to greatly increase the capacity of the adult for physical exertion. Greater exercise tolerance combined with superior strength permits individuals of both sexes to outperform vastly in physical effort their prepubescent selves.

Physical ability is both a function of biological endowment and the effect of exercise and training. There are no striking differences in the physical abilities of boys and girls before puberty. After puberty males have an advantage in overall muscular strength as well as in heart and lung functions. Nevertheless, the differences in physical ability we commonly observe between men and women are also to a significant extent the result of men typically using their bodies more strenuously in work and play. When women exercise to the same extent that men do, the difference in physical performance narrows considerably. This is true for ordinary activities as well as for peak performances by competitive athletes. For example, Olympic records in the 400-meter freestyle swimming events reveal that men were 16 percent faster than women in 1924, 11 percent faster in 1948, and only 7 percent faster in 1972. Both women and men have been improving their times throughout this period, but women have been improving faster. The female record in 1970 was faster than the male record in the mid-1950s (Wilmore, 1975, 1977).

Reproductive Maturation

The maturation of the reproductive system in both sexes is the primary mark of puberty. It involves the accelerated growth of the internal sex organs and the genitals, which we call the *primary sexual characteristics*. It also involves the development of *secondary sexual characteristics*— the female breasts, male facial hair, pubic and axillary hair in both sexes, and voice changes (Savage and Evans, 1984).

Female Maturation Breast development is usually the first visible sign of female puberty (Figure 4.4). It usually starts between the ages of 8 and 13 and is completed between 13 and 18. Sometimes the two breasts develop at different rates. This unevenness need not be a source of concern: the breast growing more slowly will eventually catch up, and the asymmetry is usually corrected by the end of adolescence.

Pubic hair usually appears next, attaining the adult pattern by about age 18. It precedes the growth of hair under the armpits (axillary hair) and on the legs by about one year. Both breast and pubic hair growth follow predictable patterns, which are useful for monitoring the rate of pubertal development (Figure 4.5).

The musculature of the uterus develops markedly during puberty. The vagina enlarges

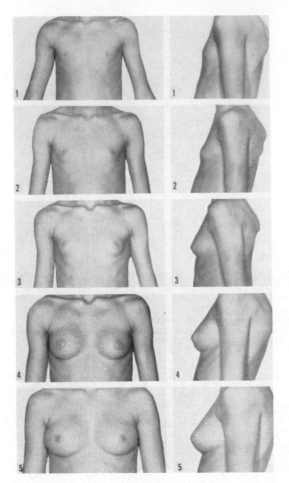

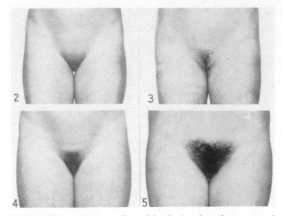

Figure 4.5   Stages of pubic hair development in adolescent girls: (1) prepubertal stage (not shown), with no true pubic hair; (2) sparse growth of downy hair, mainly at sides of labia; (3) pigmentation, coarsening, and curling with an increase in the amount of hair; (4) adult hair, but limited in area; (5) adult hair with horizontal upper border.

Figure 4.4   Stages of breast development in adolescent girls: (1) prepubertal flat·appearance, like that of a child; (2) small, raised breast bud; (3) general enlargement and raising of breast and areola; (4) areola and papilla (nipple) form contour separate from that of breast; (5) adult breast with areola and breast in same contour.

and its inner walls become thicker and more furrowed. The external genitals, including the clitoris, become enlarged and their erotic sensitivity heightened. The most fundamental change of all involves the ovaries, where the ovulatory cycle becomes activated, beginning the menstrual cycle, which we shall discuss later.

Male Maturation   The onset of puberty in males is marked by the enlargement of the testes, starting between the ages of 10 and 13 and continuing until the ages of 14 to 18. The development of pubic hair that occurs between 12 and 16 anticipates by two years or so the growth of hair on the face and the armpits. The first ejaculation usually occurs at about the age of 11 or 12, but mature sperm usually take a few more years to appear. Like girls, pubescent boys are usually not fully fertile, but that does not provide them with contraceptive security if they engage in sex.

The penis begins begins to grow markedly about a year after the onset of testicular and pubic hair development (Figure 4.6). Deepening of the voice, which results from the enlargement of the larynx, is a late event, but an important one for the adolescent boy's sense of masculinity. Girls experience a similar change, but it is much less marked. Similarly, boys experience some breast enlargement, which may alarm them, but usually it is slight and eventually it disappears.

Another distressing event of puberty may

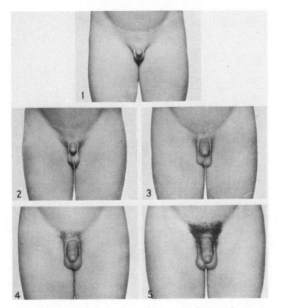

Figure 4.6   Stages of male genital development: (1) prepubertal stage, in which the size of the testes and penis is similar to that in early childhood; (2) testes become larger and scrotal skin reddens and coarsens; (3) continuation of stage 2, with lengthening of penis; (4) penis enlarges in general size, and scrotal skin becomes pigmented; (5) adult genitalia.

be the appearance of acne, a transient skin condition that is more common among boys than girls and is related to the effects of androgen. Though usually no more than a minor cosmetic problem, acne can be severe enough to require medical treatment (Johnson, 1985).

As the reproductive system matures, pubescent boys begin to experience erections (nocturnal penile tumescence) and orgasm (nocturnal emission) in their sleep, and girls may manifest vaginal lubrication under similar conditions.

Neuroendocrine Control

All these changes of puberty are triggered and regulated by nervous and hormonal mechanisms. Because they are closely integrated, we refer to them as *neuroendocrine mechanisms*.

They involve the interaction of hormones from the hypothalamus, the anterior pituitary, the gonads, and the adrenal cortex. Their effects are illustrated in Figure 4.7.

Although the nature and actions of these hormones are now well known, the precise mechanism that initiates puberty remains unclear. The hypothalamic-pituitary-gonadal system can function before puberty. The pituitary, the gonads, and the target tissues, such as the breasts, can be stimulated to develop at any time if the person is given the appropriate hormones. The onset of puberty must be restrained by the brain, either at the hypothalamus or at some higher level (Guyton, 1986).

Before puberty, low levels of gonadal hormones are already circulating in the bloodstream. It is assumed that a hypothalamic center operates like a *gonadostat* (a term coined in analogy to a thermostat). This gonadostat is highly sensitive in childhood, so even the low levels of steroid hormones in circulation would keep it turned off through negative feedback and inhibit the production of GnRH. As the child matures, the hypothalamic gonadostat becomes less sensitive. It is no longer turned off by the low levels of circulating gonadal hormones. As a result, the hypothalamus produces more GnRH, prompting the pituitary to produce larger amounts of gonadotropin, which in turn increases gonadal hormonal output. Mounting levels of gonadal hormones finally reach a level that induces the tissues of the body to respond, and sets the physical changes of puberty in motion (Grumbach et al., 1974; Grumbach, 1980).

Other theories propose alternative processes. For instance, Frisch (1974) believes that puberty is triggered by reaching a critical body weight. Because children grow at somewhat different rates, they attain the critical weight level and enter puberty at different ages. Others ascribe the initiating function to the adrenal gland.

THE MENSTRUAL CYCLE

The *menstrual cycle* is one of the key physiological functions of the female body. An ovarian

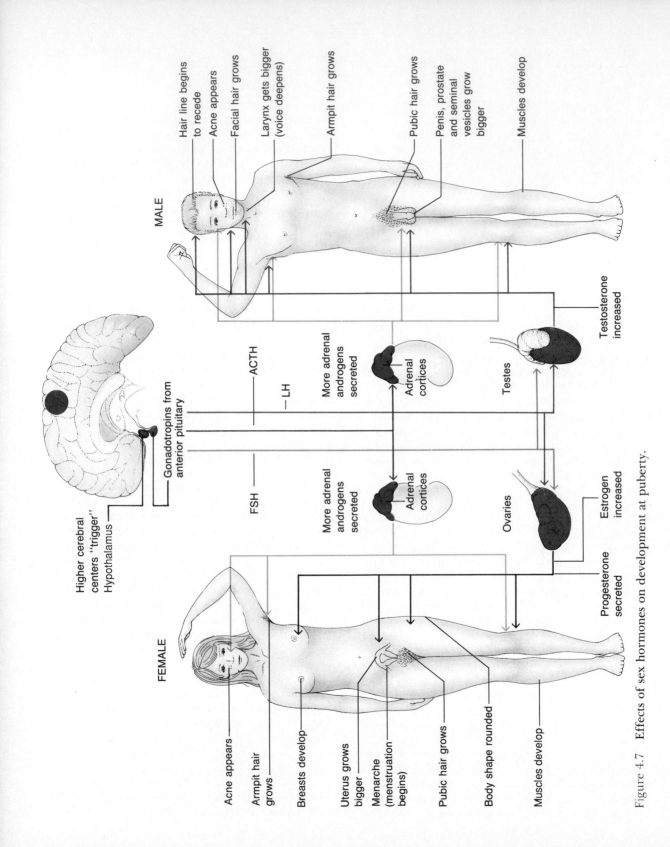

Figure 4.7   Effects of sex hormones on development at puberty.

cycle, or *estrus cycle*, is characteristic of all mammals, but *menstruation*, the periodic shedding of the uterine endometrium—"bleeding"—that accompanies the ovarian cycle, exists only in women, female apes, and some monkeys. The significance of menstruation in women goes beyond physiology; it has important psychological and social ramifications.

Menarche

The onset of menstrual cycles in puberty is called *menarche;* the cessation of menstrual cycles in midlife is the *menopause* (Box 4.1). Both are highly significant biological landmarks in a woman's life, with many psychological and social implications, which are determined largely by cultural attitudes.

The onset of menarche is determined by a variety of genetic and environmental factors. In the United States, menarche now occurs at the average age of 12.8 years, with a normal range of 9 to 18 years (Zacharias et al., 1976). Information from around the world shows considerable variation in this regard. For instance, among girls in Cuba the median age of menarche is 12.4 years, whereas among the Bundi tribes of New Guinea it is 18.8 years (Hiernaux, 1968).

Among industrialized societies the average age at menarche has gradually declined from 17 years in 1840 to the current levels, which stabilized a few decades ago. Teenage girls therefore now reach menarche at about the same ages that their mothers did.

The decline in the age of menarche, as well as cross-cultural differences, are generally attributed to environmental factors, such as better nutrition and improved health care. Nonetheless, the force of genetic factors is evident in family tendencies for the onset of menarche. For instance, randomly chosen girls reach menarche differing on the average by 19 months; for sisters who are not twins, the difference is 13 months; for nonidentical twins, it is 10 months; for identical twins, 2.8 months (Tanner, 1978).

Following menarche, it usually takes a few years before menstrual cycles become regular-

ized (they become irregular again during the menopause before stopping altogether). During puberty, when the menstrual cycle is becoming established, ovulation tends to be inconsistent. *Anovulatory cycles* (cycles without ovulation) make it less likely that a teenager will get pregnant, but this relative adolescent sterility is highly unreliable as a birth control method. Millions of teenagers do get pregnant.

Phases of the Menstrual Cycle

The length of the ovarian cycle is specific for each species. It is approximately 36 days in the chimpanzee, 20 days in the cow, 16 days in sheep, and 5 days in mice. Dogs and cats are seasonal breeders and usually ovulate only twice a year.

The average length of the human menstrual cycle is 28 days (hence its association with the lunar month and the derivation of "menstrual" from the Latin for "monthly"). Cycles that are shorter or longer by several days are also perfectly normal. Because the onset of menstrual bleeding is more abrupt than its gradual end, the time that bleeding starts is counted as day 1 of the menstrual cycle.

The menstrual cycle is a continuous process, one cycle following another. Once the cycles have become regularized, most women go through their menstrual periods on a fairly predictable rhythm. However, this rhythm is often influenced by physiological and psychological fators, so it is common for a woman's period to come a few days early or late.

The menstrual cycle is controlled by the hypothalamic, pituitary, and gonadal hormones. It involves the entire reproductive tract, especially the ovaries, uterus, and vagina. Though the menstrual cycle is one continuous process, we describe it in four phases: the *preovulatory phase, ovulation,* the *postovulatory phase,* and the *menstrual phase.*

Preovulatory Phase   The *preovulatory phase* starts as soon as menstrual bleeding from the previous cycle has ended. This period is also known as the *follicular phase,* because the ovarian follicles in the ovary develop at this time, and as the *proliferative phase,* because of the changes in the uterine lining.

As illustrated in Figure 4.8, while menstrual bleeding is in progress, the anterior pituitary increases its production of FSH. As a result, the production of estrogens from the ovaries begins to increase sharply. Higher levels of estrogen cause thickening of the endometrium with proliferation of its superficial blood vessels and uterine glands. The cervical glands produce a thick and cloudy mucous discharge, which gradually takes on a watery character. Simultaneously the lining of the vagina thickens, and its cells undergo characteristic changes. These changes are so distinctive that examination of vaginal cells, cervical mucus, and the endometrial lining is used in tests of ovarian hormonal function.

Increasing levels of estrogen in the bloodstream gradually reduce the producton of FSH through negative feedback. The production of LH and of progesterone remain low throughout the cycle's preovulatory phase.

Ovulation   The central event in the ovarian cycle is *ovulation,* which takes place at about the midpoint of a 28-day cycle. Ovulation always occurs approximately 14 days before the onset of menstruation, no matter how long the cycle. Differences in the length of menstrual cycles are due to the length of the preovulatory phase only.

Ovulation is triggered by an upsurge in LH secretion and to a lesser extent by a rise in FSH secretion. Both changes are thought to be caused by the rapid increase in the levels of estrogen, which act in a "positive feedback" relationship. Unlike the negative feedback model discussed earlier, in this case increasing levels of estrogen do not inhibit but enhance the production of the pituitary hormones. FSH and LH act synergistically (that is, they enhance each other's actions) to cause rapid swelling of the ovarian follicle. Simultaneously, there is a reduction in the secretion of estrogens, and an increase in the production of progesterone. All of these changes jointly trigger ovulation (Guyton, 1986).

Some women experience mild pain in midcycle during ovulation (called *Mittelschmerz,* German for "middle pain"). This symptom is most common in young women and consists of intermittent cramping pains on one or both sides of the lower abdomen that last for about a day.

Postovulatory Phase   As the dischanged ovum starts on its journey through the fallopian tube, the menstrual cycle enters its *postovulatory phase.* Under the influence of high levels of LH and facilitated by the pituitary hormone prolactin, the corpus luteum is formed, leading to a sharp increase in the secretion of progestins. The production of estrogens, which had dropped sharply at the time of ovulation, now begins to climb again to a higher level. Because it is now the corpus luteum rather than the developing follicle that is the main source of hormones, the postovulatory phase is also known as the *luteal phase.*

The sustained level of estrogens continues to act on the uterine endometrium, making it progressively thicker. Under the influence of progestins the uterine glands now become active and secrete a nutrient fluid; hence *secretory phase* is another name for the postovulatory period. Secretions of cervical mucus gradually thicken and regain their cloudy and sticky consistency. Since the changes in cervical mucus are linked to the phases of the menstrual cycle, they are relied on in the "rhythm" method of birth control (Chapter 6).

High levels of estrogens and progestins in the bloodstream inhibit gonadotropin production by the anterior pituitary, which results in gradually declining levels of FSH and LH. In addition, the corpus luteum secretes moderate amounts of the hormone *inhibin* (the same as the inhibin secreted by the Sertoli cells of the testes); inhibin further reduces the secretion of FSH and LH. This reduction in turn means less stimulation of the corpus luteum and less production of estrogens and progestins. These hormones are being constantly used up, so their blood levels gradually decline as well. Without the sustenance provided by these hor-

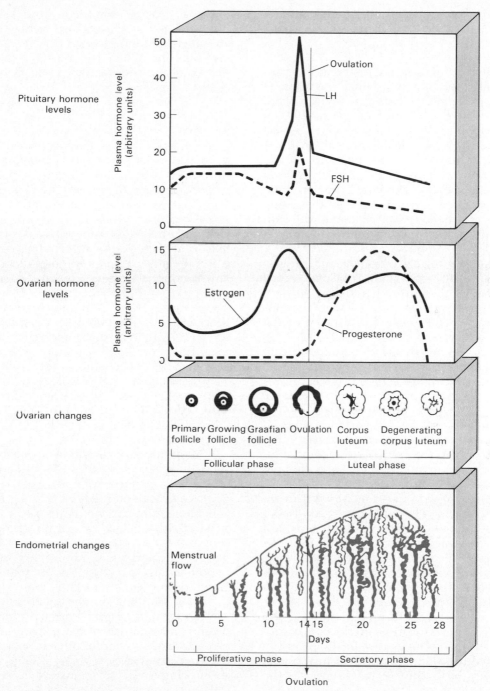

The menstral cycle.

mones, the endometrial lining degenerates and sloughs off. This material is the menstrual discharge.

Menstrual Phase   Discharge of the uterine lining, which appears as vaginal bleeding, is called *menstruation*. It lasts four or five days, during which a woman loses 50 to 200 ml (about half a cup) of blood; however, the amount varies a good deal between individuals and in different cycles. Normally this blood loss is rapidly replenished, and there are no ill effects whatsoever. However, women who menstruate need sufficient iron in their diet, and some may need to supplement it with iron compounds to help the production of red blood cells.

As blood levels of estrogens and progestins go down during the menstrual phase, they stop inhibiting the anterior pituitary. As a result, FSH production rapidly picks up, heralding the start of the proliferative phase of the next cycle.

The foregoing description supposes that the woman has not become pregnant. If she has, the cells of the developing placenta produce *chorionic gonadotropin*, which has the same effect as pituitary LH on the corpus luteum in maintaining high levels of hormone production. Under these conditions the uterine lining (which now holds the embryo) does not slough off, and the pregnant woman misses her period for the first time (Chapter 6).

A similar process explains how birth control pills function. The pill typically consists of some combination of synthetic estrogens and progestins. These hormones act the same way as their natural counterparts produced by the ovary. Taken on a daily basis, they maintain high blood levels of these hormones. The anterior pituitary has no way of telling if these hormones are coming from the ovary or the pharmacy. They suppress production of gonadotropins through negative feedback, just as they do naturally in the postovulatory phase. Because the birth control pill suppresses the pituitary before ovulation, there is no LH surge, no ovulation, and no chance to become

pregnant. When a woman stops taking the hormone pill at the end of each month, the endometrial lining sloughs off and menstruation follows. When she resumes taking the pill, another anovulatory cycle is repeated (Chapter 7).

Menstrual Discomfort

The menstrual cycle is subject to a variety of disturbances beyond normal fluctuations in its timing, duration, and amount of flow. They include absence of menstruation (*amenorrhea*) and increased amount or duration of menstrual bleeding (*menorrhagia*). Also, for a significant number of women, menstruation entails some pain or discomfort. This discomfort is usually mild, but in a number of cases (usually among younger women) it is severe enough to necessitate bed rest.

Some of these conditions are clear-cut: they have specific causes and symptoms and interfere with normal reproductive functioning. Other conditions are more ambiguous. A case in point is the wide variety of sensations some women experience before or during the menstrual period. These women are not ill. The process they are going through is physiologically normal; yet some of them feel acutely uncomfortable.

Many physical conditions can account for such menstrual disorders. So can emotional factors. Sometimes even an experience like going to college or going back home from college is enough to disrupt temporarily the menstrual rhythm. Following unprotected intercourse the fear of pregnancy may also cause delayed menstruation. The reason is that the hypothalamus takes part in the regulation of emotions and endocrine functions.

The symptoms of menstrual-cycle distress are categorized as premenstrual tension syndrome and dysmenorrhea (painful menstruation). Some 150 symptoms have been linked to the menstrual cycle (Moos, 1969). Premenstrual symptoms are manisfested during the week preceding the menstrual flow; the symptoms of dysmenorrhea accompany menstruation itself. Though there may be some overlap in discomfort, these two conditions should be dealt with separately. Their manifestations, probable causes, and treatments differ.

Dysmenorrhea   Menstrual discomfort of varying severity affects half of all women in early adult life. Typically they have cramps in the lower abdomen, backache, and aches in the thighs. Less often, there may be nausea, vomiting, diarrhea, headache, and loss of appetite. When these symptoms become severe enough to interfere with work or school, and last two days or more, the woman can be said to suffer from *dysmenorrhea.*

If menstrual pain is due to pelvic disease, it is called *secondary* dysmenorrhea. Where there is no apparent illness causing it, then it is *primary* dysmenorrhea.

The menstrual cramps of primary dysmenorrhea come from spasms of the uterine muscles. Chemical substances called *prostaglandins* cause the spasms. In women they are normally released from the endometrial lining, shed during menstruation. Menstrual blood has four times the level of prostaglandins in ordinary uterine blood.

Prostaglandins were first isolated from prostatic fluid (Chapter 2), but they are a family of compounds, widely distributed in various tissues of the body, with a broad range of actions. These actions include contraction of the smooth muscles of the uterus (hence their use in inducing labor); various effects on the gastrointestinal tract (causing diarrhea, cramps, nausea, and vomiting); dilation of blood vessels (causing hot flashes); and other changes in the endocrine and nervous systems of the body.

The time-honored, though not the best, remedy for dysmenorrhea is aspirin with bed rest and warm fluids. Aspirin is an *analgesic* (pain reliever). It also belongs to a class of compounds called *antiprostaglandins,* which can prevent as well as treat dysmenorrhea by inhibiting the synthesis and the actions of the prostaglandins. Other compounds are even more effective antiprostaglandins. For instance, among over-the-counter drugs ibupro-

fen (marketed as Advil, Nuprin, and other labels) is most effective, especially in its stronger form (Motrin), which must be prescribed. By contrast, pain killers such as acetominophen (Tylenol) do not have an antiprostaglandin action; so they are less useful for treating dysmenorrhea. Other substances like vitamin B6 may diminish the reaction to prostaglandins by relaxing the uterine muscles.

Women on the pill are also likely to get relief. Steroid hormones reduce the amount of endometrial sloughing and hence the level of prostaglandins released in the process. However, a woman should not put herself on the pill for this purpose only.

A similar caution applies to *menstrual extraction,* especially when practiced by self-help groups. This procedure is the same as the vacuum aspiration method of abortion (Chapter 7)—suctioning out the uterine contents. This so-called "five-minute period" may not be a safe procedure, particularly for women with a history of uterine infections, tumors, and a number of other conditions.

Premenstrual Syndrome   The symptoms of the *premenstrual syndrome* (PMS) are more varied and less distinct than those of dysmenorrhea. No explanation of them has been widely accepted.

The symptoms of premenstrual tension appear two to ten days before the onset of menstrual flow and usually subside by the time the menses start. In some cases, these symptoms merge with dysmenorrhea, but either can occur alone.

The more common symptoms of premenstrual tension can be grouped in three categories. First are symptoms associated with *edema* (swelling): a bloated feeling in the abdomen, swelling of the fingers and legs, swelling and tenderness of the breasts, and weight gain. Second is headache. Third is emotional instability, including anxiety, irritability, outbursts of anger, depression, lethargy, insomnia, and cognitive changes such as difficulty concentrating and forgetfulness. More idiosyncratic reactions include changes in eating habits (such as a craving for sweets), excessive thirst, and shifts in sex drive (increase for some, decrease for others) (Rubinow, 1984). The implications of these changes are discussed in Box 4.2.

There has been much confusion in the assessment of these symptoms and their causes. To begin with, it is unclear what proportion of women actually suffer from such symptoms: estimates range from 30 percent to 90 percent. Moreover, these symptoms tend to be highly subjective: though most women perceive them as unpleasant, other women experience bursts of energy and creative activity at this time. The symptoms of edema are ascribed to fluid retention, but recent studies have failed to document significant weight gain in the premenstrual period. The swelling must be explained by internal shifts of fluid from one part of the body to another. Finally, in studies where women with premenstrual tension believed that their mental or physical performance was impaired, tests and other objective measures have failed to substantiate these self-perceptions.

Does this mean that PMS is a myth? Do women imagine these symptoms? The experiences of countless women testify to the contrary. The symptoms can be considered so far to be the standard criterion for defining PMS (Abplanalp et al., 1980).

A variety of physiological causes have been proposed for premenstrual symptoms: the drop in *progesterone* level late in the postovulatory phase; *vitamin B-complex* deficiencies causing decreased liver metabolism, hence higher levels of estrogen; increased *aldosterone* (a steroid hormone secreted by the adrenal cortex); and abnormalities in *endorphin* production, causing estrogen-progesterone imbalances, and disturbances in the synthesis of neurotransmitters (Debrovner, 1983; Sondheimer, 1985).

Futher evidence pointing to the physiological roots of the syndrome comes from the study of female baboons, who are reported to become less social during the premenstrual period and eat more (Hausfater and Skoblick, 1985).

The likelihood of experiencing menstrual distress also seems to be linked to psychological or cultural factors, such as religious background. Among college women, Catholics and Orthodox Jews tend to have a higher prevalence of menstrual distress (Paige, 1973). Negative attitudes toward menstruation and the expectation of menstrual discomfort may also create a self-fulfilling prophecy. For instance, in one study, women who were led to believe that their menstrual periods were due in a few days were more likely to experience premenstrual distress than other women who had been persuaded by the experimenter not to expect their period until quite a bit later (Ruble, 1977). Other attempts to establish a psychological origin for premenstrual tension have linked it to personality types and life circumstances (Kinch, 1979). Negative associations that society or an individual attaches to menstruation have received special attention. High levels of stress, such as family conflicts or even college exams, tend to accentuate premenstrual tension (Rubin et al., 1981).

Premenstrual tension may be helped by a low-salt diet during the week before the period, or by taking diuretics, both of which can counter fluid retention. Coffee, tea, cola drinks that contain caffeine, and sweets tend to worsen the condition. If symptoms of physical discomfort and headache are severe enough, analgesics may be helpful. One must be careful, however, not to substitute one problem for another by becoming dependent on drugs in dealing with the ordinary problems of menstrual distress.

There have also been claims of relief through the administration of progesterone and vitamin B-complex, to correct a deficiency that presumably causes the premenstrual symptoms (Freeman, 1985). On the other hand, these symptoms seem to be also relieved in a considerable proportion of cases by *placebos*, chemically inert substances which work because patients expect them to. Current treatments of PMS therefore emphasize stress reduction and emotional support. Finally, some women find that orgasm (through any means) relieves both premenstrual tension and men-

Box 4.2

BEHAVIOR AND PMS

Some women claim that a variety of changes affect their thoughts, feelings, and actions in the premenstrual period. Cognitive changes include forgetfulness and difficulty in concentration; emotional changes include moodiness (especially depression and apathy) and instability (anxiety, irritability, anger) (Goldstein et al., 1983; Andersch et al., 1986). Long before PMS was acknowledged, it was observed that some women acted out of character about the time of their menstrual periods. Queen Victoria reportedly suffered from fierce tempers during her premenstrual period. "Even (Lord) Melbourne, a past master at dealing with women, had on one occasion quavered and feared to sit down as the fire blazed in the eyes of the eighteen-year-old queen" (quoted in Dalton, 1979, p. xiii).

Premenstrual distress is thought to be so overpowering that in some countries, such as France and England, it constitutes a mitigating circumstance in violent crimes. In 1981, a 33-year-old woman in Britain was found guilty of murdering her lover following a quarrel. The verdict: manslaughter due to diminished responsibility owing to the premenstrual syndrome. The woman was placed on three years' probation and released (Laws, 1983).

Do a significant proportion of women show significant behavioral changes at these periods? How should they be dealt with? At present, there are no convincing answers, but sharply divided opinions.

In Britain, Dalton (1969, 1979) has been among the chief proponents of the view that PMS is a clinical entity with distinct behavioral manifestations. She considers the four days before and the four days after the onset of menses (the *paramenstruum*) to be a particularly stressful time for women (Dalton, 1972). Statistics show higher rates of accidents, psychiatric admissions for acute illness, and attempted suicides during the paramenstruum. Similarly, this period has been associated with significantly higher rates of behaviors leading to imprisonment and absenteeism from work (Dalton,

1979). It follows from this perspective that the paramenstruum should be recognized as a time of higher risk and distress for women. Therefore, although it is part of a normal physiological process, serious attention should be paid to the anticipation of paramenstrual stress and the relief and management of its symptoms. Furthermore, it is argued that allowances must be made for the behaviors of women during these periods when their control of their actions is diminished.

The opposing view questions the very basis of the reported association of behavioral symptoms with the menstrual cycle (Rubinow et al., 1986; Ghadirian and Kamaraju, 1987). For one thing, the studies that relate abnormal behaviors to PMS are full of methodological pitfalls. They draw conclusions about hormonal effects without even measuring hormone levels. Systematic attempts to relate levels of estrogen and progestins to moods and daily activities have failed to show any correlation (Abplanalp et al., 1980). Although some women may well be adversely affected during the paramenstruum, their behavioral aberrations may have been highly exaggerated. There is no good evidence that the mental and physical capabilities of women are generally compromised during the paramenstruum. To make allowances or to recognize diminished responsibility for actions committed by women during this period casts an aura of unreliability on all women, and creates one more excuse to discriminate against them in various occupational and social settings (Reid, 1986).

To resolve this dilemma we need to rethink the concept of PMS and to reexamine its purported manifestations (Fausto-Sterling, 1985). Unless there is some agreement of what it is we are talking about, we can hardly make progress about its management. Meanwhile, women who experience genuine distress must not be burdened with guilt, and women who are undisturbed by PMS must not be assumed to be in distress.

strual cramps (Budoff, 1980; Holt and Weber, 1982).

ATYPICAL SEXUAL DEVELOPMENT

We have been looking at normal sexual development, but a lot can be learned from abnormalities. A good deal of research on gender identity and gender-related behavior has been based on conditions caused by hormonal disturbances. Let us consider some representative examples.

Normally the reproductive system differentiates in the embryo (Chapter 2) and matures at puberty. Rarely, abnormalities in sexual differentiation occur at the level of the chromosomes, the hormones (hypothalamic, pituitary, or gonadal), or the target tissues.

The developmental variation may be unusual timing or faulty differentiation. When sexual development occurs unusually early, the result is *precocious puberty* (Box 4.3); if late, it leads to *delayed puberty*. Failure of puberty to occur altogether results in the inadequate development of sex organs; it is called *sexual infantilism*. New procedures now make the early diagnosis and treatment of these conditions more likely (Ortner et al., 1987).

Faulty sexual differentiation creates incongruous combinations of male and female structures. Individuals whose genitals are mismatched with the rest of their body have long been known since antiquity as *hermaphrodites* (a term derived from the Greek gods *Hermes* and *Aphrodite*).

The true hermaphrodite has both male and female gonads, or a mixture of ovarian and testicular tissue in the gonads. Though usually a genetic female (XX), such a person has external genitals that look predominantly male or female (Figure 4.9) or may combine features of both sexes, with one form predominating. Where the sex chromosomes are correctly matched with male or female gonads but mismatched with external genitals, the term *pseudohermaphrodite* is applied. A female pseudohermaphrodite will have female sex chromosomes (XX), ovaries, fallopian tubes, and

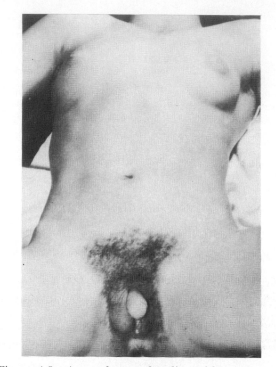

Figure 4.9   A true hermaphrodite, with one ovary and one testis. A genetic female who has always lived as a male.

uterus, but external genitals that appear to be male. A male pseudohermaphrodite will have male sex chromosomes (XY), testes, and other male structures, but female external genitals and even a feminine body build (Figure 4.10). There are many variants of these abnormalities and numerous causes (Imperato-McGinley, 1985).

Chromosome Disorders

Instead of the normal complement of 46 chromosomes, including XX or XY chromosomes, some individuals are born with extra or missing sex chromosomes. *Klinefelter's syndrome* results when a male has an extra X chromosome (XXY); *Turner's syndrome* results when a female is missing one X chromosome (XO) (Grumbach and Conte, 1985).

In Klinefelter's syndrome masculinization is incomplete, resulting in a small penis, small

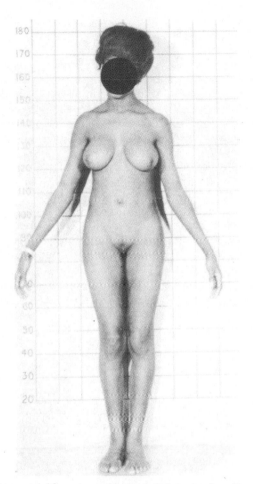

Figure 4.10   A male pseudohermaphrodite. A genetic male with the androgen insensitivity syndrome.

testes, low testosterone production, and therefore incomplete development of secondary sex traits. Some men show partial breast development at puberty. These men are infertile. Some have problems with social adaptation, perhaps because of their lower average intelligence (Federman, 1968), but others lead full lives.

Individuals with Turner's syndrome have a female body build, but ovaries are absent or rudimentary, incapable of producing ova or female hormones. Therefore, the XO female is infertile and does not undergo puberty unless treated with female sex hormones. These women are of short stature and may have congenital organ defects, and webbing between the fingers and toes or between the neck and shoulders. Despite incomplete development of the female organs, XO females usually have no gender problems during psychosexual development.

Another chromosomal abnormality with intriguing behavioral consequences is the presence of an extra Y chromosome in males (Hamerton, 1988). XYY individuals tend to be tall, have lower than average intelligence, and suffer from severe acne. The XYY syndrome has attained some notoriety because of its higher frequency among prisoners, implying greater aggressivity (Jacobs et al., 1965). Yet when a random sample of men outside of prison was examined, violence did not appear to be higher in XYY than in XY males. The validity of the association between an extra Y chromosome and antisocial sexual behavior is therefore doubtful (Witkin et al., 1976).

Hormone Disorders

Disorders of differentiation due to chromosomes usually affect people by causing hormonal abnormalities. Such abnormalities can occur too even when sex chromosomes are normal. Hormones may be overactive or underactive at any level of the hypothalamic-pituitary-gonadal system of the embryo; or abnormal sexual differentiation may be caused by hormones produced or ingested by the mother during pregnancy.

In the female, hormones cause pseudohermaphroditism most commonly in *congenital adrenal hyperplasia* (CAH). This condition (also known as the *adrenogenital syndrome*) results from a genetic defect. Too much androgen is produced by the fetal adrenal cortex (Behrman and Vaughan, 1983). In a male child, this defect causes precocious puberty. The female infant is born with her external genitalia masculinized to various degrees. The clitoris may look like a penis, and the labial folds may be fused together, giving the appearance of a

Box 4.3

PRECOCIOUS PUBERTY

When puberty begins before age eight in girls and age ten in boys, it is considered precocious. Girls are twice as likely to have precocious puberty as boys. There are cases of menstruation beginning in the first year of life. The youngest known mother was a Peruvian girl who began menstruating at three years and gave birth (by Caesarean section) to a baby boy at the age of 5 years 7 months (see the figure) (Wilkins et al., 1965). Among precocious boys, penile development may begin at five months and spermatogenesis at five years (see figure). A seven-year-old boy was reported in the 19th century to have fathered a child (Reichlin, 1963).

In the majority of cases, especially in girls, precocious puberty does not reflect any underlying pathology; it is simply a natural variation in the body's timing mechanisms. However, in 20 percent of girls and in 60 percent of boys precocious puberty is caused by a serious underlying disease. For instance, tumors in the hypothalamic region of the brain, the gonads, or the adrenals can trigger the precocious production of gonadotropins or sex hormones, which in turn lead to the early sexual maturation of the body. These cases clearly show that puberty depends on the maturation of the hormonal system and not of the body tissues; the reproductive system is ready to mature at any time.

In some cases, a child may undergo incomplete precocious puberty, with early breast development or early growth of pubic hair, but no other changes. These cases are less likely to be related to an underlying illness. Precocious puberty is not accompanied by the heightened erotic and romantic interest typical of normal puberty. In other words "erotic age" does not automatically match physical development (Money and Ehrhardt, 1972).

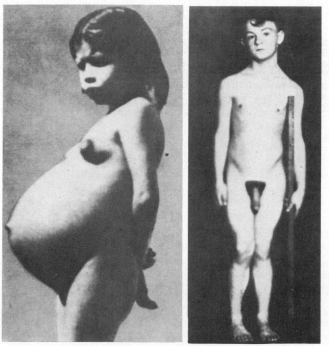

(Left) Linda Medina is the youngest known mother in the medical literature. She gave birth in 1939 at the age of 5 years 7 months. (Right) A five-year old boy with unusual adrenal development has the height of an 11.5-year-old and precocious penile development. (From Wilkins, Blizzard and Midgeon 1965 *The Diagnosis and Treatment of Endocrine Disorders in Childhood and Adolescence*. Courtesy of Charles C. Thomas, Publisher, Springfield, Illinois.)

scrotum. The condition is treated by suppressing the excessive androgen production and surgically correcting the genitals within the first few weeks of life.

The same end result will occur in a perfectly normal female embryo if the pregnant mother produces abnormal amounts of androgen (for example, due to an androgen-producing tumor) or takes androgenic hormones. Some years ago pregnant women were given certain synthetic steroids to prevent miscarriage (before the masculinizing effects of these compounds were known), and they gave birth to female babies with masculinized genitals.

Male pseudohermaphroditism results if there is not enough androgen before birth, or if the hormones produced are not biologically active. In the absence of normal male hormone, genetically male embryos will develop female genitals.

Tissue Disorders

Even where the hormonal system is quite normal, abnormalities will result if the bodily tissues do not respond normally to hormonal stimulation. Such a condition, known as *androgen insensitivity* or the *testicular feminization syndrome*, is the most common cause of male pseudohermaphroditism. In this case, a genetic male with androgen-producing normal testes will develop a female appearance with female genitals and breasts, because body tissues fail to respond to testosterone (Figure 4.10). The net effect is therefore the same as if no androgen were present. Because the body does respond to the Mullerian duct-inhibiting substance, no uterus or fallopian tubes develop. Genetically male infants with this syndrome look like normal baby girls. The condition is usually diagnosed at puberty when menarche fails to occur.

Unlike other forms of pseudohermaphroditism, which can be treated with surgery and hormones if detected early, people with complete testicular feminization must be reared as female. Although sex-change surgery and hormonal treatment could reverse the established

female apearance, these individuals have female gender identities (think of themselves as women) and do not desire such change.

In conditions like *transsexualism,* the person's sense of gender identity (being male or female) is in conflict with his or her anatomical sex without demonstrable genetic or hormonal abnormalities.

HORMONES AND SEXUAL BEHAVIOR

How much of sexual behavior is controlled by hormones? Over the past half century a great deal of evidence has been gathered to demonstrate the role that hormones play in shaping and sustaining mammalian sexual behavior (Davidson et al., 1982). The extent to which this influence persists among humans is not yet fully established, but it seems certain that sex hormones are also significantly linked to some aspects of human sexual behavior.

Mammalian Behavior

Our basic knowledge of the effects of steroid hormones on sexual behavior comes from research with rodents like the rat, and to a lesser extent with nonhuman primates such as chimpanzees, gorillas, and orangutans (Beach, 1971). These animals have predictable and gender-specific sexual behaviors that are mainly under the control of sex hormones.

The sexual behavior of the female animal is dependent on ovarian hormones and linked to the period of *estrus* ("frenzy") that coincides with ovulation, during which the animal is said to be "in heat." The removal of the ovaries stops sexual activity; the administration of ovarian hormones restores it. Female rodents are dependent on both estrogens and progesterone; dogs and cats require only estrogens for normal sexual function.

The estrous female arouses sexual interest in the male by physical changes in her genital region and the production of potent scent signals conveyed by pheromones (Box 4.4). Whereas hormones affect the individual producing them, pheromones influence the behavior of other animals of the same species.

Box 4.4

PHEROMONES

The term *pheromone* ("to transfer excitement") was coined in 1959 to describe chemical sex attractants in insects; its existence had been known since the 19th century. These substances are remarkably potent; the minute amount of pheromone in a single female gypsy moth is enough to sexually excite more than one billion males from as far as two miles away. In a number of insect species, the males in turn produce pheromones to induce the female to copulate. In the male cockroach this is an oily substance, the consumption of which induces the female cockroach to take the coital posture (Hopson, 1979).

There are numerous examples of the influence of pheromones in mammals' reproductive and social behavior: housing female mice together inhibits their ovarian cycles; exposure to a male or just his urine will revive the cycles; the odor of male mice will accelerate puberty of young female mice; the introduction of a male from a foreign colony will suppress the pregnancy of mice who have mated with their own males. When female rats live together (or just breathe the same air) they tend to ovulate and come into heat the same day (McClintock, 1983).

The menstrual cycles of women show a similar tendency to be influenced by the odors of other women. For example, the menstrual periods of women who live together in college dormitories, or other close quarters, often become synchronized, so they menstruate close to the same time (McClintock, 1971). In experiments on the effect of scents, when a "donor" woman wears cotton pads under her arms for a 24-hour period and other women are then exposed to her armpit odor, the menstrual periods of the recipient women shift to become closer to the periods of the donor (Russell et al., 1977).

There is also some correlation, if not a causal link, with the influence of male odors. In one study, women who seldom dated had longer cycles; those who dated more often had shorter and more regular cycles. Similarly, women who had slept with men once a week were more likely to have regular menstrual cycles and fewer fertility problems. Sexual activity is not the key factor in these studies (mastur-bation makes no difference); it is the presence of the male, or to be more precise, his odor that makes the difference. Under-arm secretions from men mixed with alcohol and dabbed on the upper lip of women will also regularize their periods almost as effectively (Cutler et al., 1985).

What are the substances that exert such a curious influence? Some investigators have reported the presence of volatile fatty acids ("copulins") in vaginal secretions of rhesus monkeys in mid-cycle that stimulate male sexual interest, mounting, and ejaculation. Similar compounds have been identified in human vaginal secretions; they peak at mid-cycle, unless women are on the pill (Michael and Keverne, 1968; Michael et al., 1974, 1976; Bonsall and Michael, 1978). Other investigators have failed to confirm these findings. The role of such human "copulins" is unclear (Goldfoot et al., 1976).

People are apparently able to detect differences in vaginal odors (without being told the source) under experimental conditions; most men and women find them rather unpleasant, but less so for mid-cycle secretions (Doty et al., 1975). On the other hand, erotic literature is full of testimonials to the arousing smell of genital secretions; perhaps when sexually excited, people react to odors in quite different ways.

Vaginal secretions, moreover, need not be the primary source of erotic scent signals. Our body is studded with odor-producing glands, and despite our scrupulous efforts to eliminate and conceal them, they may still be part of the "silent language" of sex. One perfume company has marketed a fragrance that contains *alpha androstenol*, a synthetic chemical similar to a substance in human perspiration. This substance, when produced by boars, makes the sow in heat adopt the mating posture. However, there is no scientific evidence of its having any sexual effect on humans (Benton and Wastell, 1986). Even if human pheromones were to be identified—and they may well exist—it is unlikely that they would override all of the other characteristics that make people attractive to each other. Rather, pheromones would be one more element in this complex interaction.

The female animal's sexual interest is cyclical; the male is more or less ready to copulate anytime, provided there is a receptive female available. In effect, then, male sexual behavior is mainly controlled by female receptivity.

The male response is also dependent on an adequate level of testosterone. Castration of adult rats is followed by decline in sexual behavior; administration of testosterone restores the sexual behavior to its earlier levels (Goy and McEwen, 1980; Bermant and Davidson, 1974).

Equally important is the presence of testosterone during early life for adequate behavioral responses to testosterone during adulthood; reduced testosterone stimulation in infancy impairs sexual performance in adulthood. Based on animal research, the following model has been developed linking hormones to sexual behavior.

Organization and Activation   The influence of sex hormones on behavior occurs in two stages. The first stage is *organizational*. Hormones influence the development and differentiation of those portions of the brain that deal with sexual behavior. The second stage is *activational*. Hormones go on to initiate and maintain sexual behavior (Ehrhardt and Meyer-Bahlburg, 1981).

The organizational effects typically occur during a limited period of greater susceptibility, or *critical phase*, either before birth (the prenatal period) or around the time of birth (the perinatal period). Their influence tends to be long-term and permanent. By contrast, the activational effects of hormones are reversible, and not limited to a critical phase of development. The organizational effects of hormones are like the exposure of photographic film to light (which captures the image); the activational effect, like the effect of developing fluid on exposed film (which brings out the image) (Wilson, 1978). Through this double influence, hormones first develop portions of the brain that deal with sexual functions and then activate and maintain those functions.

This model suggests that the male and female brains are in some respects different. Just as the reproductive system becomes male or female before birth through the influence of sex hormones (Chapter 2), presumably the brain develops male or female structure, resulting in distinctive sexual behaviors. These differences are called *brain dimorphism*.

Sex Differences in the Brain   Evidence for sexual dimorphism in the mammal's brain comes from three areas. The first evidence is behavioral. Mammalian sexual behaviors are typically sex-specific. For instance, the sexually receptive female rat will arch its back (lordosis) while the male rat will mount and rhythmically thrust its pelvis during copulation. Because such behaviors are not acquired by learning, we infer that they reflect brain differences.

More direct evidence comes from physiological functions. In both sexes, the production of gonadal hormones is controlled by the same pituitary hormones (FSH and LH). The reason that the female produces gonadal hormones in a cyclical fashion and the male does not must reflect differences in brain function.

The third type of evidence is the demonstration of anatomical differences between the male and female mammalian brains (Gorski et al., 1978). In the male rat, the *medial preoptic area* of the hypothalamus is strikingly larger than in the female, a structural change that becomes apparent shortly after birth under the influence of testosterone (Raisman and Field, 1971; Goy and McEwen, 1980). The preoptic area is essential for the preovulatory release of LH that occurs in the female only; it is also important for both male and female sexual behavior (Lisk, 1967). Such a structural sex difference in the brain is of considerable importance because it could cause differential sexual behaviors in the adult, even in the absence of immediate hormonal action (Thompson, 1985).

The significance of such animal findings for human sexual behavior remains to be demonstrated (Fausto-Sterling, 1985; Bleier, 1984). However, evidence is beginning to emerge that brain sex differences, such as the size of the medial preoptic area, may also exist in humans. Debate on the issue is lively.

Experimental Evidence   A good deal of evidence points to androgen as the key hormone in the sexual organization of the brain, as it is in the sexual differentiation of the reproductive system. How do we know? Let us follow a typical experiment.

First an adult male rat is castrated. He no longer reacts sexually. Treatment with estrogens and progestins has no effect in either restoring male activity (mounting) or instituting a female pattern of response (lordosis), but treatment with testosterone does restore normal male behavior. Next, a rat is castrated soon after birth. As an adult, he shows no sexual behavior. When given estrogen and progestins, he develops a female pattern of response (lordosis). Treatment with testosterone fails to set up a male pattern. Finally, the same experiment is repeated, but this time the rat is given an injection of testosterone right after being castrated in infancy. As an adult, his sexual responses are identical to that of the male rat castrated in adulthood: female hormones have no effect; testosterone restores normal male sexual function.

From this experiment two conclusions are clear. First, male sexual behavior is dependent on the presence of testosterone, both in infancy and in adulthood. Second, the presence of testosterone in infancy stamps the brain in the male pattern, following which female hormones have no effect. In the absence of testosterone in infancy, estrogen and progesterone induce female response in a genetic male.

Repeating the same experiment on a female rat, we get comparable results with a key difference. The adult female when castrated loses sexual function: estrogen and progesterone treatment restore normal female functions; testosterone has little or no effect. A female rat whose ovaries are removed in infancy behaves exactly like her castrated male counterpart: female hormones given in adulthood induce a female pattern of sexual behavior; testosterone has no effect. Finally—and this is the key point—a female rat whose ovaries are removed in infancy and who is given testosterone will also respond like her male counterpart. Estrogens and progestins then have no

effect in adulthood; testosterone induces a male pattern of response. This female will mount another female and go through the ejaculatory motions, even though she cannot ejaculate (Daly and Wilson, 1978).

What determines the pattern of sexual response is clearly not genetic sex but sex hormones. The development of sexuality in both male and female mammals depends on the organizational effect of androgen or its absence in early life, complemented by the activational effects of male or female hormones in adulthood.

Many questions still need to be answered about these issues, and there are many complexities we have not dealt with here. For instance, androgens may not be the sole hormones with an organizational effect; progesterone may play a protective role against the effects of androgens on the brain; estrogens too, in low concentration, may exert an influence (MacLusky and Naftolin, 1981).

Further complicating the story, testosterone does not act directly on the brain cells. It must be first converted to estradiol in order to combine with the receptors on brain cells. Thus, ironically, it is the "female" hormone that is ultimately responsible for masculinizing the brain (Ehrhardt and Meyer-Bahlburg, 1981).

It would be misleading to think of hormones as a form of "sex fuel" that keeps the sexual drive going. The effects of hormones are highly influenced by the condition of the body on which they are acting. For example, age, conditions of rearing, sexual experience, nutrition, and the testing situation all influence how an animal will behave under the stimulation of sex hormones.

Primate Behavior   Sex hormones exert an important influence on the sexual behavior of nonhuman primates, but their role is not as definitive as in other mammals.

One way of assessing this role is to look at estrus, the key event that triggers the sexual activity of the female and the response of the male. Primates who evolved earlier, such as lemurs, show fairly distinct estrous cycles. In the more developed primates, however, such cycles tend to be influenced by social and environmental factors–in particular, the social structure of the species and the sexual drive of the male (Rowell, 1972).

For example, although the male gorilla lives with a group of females, he has a low sexual drive, and sexual activity in the group is sporadic, mainly initiated by the female. In this situation it is essential that sexual activity occur around ovulation to ensure reproduction. Indeed, female gorillas do show definite estrous patterns–behavior that invites copulation (Short, 1980).

The male orangutan is much more highly sexed. These apes are more isolated from each other, so a male will attempt to copulate with any female he encounters. In joint captivity this fact leads to a good deal of sexual activity, because the females cannot get away from the males. This setup makes it appear as if the females had no clear estrous period. However, if the males are caged separately with an opening to the females' cage that is too small for them but large enough for a female, then a clearer pattern emerges. Copulation is initiated by the female at the time of estrus but not otherwise (Nadler, 1977).

Chimpanzees live in small social groups. The female is sexually receptive almost anytime, but more so during her period of estrus, when her external genitals visibly swell. During this time the female mates actively with most of the males in the group, but because ovulation occurs toward the end of this period, this sexual activity usually does not lead to impregnation. Typically, the female will pair off with a single male at the end of this period, and conception is most likely to result from their temporary consortship (Tutin, 1980).

It is clear from these examples that in primates sexual interactions are influenced by hormones, but not in the lockstep fashion characteristic of rodents. Furthermore, as shown by chimpanzees, sex now serves needs that are not purely reproductive; the mating of the female prior to ovulation possibly fosters group cohesiveness by providing sexual access to all of the males, even though one of them is eventually selected by the female as consort. In other words, the stage appears to be set for the more complicated and socially influenced sexual interaction we encounter among humans.

Even at the level of primates, there is no longer a single answer to the question of whether or not hormones influence sexual behavior. They do and they do not, depending on what hormone we are looking at, which aspect of sexual behavior is involved, and under what circumstances it takes place. These considerations apply not only to different species and to differences in males and females, but even to the various aspects of sexual responsiveness. For instance, Beach (1976) distinguishes three aspects of female sexuality: *attractiveness* is some feature of the female that arouses the male's sexual interest; *receptivity* is the extent to which the female accepts the male's sexual advances; *proceptivity* is the extent to which the female takes the initiative in approaching the male (presumably a measure of the male's attractiveness). In the rhesus monkey female attractiveness is dependent on estrogens (which bring about swelling of the vulva and color changes during estrus); proceptivity depends on androgen; and receptivity seems to be less dependent on hormones (Bancroft, 1983).

Human Behavior

The role of sex hormones in behavior is far more difficult to determine among humans than among other animals, for several reasons.

First, human sexual behavior is incomparably more complex than its animal counterpart. Cultural influences are so pervasive and infinitely varied that the subtlety and range of human sexual behavior is bewilderingly diverse (Gregersen, 1983). In studying the relationship of hormones to sexual behavior we have no simple measure like mounting or lordosis to go by.

Second, the key experiments conducted with animals cannot be replicated among humans. We are not about to castrate baby boys or give androgens to baby girls under experimentally controlled conditions to see what happens to their sexual development. Instead we must rely on studying conditions such as the androgenital syndrome.

Third, it may be argued that we have difficulty finding convincing evidence for the influence of hormones on human sexual behavior because there is no such influence to be found. Perhaps during the course of evolution, there has been a progressive "emancipation" of sexual behavior from hormonal control (Beach, 1947). Instead, social learning and individual experience have taken over as the forces that shape human sexual behavior.

The great role of social learning notwithstanding, there is now evidence that biological factors do influence human psychosexual differentiation. As we have seen, anatomical evidence is beginning to point to sex differences in the human brain, specifically in the medial preoptic area of the hypothalamus.

The cycles of gonadal hormone secretion in women, but not in men, presuppose differences in brain function, just as they do among other mammals. Furthermore, the surge in LH secretion in response to estrogen stimulation is just a female characteristic. Not only is this reaction present normally during the menstrual cycle, but it can be induced at other times with injections of estrogen. By contrast, men do not respond to estrogen injections with an increase in LH production. This difference too points to a difference in the brain (Gladue et al., 1984).

We need to ask more sophisticated questions than whether or not hormones have anything to do with sex. Hormones may have a significant impact on a particular facet of sexuality but not on another, under certain conditions but not others. In short, hormonal influences only make sense in the broader context of human experience, as one factor among many that predispose us to act. They are not some irresistible force that drives us into sexual activity willy-nilly. They encourage, but they do not compel us (Money, 1987).

Male Sexual Drive   Erotic desire is usually taken for granted as part of "human nature,"

but it is still a mystery. To what extent do biological factors—hormones in particular—generate and sustain sexual drive (Chapter 3)?

The most presuasive case for the effects of hormones on sexual drive is based on the role of androgens in the male. Let us first consider the effects of the lack of testosterone. It is now generally agreed that castration prevents normal sexual function in men (Box 4.5), but a lot of confusion has existed in this regard. Based on a review of the literature Kinsey concluded: "Human males who are castrated as adults are, in many but not in all cases, still capable of being aroused by tactile or psychogenic stimuli" (Kinsey et al., 1953, p. 744). Subsequent surveys also reported that some castrates remain responsive for years. These early studies were plagued with inconsistent methods and the failure to differentiate between changes in sexual desire and loss of erectile ability, or between response to various forms of sexual stimulation (Heim and Hursch, 1979; Heim, 1981). At present, there is "little doubt that after castration sexual behavior is . . . drastically reduced or completely suppressed in a high percentage of men" (Davidson et al., 1982). Although the decline in sexual drive usually follows castration quite rapidly (Davidson, 1980), the timing varies from one individual to another (Bancroft, 1983). Moreover, up to 63 percent (Sturup, 1979) of castrates may retain some degree of sexual drive for years, although how effective it is remains unclear.

Further confirmation of the importance of androgen for male sexual drive comes from carefully controlled studies of *hypogonadal* men, who need hormone injections. If treatment stops, within a month there is a decline of sexual interest. Soon after, they lose capacity for seminal emission, but not necessarily for orgasm (in other words, a man may be able to have an orgasm but without the ejaculation of semen). As a result of reduced sexual desire, sexual activity declines. All of these changes are reversed with one to two weeks of giving hormones again (Davidson et al., 1979; Skaakeback et al., 1981).

The relationship between androgens and

Box 4.5

EUNUCHS

Of the many causes of androgen insufficiency, the most dramatic is the removal of the testes by castration. Before puberty, castration results in inadequate genital development and the absence of the secondary sexual characteristics. The person, known as a *eunuch*, will have bypassed most of the changes of puberty (see figure). He will have a high-pitched voice, poor muscular development, underdeveloped genitals, no beard, no pubic or axillary hair, and female-type subcutaneous fat deposits with partially developed breasts. He will be of normal height or taller, because the long bones of the extremities continue to grow in the absence of androgen. Eunuchs have a low sexual drive though they are not necessarily impotent. Those castrated after puberty do not lose their secondary characteristics.

Castration has been practiced in many cultures for various reasons, most notably to provide "safe" guardians for women (*eunoukhos* means "guardian of the bed" in Greek). Best known in connection with Islamic and Chinese harems, the practice goes back to remote antiquity.

Some eunuchs wielded great power in Islamic and Chinese courts. With no family ties or offspring, their sole loyalty was to the ruler, and they often acted as members of his personal staff. At the peak of their influence, there were over 70,000 eunuchs in the service of the Ming dynasty. The eunuch system in China was not completely abolished until 1924, when the final 470 eunuchs were driven from the last emperor's palace (Mitamura, 1970).

In Europe as late as the turn of the 19th century, boys were castrated to maintain their soprano voices. These *castrati* sang in church choirs and performed female roles in opera, which excluded women. During our own century, castration has been used in Europe and the United States in efforts to modify sexual behaviors such as "chronic" masturbation, homosexuality, exhibitionism, and child molestation. Surgical castration is no longer

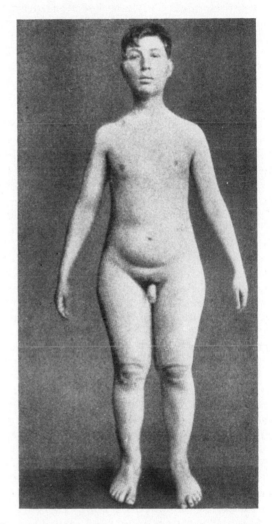

Twenty-two-year-old eunuch castrated before puberty.

practiced for punitive reasons or to alter behavior. But "chemical castration" through the use of antiandrogenic drugs is used occasionally in the treatment of sex offenders.

the ability to have an erection is more complex. Hypogonadal men have impaired nocturnal penile tumescence (NPT) (Chapter 3), and treatment with androgen significantly improves their ability to have erections during sleep. These men seem unaffected in their ability to have erections in response to erotic films (Bancroft and Wu, 1983; Kwan et al., 1983). These findings suggest that some aspects of sexual arousal and erectile response are androgen-dependent, and other aspects are not (Bancroft, 1986a).

Drugs that have an *antiandrogenic* effect will interfere with sexual function, particularly sexual drive. Most notable are cyproterone and medroxyprogesterone, which are used in Europe and the United States, respectively, for the treatment of sexual offenders.

Among normal males with no testosterone deficiency, levels of androgen do not relate significantly to sexual desire, excitement, or frequency of coitus (Persky, 1983). In other words, if a man has normal levels of androgen, giving him more androgen has no effect on his sexual behavior. As with a glass full of water, adding more water does not fill it up any further; the excess merely spills over.

This issue has an interesting corollary. Testosterone may increase aggression, which in turn may affect sexual activity. Ordinarily there are no discernible differences in testosterone levels between less and more aggressive men. It is possible, though, that for men already predisposed to violence, testosterone might facilitate it.

Female Sexual Drive   Unlike nonhuman primates and other mammals, women have no estrus. They are potentially sexually receptive at any time. Then is a woman's sexual drive completely free from hormonal influences? The question remains open. Studies in this regard have focused on the role of estrogens and progestins, as well as androgen.

The most obvious approach would be to look for fluctuations in sexual interest during the menstrual cycle, matching the clear and predictable shifts in hormone levels. Such attempts have led so far to inconclusive results (Persky, 1983).

Women have no estrus. Does that mean that they have no cyclic fluctuations whatsoever in the level of their sexual desire? Numerous studies have failed to answer that question with certainty. Several studies have shown an increase in the likelihood of intercourse at or shortly after ovulation at mid-cycle (Udry and Morris, 1968; Harvey, 1987). A similar pattern has been found based on self-ratings of sexual arousal among young women during various menstrual phases (McCance et al., 1952; Moos, 1969; James, 1971; Gold and Adams, 1978). However, many other studies have failed to support a mid-cycle peak of sexual interest (Bancroft, 1983).

These findings may be influenced by the fluctuations of sexual interest in the male partner. Whether or not a woman has coitus is not determined by her own level of interest alone but influenced by her partner. This problem may be less in lesbian relationships, where gender conflict, fear of pregnancy, and contraceptive usage are not an issue. In such couples, the significant peaks in sexual encounters and orgasm are found to be at mid-cycle (Matteo and Rissman, 1984).

Other studies show evidence of coitus to peak after menstruation with another peak at mid-cycle (McCance et al., 1952). The peak after menstruation may be compensating for the period of abstinence during menses. Women who engage in coitus during their period do not show such a rebound reaction.

The mid-cycle peak, when it occurs, may be in response to increased levels of estrogen, or to higher levels of androgen, which are also present at this time. The decline in sexual activity during the postovulatory period is consistent with the higher levels of progestogens, which are known to inhibit libido.

Another approach is to look at women whose ovaries do not function because of surgery or natural menopause. Traditionally, the menopause has supposedly meant not only the end of a woman's ability to bear children but a lessening of her sexual attractiveness and interest. These attitudes have recently changed markedly (Box 4.1). Nevertheless, several studies have shown that in a majority of women there is a definite decline in sexual interest and orgasmic response following the menopause (Hälström, 1973). A similar decline has been shown more generally for women between the ages of 45 and 55 (Pfeiffer, Verwoerdt, and Davis, 1972).

The fact that these changes correlate with the decline of ovarian hormones does not prove that they are caused by it. First, there appears to be a great deal of variation in how women respond to the menopause, both in the severity of symptoms and in sexual function. Some women experience a markedly negative change, others do not, and still others experience a positive enhancement of sexual interest and responsiveness in midlife (Masters and Johnson, 1970).

Second, the menopause brings about distinctive structural changes, with thinning of the vaginal tissues and reduction in the lubricatory response. If untreated, these changes may make intercourse painful for menopausal women. The loss of sexual interest may be a secondary effect of physical discomfort. Menopausal women must also still contend with the cultural prejudices with regard to the sexuality of older women, which cannot fail to influence their own self-perceptions. Finally, hormonal replacement therapy does not seem to have an effect on female sexual drive, although it distinctly improves vaginal dryness; this fact casts doubt on a direct relationship between the level of ovarian hormones and female sexual behavior (Bancroft, 1983).

For some reason, women who go through menopause before midlife because of surgical removal of the ovaries seem more likely to suffer sexually negative effects than those who become menopausal through normal aging, and they are more likely to respond to estrogen replacement (Dennerstein et al., 1980).

The behavioral effects of progestins are even less clear. Progesterone has been shown to inhibit sexual desire (Bancroft, 1980); yet young women who exhibit higher levels of sexual activity in the luteal phase of their cycle also have higher progestin levels than women who do not.

Studies of oral contraceptives have produced conflicting results about the effect of steroid hormones on female sex drive. When there is an impact, it seems that the birth control pill inhibits sexual desire. On the other hand, such a physiological effect can be more than counteracted by the freedom from fear of pregnancy, which promotes greater sexual interest and responsiveness. Futhermore, giving sex hormones to individuals who already have normal levels (such as women on birth control pills) is not a good test of whether or not hormones sustain the sexual drive. Recall that testosterone has no effect on men who have no hormonal deficiency.

Based on clinical studies, it has been suggested that androgens may be the hormones responsible for the female sexual drive. For example, removal of the adrenal glands (the main source of female androgen) reduces a woman's sexual desire, responsiveness, and activity. However, such studies have involved women who were very ill and whose adrenals were surgically removed as part of their treatment; under such circumstances an objective assessment of sexual activity is hardly possible (Waxenburg et al., 1959). In other cases women receiving large doses of testosterone as part of the treatment for breast cancer have reported enhanced sex drive. Improvement here may be the result of the more general anabolic effects of testosterone on body tissue, or because androgen enlarges the clitoris, it may enhance its sensitivity as well (Gray and Gorzalka, 1980). Nonetheless, when, under controlled circumstances, androgens are given to menopausal women whose ovaries have been removed for medical reasons, there is an increase in the intensity of sexual desire and arousal, as well as in the frequency of sexual fantasies (Sherwin et al., 1985). Although the relationship of female sex drive to sex hormones remains unsettled, it is probable at least that arousal is in part dependent on the level of testosterone.

A great deal more needs to be learned about how hormones influence sexual behavior under ordinary circumstances, as well as in the laboratory. Nonetheless, we know gonadal hormones play an undisputable role in the sexual behavior of animals. We are sure they play a similar role in men; and although the issue is still controversial, we now question the nature and extent rather than the existence of such a role in women as well. More uncertain are the possible relationships between hor- mones and gender identity (masculinity and femininity) and the development of heterosex- ual or homosexual orientation.

REVIEW QUESTIONS

1. Which hypothalamic, pituitary, and gonadal hor- mones are involved in sexual functions?

2. What changes happen to the body at puberty?

3. Give examples of disorders of sexual develop- ment in males and females caused by sex chro- mosomes, sex hormones, and tissue response.

4. Describe the changes during the menstrual cycle involving pituitary hormones, the ovary, ovarian hormones, and the uterine endometrium.

5. What are the symptoms of premenstrual syn- drome and dysmenorrhea?

6. Compare and contrast the organizational and activational effects of hormones on sexual behavior.

THOUGHT QUESTIONS

1. How can abnormal patterns of sexual develop- ment help you to understand sexual behavior?

2. Why are there menstrual taboos? Why do some modern women feel self-conscious about men- struction?

3. What would be the psychological and social con- sequences if an effective human pheromone were marketed?

4. Should PMS be a mitigating circumstance in crimes committed by women? Argue both ways.

SUGGESTED READINGS

Crapo, L. (1985). *Hormones*. San Francisco: Free- man. An authoritative yet simply written introduc- tion to hormones.

Delaney, J., Lufton, M. J., and Toth, E. (1988). *The Curse*. Urbana, Ill.: University of Chicago Press. A fascinating cultural history of various facets of menstrucation.

Hopson, J. S. (1979). *Scent signals: The silent language of sex*. New York: Morrow. An informative and en- tertaining account of pheromones.

Katchadourian, H. (1977). *The biology of adolescence*. San Francisco: Freeman. Introductory-level text on aspects of puberty.

Kelley, K. (Ed.). (1987). *Females, males, and sexuality*. Albany: State University of New York Press. Chap- ter 2 provides an overview of hormones and sex- related behavior. Chapter 4 deals with the pre- menstrual syndrome.

Sexual Health and Illness

When I first realized that I had a sexually transmitted disease, I felt like a monster. I thought all my friends would be horrified if they knew I had such an affliction .... I wanted to crawl into a hole and die ...."

ANONYMOUS
COLLEGE STUDENT
Miller, Rich, and Steinberg
(1987)

Sex is wonderful, but it can be dangerous. Currently, people are most worried about AIDS. It is the newest of many sexually transmitted diseases (STDs) people can get through sexual contact. Can it happen to you? Knowing the truth about these illnesses will keep your concerns realistic. It can keep you, your partner, and someday even your children healthy.

This chapter looks first at how to maintain a healthy reproductive system, then at ailments that system shares with the rest of the body, and finally at the STDs. In each case you can learn how to take care of yourself and when to go for help.

There are many facets to the STDs. Their clinical signs and symptoms, or patterns of transmission, are primarily medical concerns. But the STDs also have an important impact on how individuals behave sexually and relate to their sexual partners. By affecting large numbers of people, the STDs have serious economic and political consequences, and they raise difficult ethical and legal issues. These are some of the most difficult problems confronting our society.

Diseases of sexual organs do not always interfere with sexual functions. A man may have herpes but no difficulty in having an erection; a woman may have gonorrhea but no problem reaching orgasm. In other circumstances, physical ailments interfere with sexual function, as do psychological and interpersonal problems. Disturbances in sexual performance or satisfaction are referred to as *sexual dysfunction*; we shall deal with them separately, in Chapter 8.

MAINTAINING SEXUAL HEALTH

The human body is superbly designed, but it requires proper care and maintenance to work well. Prevention is the key to good health. That means knowing your own body, keeping fit, and getting medical attention through checkups, as well as at the first signs of illness.

Keeping Clean

One key to sexual health for men and women is cleanliness. Apart from hygienic considera-

tions, a clean and fresh body enhances sexual attractiveness. Cleanliness does not mean eliminating or covering up all natural odors. The natural scents of the body can be erotic; stale and offensive smells result from the action of skin bacteria and other microorganisms on accumulated body secretions.

Although the vagina is a self-cleansing organ, some women like to wash it by *douching* after menstruation or coitus. Douching is not necessary to keep clean. It may predispose a woman to yeast infections, because it disturbs the bacteria that normally live in the vagina.

For effective use of spermicides (foams, diaphragm jelly, and so on) it is important *not* to douche for at least six hours after intercourse. Beyond that time a woman may douche to rinse off the residues of the contraceptive substances. Disposable douching kits are convenient, but they are a needless expense, and the chemicals may be irritating. A woman can instead buy a douche bag and use plain lukewarm tap water, or make it mildly acidic by adding two tablespoons of vinegar to one quart of water. Doctors recommend douching for certain infections, to convey the medication to the vagina.

Toxic Shock Syndrome

To absorb the menstrual flow women use *tampons*, absorbent cylinders inserted into the vagina, or *sanitary napkins*, absorbent pads placed against the vaginal opening. Tampons come in various sizes and can be inserted into the vagina without damage to the hymen. It is preferable that the size of the tampon used on a given day be matched with the amount of menstrual flow expected. Small tampons will not be able to handle heavy flow; large ones will be insufficiently saturated by scanty flow to be taken out easily and may cause vaginal irritation.

A great deal of concern has been generated since 1980 by a newly recognized condition associated with tampon use, called *toxic shock syndrome (TSS)*. It usually occurs in younger women during or immediately following menstruation. It is caused by toxins produced by certain bacteria (*Staphylococcus aureus*) that are normally present in many women but tend to multiply rapidly in and around absorbent tampons. A bacterial virus may be responsible for triggering these bacteria to produce their toxin.

The early symptoms of toxic shock are sudden high fever accompanied by sore throat, rash, vomiting, diarrhea, dizziness, and abdominal pain. The condition is very rare, occurring in less than 1 per 100,000 menstruating women (of whom there are 50 million in the United States), but it can be fatal; 88 deaths were ascribed to it between 1978 and mid-1982.

The more absorbent a tampon, the higher the risk of its causing toxic shock; the likelihood is 60 times higher for the most absorbent type (Berkeley et al., 1987). The particular brand of tampon (*Rely*) that was found to be most frequently associated with this condition has been taken off the market (Paige, 1978), but there has been no significant drop in the number of cases—still about 40 to 65 per month in the United States.

Though there are many other causes for the TSS-like symptoms, if a woman experiences them while wearing tampons, she should seek immediate medical care. To help reduce the risk, tampons should be changed at least every six hours, hands washed prior to insertion, sanitary napkins substituted at night, extremely absorbent types ("super" or "superplus") used only with heavy flow, and the use of superabsorbent tampons avoided by adolescents.

Sex during Menstruation

Some women refrain from sexual intercourse while menstruating; others do not mind the presence of menstrual blood, which may be washed away before coitus or held back with a diaphragm. Until recently, personal rather than medical considerations determined these decisions. The situation has now changed because certain health risks have become linked with having sexual intercourse while menstruating.

Intercourse during menstruation has been associated with increased risk of *pelvic inflammatory disease* (PID), infections of the uterus and fallopian tubes. During menstruation, the cervix is more open, and the normal mucous plug is absent, while the free blood in the vagina and uterus provides a growth medium for bacteria (Hatcher et al., 1988). The risk of acquiring PID from a single act of intercourse is three to six times greater right around menstruation.

In a long-term mutually monogamous relationship, or if a condom is used, the risk of pelvic infection for the woman is negligible. However, menstrual blood is also a possible source of AIDS virus and hepatitis B virus, so a male who has intercourse with a menstruating woman may be taking a higher risk of exposing himself to these diseases. The risk is less if he uses a condom. Orgasm during menstruation through any means may relieve or sometimes increase menstrual cramping (Hatcher et al., 1988).

COMMON AILMENTS OF THE REPRODUCTIVE SYSTEM

Genitourinary Infections

The most serious infections of the genital and urinary tract are due to sexually transmitted diseases, which we discuss separately. A number of common, milder conditions should not be mistaken for them.

Genital Discharge   Normally, men should pass only urine and semen through the urethra; any other discharge, or the presence of blood in urine or semen, is abnormal and is probably due to infection.

Women, in addition to menstrual bleeding, secrete vaginal mucus during their monthly cycle (apart from vaginal fluid generated during sexual excitement). When vaginal discharge is excessive and contains pus cells it is called *leukorrhea* (Barclay, 1987). Almost every woman experiences leukorrhea at some time in her life. It is not a disease but a condition that can be caused by infections, chemicals, and physical changes. For example, irritating chemicals in douche preparations, foreign bodies such as contraceptive devices, and alterations in hormone balance (as during pregnancy or menopause) may cause it (Capraro et al., 1983). If excess discharge is associated with itching, pain, a bad odor, or an unusual color the woman should see her doctor to rule out a possible infection (Hatcher et al., 1988).

Vaginitis   Vaginitis is an inflammation of the vagina that causes itching, pain, discharge, and discomfort with intercourse (Eschenbach, 1986). Several organisms can cause vaginitis; only some of them are sexually transmitted. *Hemophilus vaginalis* is a frequent cause. Because it can often be found in the sexual partners of infected women, it is possible that the infection is sexually transmitted; but it is also found in people who have not had sexual intercourse.

A common vaginal infection is caused by a protozoan called *Trichomonas* ("Trich"). Close to one million women per year are seen by doctors for this condition. It is characterized by a smelly and foamy yellowish or greenish discharge that irritates the vulva, producing itching and burning. A man may harbor this organism in his urethra or prostate gland without symptoms, or he may have a slight urethral discharge. Because sexual partners usually infect each other, both partners are treated simultaneously with metronidazole (Flagyl) to prevent reinfection. Although usually sexually transmitted, trichomonas infections occasionally have been spread by genital contact with a wet bathing suit or washcloth. There are no known long-term consequences to men or women from this infection.

Candidiasis   Another common vaginal infection is *candidiasis* (or *moniliasis*). It is caused by a yeast-like fungus called *Candida albicans*. The thick white discharge it produces causes itching and discomfort, which may be severe. The organism is present in the vagina of a substantial number of women, but it produces problems only when it multiplies excessively.

Candidiasis is more commonly seen in women who are using oral contraceptives, are diabetic, pregnant, or on prolonged antibiotic therapy. The condition responds to treatment with nystatin (Mycostatin) suppositories or cream but is difficult to completely eradicate.

Though much less common, the same infection occurs in about 15 percent of the male sexual partners of women with moniliasis. They usually have no symptoms, but there may be marked inflammation of the glans, especially under the foreskin. It is also treated with nystatin cream.

Cystitis   The closeness of the urethral opening to the vagina and anus predisposes women to urinary bladder infections (cystitis). Sometimes bladder irritation is caused by frequent or vigorous coitus ("honeymoon cystitis").

Cystitis is most commonly found in young, sexually active women and in older men with an enlarged prostate. The cystitis of young women and older men is usually caused by bacterial infection irritating the bladder wall. Similar symptoms may be experienced by menopausal women whose bladders become irritated by pressure through thinned vaginal walls. Women who are sexually active are more likely to have urinary tract infections than those who are not (Leibovici et al., 1987).

The primary symptom of cystitis is frequent urination accompanied by pain and a burning sensation in the urethra. Although these symptoms may disappear spontaneously in a few days, it is advisable to receive proper treatment, because untreated infections may spread from the bladder to the kidneys, with more serious consequences.

Measures that help prevent cystitis are drinking lots of fluids, urinating after coitus, wearing cotton underpants, and maintaining good general hygiene. The use of condoms can also reduce the chances of infection (Hatcher et al., 1988).

Prostatitis   A common problem among men is inflammation of the prostate gland (*prostatitis*), manifested by increased frequency and burning on urination, and painful ejaculation.

Sometimes there is no apparent bacterial cause. A curious association exists between prostatitis and irregular sexual activity—long periods of abstinence followed by bouts of intensive sex (hence, it is called the "sailor's disease") (Silber, 1981). However, in healthy young men prostatitis is usually (but not always) caused by one of the sexually transmitted diseases (Holmes et al., 1984).

Cancer

The reproductive systems of both sexes, especially the female sex organs, are some of the most common sites for the development of cancer (Rutledge, 1986). The early detection of these cancers is of great importance in treating them. It is essential that everyone know enough about the early signs and symptoms to seek medical help promptly.

Cancer of the Breast

Cancers are malignant tumors which grow and spread in the body. Cancer of the breast is the most common form of cancer in women, accounting for 25 percent of all female cancers. In the United States about 5 percent of women develop breast cancer (100,000 new cases) a year; one out of ten women develops it in her lifetime. Though rare before age 25, it increases steadily in each decade thereafter. For women 40 to 44 years old it is the most common cause of death (Giuliano, 1987).

Women who are at higher risk include those over 50, those who have a family history of breast cancer, those who experience a late menopause, and those who have never had children. Men rarely develop breast cancer; only about 1 percent of breast cancers do occur in men.

The primary symptom of breast cancer is a painless mass in the breast; much less common are dimples on the breast surface or discharge from the nipple. The early cancerous lump will not show or make itself felt; so regular and systematic breast self-examination is of great importance. Examinations should be done by every woman, once a month, about a week after the end of the menses when the breasts are not likely to be tender. Detecting change requires having a sense of what the breast feels like normally. Doing it on a set day each month makes it easier to remember, and because breast tissues will be in a similar hormonal state, it is easier to feel changes (Box 5.1).

The most important factor that determines a woman's chance of surviving breast cancer is how early and how small the cancer is when detected and treated. Women who regularly examine their own breasts can detect a mass smaller and sooner than their physicians can at annual visits. Most breast masses, especially in younger women, are due to other causes, such as benign cysts, not cancer. A woman who feels a mass should not panic, but should seek prompt medical attention.

The basic treatment for breast cancer is either surgical removal of the lump, followed by radiation treatments, or the surgical removal of the breast (mastectomy). Though current surgical procedures cause as little disfigurement as possible, and breast reconstruction is possible through plastic surgery, women still worry that they will lose sexual attractiveness. Some women become so self-conscious following such surgery that they give up sex altogether if it involves nudity. They need reassurance that they are still sexually attractive. Counseling and support groups can help. Prosthetics can be worn to give the breast a normal contour.

Radiation treatments have the same success rate, without causing a dramatic change in body shape, but they do have side effects. The treatments can only be carried out successfully in patients who fit into certain guidelines, such as not being obese or having extremely large breasts. Radiation therapy requires almost daily visits to a specialist at a hospital with special equipment, for one or two months (Rubin, 1987).

Cancer of the breast can be rapidly fatal if it spreads to vital organs, but with early diagnosis and treatment the prognosis is much more favorable. About 65 percent of patients with cancer of the breast now remain alive for at least five years after the initial diagnosis.

Box 5.1

HOW TO DO A BREAST SELF-EXAM

1. *In the shower*. Examine your breasts during your bath or shower. Hands glide more easily over wet skin. Hold your fingers flat and move them gently over every part of each breast. Use the right hand to examine the left breast and the left hand for the right breast. Check for any lump, hard knot, or thickening.

2. *Before a mirror*. Inspect your breasts with arms at your sides. Next, raise your arms high overhead. Look for any changes in the contours of each breast: swelling, dimpling of skin, or changes in the nipple.

Then rest your palms on your hips and press down firmly to flex your chest muscles. Left and right breast may not exactly match—few women's breasts do. Again, look for changes and irregularities. Regular inspection will show what is normal for you and will give you confidence in your examination.

3. *Lying down*. To examine your right breast, put a pillow or folded towel under your right shoulder. Place your right hand behind your head; this distributes breast tissue more evenly on the chest. With the left hand, fingers flat, press gently in small circular motions around an imaginary clock face. Begin at outermost top of your right breast for 12 o'clock, then move to 1 o'clock, and so on around the circle back to 12. (A ridge of firm tissue in the lower curve of each breast is normal.) Then move 1 inch inward, toward the nipple, and repeat. Keep circling to examine every part of your breast, in-

cluding the nipple—at least three more circles. Now slowly repeat the procedure on your left breast with a pillow under your left shoulder and left hand behind your head. Notice how your breast structure feels.

Finally, squeeze the nipple of each breast gently between the thumb and index finger. Any discharge, clear or bloody, should be reported to your doctor immediately. (From an American Cancer Society pamphlet. Used with permission.)

Breasts should be examined at the same point in the menstrual cycle each month—usually early in the cycle. In addition, to the self-exam which all women over age 20 should do monthly, the following additional tests are important for early detection of breast cancer.

1. A doctor should examine the breasts every three years for women aged 20 to 40, and every year after 40.

2. A *mammogram* (breast X-ray) should be taken between the ages of 35 and 39 as a baseline for future comparisons.

3. Women with no symptoms should have the mammogram repeated every two years between age 40 and 49, and every year after 50.

4. Women of any age who have had breast cancer, have had a member of the family with breast cancer, or have other high-risk factors must set up with a physician a personal schedule of mammography (Brozan, 1987).

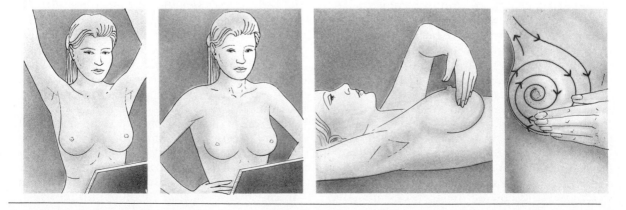

Cancer of the Cervix   Cancer of the cervix is the second most common type of cancer in women, affecting about 2 out of 100 women (60,000 new cases a year). It is rare before age 20, but the incidence rises over the next several decades. The average age of women with cancer of the cervix is 45 (Hill, 1987).

The disease is more common in women who have had large numbers of sexual contacts and who have borne children. A study of 13,000 Canadian nuns failed to reveal a single case of cervical cancer. The disease is also rare among Jewish women, which suggests that perhaps circumcision in the sexual partner is somehow linked to preventing the disease. However, in India there is no difference in rates between Muslim and Hindu women, though the husbands of the former but not the latter are circumcised (Novak et al., 1970). Three factors are known to increase a woman's risk for cervical cancer: first sexual intercourse before age 20; three or more sexual partners in her lifetime; and a male partner with many sexual contacts (Richart, 1983).

A sexually transmitted virus, human papilloma virus, the same virus that causes genital warts (discussed later), has been associated with cervical cancer. Infection with this virus appears to cause, or at least to contribute to, the development of cervical cancer. The infected areas on the cervix may be invisible to the naked eye, but may be detected with tests.

Cancer of the cervix may cause no symptoms for five or ten years, but if it is detected early, treatment is highly successful. The well-known *Pap smear* test is the best means now available for identifying cancer of the cervix in these early stages. It should be done annually beginning at age 20 (or earlier if a woman is sexually active). The procedure is simple: the cervix is scraped lightly, picking up cells from the surface; the cells are transferred to a glass slide, which is then stained and examined for the presence of abnormal cells.

As cancer of the cervix begins to invade surrounding tissues, irregular vaginal bleeding or a chronic bloody vaginal discharge may develop. Treatment is less successful when the cancer has reached this stage. If treatment by surgery, radiation, or both is instituted before the cancer spreads beyond the cervix, the five-year survival rate is about 80 percent, but it drops precipitously as the disease reaches other organs in the pelvis. The overall five-year survival rate for invasive cancer of the cervix (including all stages of the disease) is about 58 percent.

Cancer of the Endometrium   Cancer of the lining of the uterus is less common than cancer of the cervix, affecting about 1 percent of women. It usually occurs in women over 35, most commonly those between 50 to 64. Many but not all cases are detected by the yearly Pap smear test, so women over 35 also should watch for any abnormal vaginal bleeding. The five-year survival rate for endometrial cancer is 77 percent (Lacey, 1987).

Cancer of the endometrium has been linked with estrogen replacement therapy, which we discussed in connection with the menopause (Chapter 4). Birth control pills, on the contrary, seem to exert a protective influence (Chapter 6).

Cancer of the Prostate   In the male, the prostate is the organ that is the most frequent cause of disease in the reproductive system. It tends to grow larger in *benign hypertrophy* as men get older, obstructing the neck of the urethra and interfering with normal urination. As a result the man has to go to the bathroom frequently, getting up several times at night.

Cancer of the prostate is the most common cancer of male sex organs and the third most common cancer in men. About 5 percent of men will develop it (Walsh, 1985), but it is uncommon before age 60 and grows slowly. Those who have it are more likely to die of other causes, such as heart disease.

The initial symptoms of prostatic cancer are similar to those of benign enlargement of the prostate. Early in the course of the disease, sexual interest may increase because of frequent erections caused by local changes. Later on there is usually a loss of genital functioning.

A tentative diagnosis of cancer of the prostate can usually be made on the basis of a rectal examination (which involves feeling the surface of the gland with a finger inserted in the rectum), the history of symptoms, and laboratory tests. A prostate examination should therefore be part of an annual physical checkup for any man over 50. As with other cancers, the outcome is much more optimistic when it is diagnosed and treated early. The cause of prostatic cancer remains unknown, despite efforts to link it with hormones, infectious agents, sexual activity, or abstinence.

The treatment of prostatic enlargement and cancer often requires surgery. Such surgery may do damage to nerves in this region, resulting in loss of potency. This outcome is least likely (51 percent) if the prostatic tissues are removed through the urethra (Kolodny et al., 1979). Retrograde ejaculation is another possible complication (Chapter 3). Castration and estrogen therapy can cause regression of

the cancer, but it may also lead to sexual dysfunction (Chapter 8) (Walsh, 1985).

Cancer of the Testes   Unlike most other cancers, which strike later in life, cancer of the testes affects younger men. It is the most common cancer in males 29 to 35, yet it accounts for 0.7 percent of all cancers in males. Males who have undescended testes or whose testes descend after age six are at greater risk for developing testicular cancer (11–15 percent of these males will develop it).

If testicular cancer is detected early, it is curable; otherwise it can spread to other parts of the body and cause death. To check for early evidence of testicular cancer, a man should examine his testes periodically (Box 5.2). Treatment includes the removal of the affected testicle, which does not affect sexual activity or fertility. A synthetic implant is in-

Box 5.2

HOW TO EXAMINE YOUR TESTICLES

Because it is rare, much less attention is paid to early detection of cancer of the testicle than of the breast. However, the testes can be examined easily. Spending a few moments on them periodically may save your life.

The effectiveness of the testicular self-exam depends on several factors: doing it regularly, being thorough, and knowing what you are looking for. The best time is after a warm shower or bath, when the scrotal sac is relaxed. Examine one testicle at a time while you are seated or lying down. Hold the testicle between the fingers of one or both hands and roll it about, so that its surface passes under your fingertips. The point is to feel for small bumps, surface irregularities, or enlargement of the testicle itself. Do not be alarmed by the normal irregularities caused by the epididymis on the testicular surface. Get a good sense of how the testicular surface feels normally (it will take some practice). Do not become obsessed with the procedure, but whenever in doubt about a possible growth, consult a doctor promptly.

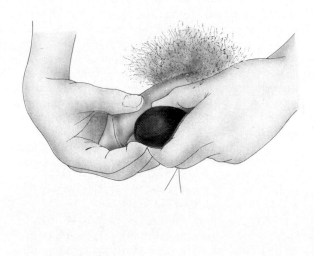

serted into the scrotum for cosmetic purposes. Some men need reassurance that lack of a testicle does not make them less masculine.

Cancer of the Penis   Cancer of the penis is rare in the United States, accounting for about 2 percent of all cancers in males. It almost never occurs among Jews, who undergo circumcision in infancy, and is also rare among Muslims, who get circumcised before puberty. In areas of the world where circumcision is not common, cancer of the penis is much more prevalent. For instance, it accounts for about 18 percent of all malignancies in Far Eastern countries. The usual explanation, though unconfirmed, is that circumcision prevents accumulation of potentially carcinogenic secretions (possibly harboring a virus) around the rim of the penis, which is the usual site of this tumor (Silber, 1981).

SEXUALLY TRANSMITTED DISEASES

Sexually transmitted diseases (STDs) are infections caused by various microorganisms (bacteria, viruses, and protozoa) that are primarily acquired through sexual contact or during birth through an infected birth canal. They were formerly called, and still are popularly known as, *venereal diseases* (in reference to Venus, the Roman goddess of love).

Prevalence

The STDs are one of the most serious public health problems. In the United States, over 10 million cases of STDs a year currently occur among the young adult population. Conservative estimates place the total cost of STDs to society at over $2 billion annually (Cates and Holmes, 1986). The spread of AIDS is likely to increase this cost greatly.

The prevalence of STDs has greatly increased in the recent past. Even rare sex-related diseases are on the rise (Leary, 1988). Younger people are particularly vulnerable; roughly two-thirds of all cases are in the 15 to 29 age group (Figures 5.1 and 5.2). It is estimated that half of the country's youth now contracts a sexually transmitted disease by age 25; this figure does not mean that every other young person you know has or will get a STD; in some groups almost everyone, in others almost none will have it (Hatcher et al., 1988).

People who are poor, those who live in urban areas, and certain ethnic minorities have the highest reported rates of STDs for each age group (Cates and Holmes, 1986). Such figures are no ground for prejudice. Part of the reason their rates appear higher is because these people more frequently use public medical services, which probably report cases more frequently than do private doctors. The inner-city poor also have higher rates of many other diseases, including most other infectious diseases, because of unsanitary living conditions, overcrowding, and inadequate medical services.

It is important to realize that *anyone* can have a sexually communicable disease, irrespective of age, sex, marital status, education, affluence, social status, or sexual orientation. In one survey, people who called an STD hotline were 83 percent white, 88 percent heterosexual, and 26 percent married. One-third of them had bachelor's degrees and a quarter earned over $25,000 a year (Hoffman, 1981). Again, the figures are deceptive. These people are more likely to use such a service and hardly representative of the general population.

Blaming a group as "carriers" of sexually transmitted diseases provokes prejudice and damages efforts to identify, treat, and control STDs (Brandt, 1987). The social consequences of some diseases can be more harmful to the infected person than the disease itself.

Types of STDs

Some two dozen microorganisms are transmitted sexually to cause a wide variety of clinical syndromes (clusters of signs or symptoms) or diseases. They include *bacteria, viruses, spirochetes, protozoa,* and *fungi* (Cates and Holmes, 1986). We shall only deal with the most common and important of the illnesses caused by these infectious agents.

A microorganism causes each illness but other conditions also affect the outcome—general health, level of immunity, the dose of the infecting organism, and so on. No one will ever develop gonorrhea, for instance, without the bacterium that causes it, but not everyone harboring the organism will develop symptoms of disease. Some people are only *carriers* of the organism, capable of infecting others although they themselves are free of symptoms. Whether a person succumbs to the illness is thus a function of the characteristics of both the *host* and the *parasite*. This model governs the way that all infectious diseases work.

There are various ways to classify STDs. For diagnosis, it is useful to categorize them by their symptoms. Some conditions primarily cause a urethral discharge (gonorrhea and chlamydia); others cause skin lesions (syphilis and herpes). The physician confronted with a patient who has a given symptom tries to sort out the underlying cause through *differential diagnosis*—a process of elimination and confirmation, using signs, symptoms, and clinical tests (Sparling, 1988).

A more fundamental method of classification, and more instructive for our purposes, is based on *etiology*, or causes. Even though herpes and hepatitis have different sets of symptoms and outcomes, they are both caused by viruses. Because viruses tend to behave similarly, we group them together. This approach provides a handle for research and treatment.

BACTERIAL STDs

Gonorrhea

Gonorrhea is an infection caused by the bacterium *Neisseria gonorrhoeae* (Hansfield, 1984). Ancient Chinese and Egyptian manuscripts refer to a contagious urethral discharge, which was probably gonorrhea. The Greek physician Galen (A.D. 130–201) is credited with coining the term *gonorrhea* from the words "seed" and "flow." This microorganism only infects humans and cannot survive for long outside the living conditions, temperature, and moisture provided by the human body.

Some 800,000 new cases of gonorrhea are reported each year in the United States, but the true incidence is estimated to be over 2 million cases a year. The rates of gonorrhea rose sharply in the 1960s and 1970s but declined somewhat in the 1980s. The rates are particularly high in 15- to 29-year-olds (Figure 5.1).

Gonorrhea is transmitted from one person to another during intimate contact with infected mucous membranes, including the membranes of the throat, genitals, or rectum. Women run a higher risk of infection following exposure than men. A woman has an approximately 50-percent chance of becoming infected with gonorrhea after one act of intercourse with an infected man; a man in the same situation runs a 25-percent chance of infection. This difference is presumably because the penis is exposed to the infectious female discharge only during coitus, whereas the infectious ejaculate is deposited deep in the vagina, where it remains after the withdrawal of the penis. Furthermore, post-coital urination can wash urethral bacteria out of the male; the vagina cleans itself much more slowly, so organisms have a greater opportunity to establish themselves. Nonetheless, post-coital urination may reduce female cystitis. Practices like post-coital washing and urination may be helpful but cannot be relied on to prevent gonorrhea (Stone et al., 1986). The use of condoms is much more effective.

Symptoms   In males, the primary symptom of gonorrhea ("clap") is a pus-filled yellow urethral discharge. The usual site of infection for men is the urethra, so this form of the disease is called *gonorrheal urethritis*. Most infected males show symptoms, but at least 10 percent do not; they are asymptomatic. A discharge from the tip of the penis usually appears within two to ten days after acquiring the infection. It is often accompanied by a burning sensation during urination and an itching feeling within the urethra. The inflammation may subside within two or three weeks without treatment, or it may persist in chronic form, in which case it may spread to the rest of the

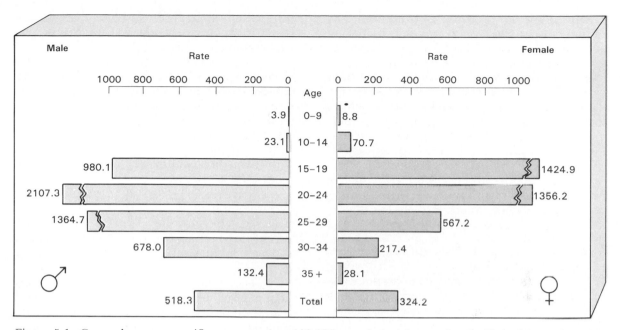

Figure 5.1   Gonorrhea: age-specific case rates (per 100,000 population) by sex, for the United States in 1982.

genital-urinary tract, involving the prostate glands, seminal vesicles, bladder, and rarely, kidneys. In some cases, the disease spreads to the joints, causing *gonorrheal arthritis* in both sexes.

The symptoms of gonorrhea in 50–80 percent of infected women are so mild or altogether absent that the woman does not realize that she is infected. This fact can lead to further complications for the woman herself, or she may act as a carrier of the disease, inadvertently spreading it to others. The primary site of infection in the woman is usually the cervix, causing *gonorrheal cervicitis.* The only early symptom may be a yellow vaginal discharge, which is difficult to distinguish from other common vaginal discharges. Untreated gonorrhea may spread up through the cervix into the uterine lining, the endometrium, and up to the fallopian tubes, causing *pelvic inflammatory disease,* which is described later.

The symptoms of nongenital gonorrhea are about the same in men as in women. *Pharyngeal gonorrhea,* an infection of the throat with gonorrhea, is transmitted most commonly during oral sex. Oral stimulation of the penis is more likely to be infectious than oral stimulation of the vulva, or kissing. The primary symptom of pharyngeal gonorrhea is a sore throat with or without fever and enlarged lymph nodes. In some cases there are no symptoms, although the person remains contagious.

Rectal gonorrhea is an infection of the rectum usually transmitted during anal intercourse. Less often, in women with gonorrheal cervicitis, the infection sometimes spreads through the vaginal discharge to the rectum. The symptoms of rectal gonorrhea are itching associated with a rectal discharge. Many cases are mild or asymptomatic.

In earlier years a common cause of blindness in children was gonorrheal infection of the eyes (*gonococcal conjunctivitis*), acquired during passage through the mother's infected organs during birth. Instilling penicillin ointment or silver nitrate drops into the eyes of all newborn babies is now compulsory and has eradicated this disease.

It takes a laboratory test to diagnose gonorrhea. The symptoms and signs may suggest it, but it must be confirmed by identifying the causative organisms in the genital discharges or in the other infected sites. To detect asymptomatic infection, cervical or urethral specimens are taken even though there are no signs or symptoms. In men, a microscopic examination of the urethral discharge may establish the diagnosis. In women's cases, and in some men's cases, the organism must be cultured—grown in a nutrient medium—to confirm its presence. There is no routine blood test that can detect gonorrhea in either sex, which is one reason why identification of asymptomatic gonorrhea has been much less successful than identification of asymptomatic syphilis.

Treatment   The usual treatment of gonorrhea is with antibiotics, usually *penicillin*. However, some rare strains that are resistant to penicillin require other antibiotics. Vaccines against gonorrhea are currently being tested (Sparling, 1988).

Chlamydia

Chlamydia is the common name for infections caused by *Chlamydia trachomatis*, bacteria that infect men, women, and babies (Stamm and Holmes, 1984).

Symptoms   The symptoms caused by chlamydia are similar to those of gonorrhea. In men, the primary infection caused by chlamydia is urethritis; it is commonly called *nongonococcal urethritis* because until recently the bacterium was difficult to identify in laboratory tests. The symptoms are painful urination, itching, and a mucoid discharge, which tends to be less profuse than that of gonorrhea. Some men have asymptomatic infections but carry the disease to others. Chlamydia does not spread to the rest of the genitourinary tract in men, nor does it cause arthritis in either sex.

In women the primary infection is cervicitis; it is called *mucopurulent cervicitis* because of a sticky yellow discharge from the cervix. Nonetheless, as with gonorrhea, the woman often cannot tell it from an ordinary vaginal discharge, and in up to 80 percent of cases she is not aware of being infected. Chlamydial infections in women can also rise through the cervix to the upper genital tract and cause pelvic inflammatory disease.

In newborns infected during passage through a mother's birth canal the sites of infection are the eyes and/or the lungs, causing *conjunctivitis* (which can lead to blindness if untreated) and *pneumonia*. The treatment given infants at birth to prevent gonorrhea does not protect against chlamydia, but most children with chlamydial conjunctivitis respond to treatment with antibiotics.

Other possible sites of infection in men and women include the pharynx and the eyes. The pharynx may be exposed during oralgenital intercourse, and the eyes contaminated through hand contact with infectious secretions.

The diagnosis of chlamydia is made by examining the likely sites of infection and noting a urethral or cervical discharge. Asymptomatic infections may be detected by growing the bacteria in culture, or by other laboratory tests that allow cheaper, more rapid diagnosis.

Treatment   The treatment of chlamydia in adults is with antibiotics, usually *tetracycline*. In infants, less potentially toxic antibiotics must be used.

Pelvic Inflammatory Disease

Pelvic inflammatory disease (PID) is not a specific disease but an inflammation of the fallopian tubes (*salpingitis*) and/or of the lining of the uterus (*endometritis*) (McGee, 1984). The inflammation is usually due to infection with various bacteria, most often gonorrhea, or chlamydia that spread "upward" from the cervix to the uterus, fallopian tubes, and eventually the lower abdominal cavity itself.

Pelvic inflammatory disease is a widespread condition, afflicting an estimated one million women each year in the United States, causing some 300,000 of them to be hospitalized (Washington et al., 1984). Over $2.6 billion are spent annually to treat this condition. PID is largely responsible for the increase in infertility among women.

Sexual activity with multiple partners and a history of PID both increase a woman's risk of acquiring a new case of PID (Eschenbach, 1986). Adolescents, young women, and black women are most susceptible; nearly 70 percent of all cases occur in women who are younger than 25. Young and black women are at higher risk for contracting sexually transmitted diseases in general, which may explain in part why their rate of PID is higher. However, the increased risk of PID in adolescents is also attributable to biological factors such as cervical and immunological immaturity (Cates and Holmes, 1986; Washington et al., 1984).

The contraceptive intrauterine device (IUD) increases risk of pelvic inflammatory disease. Whether other methods of contraception increase or decrease the risk is more controversial. Most studies have shown oral contraceptives to reduce the risk of PID, but these studies have only looked at hospitalized cases; their validity of less acute forms of PID (which are more typically caused by chlamydia) is questionable. It has also been suggested that oral contraceptives actually facilitate the spread of chlamydial PID from the cervix to the uterus (Washington et al., 1984).

Symptoms   The symptoms of pelvic inflammatory diseases may be minimal, or they may include pain and tenderness of the lower abdomen in the region of the uterus and the pelvis, chills, and fever. Gonococcal PID usually occurs in younger patients with a shorter period of pain before seeing a doctor (3 days) and higher fever; chlamydial PID is also associated with youth, but involves a longer period of pain before seeing a doctor (7–9 days), and less chance of high fever. Pelvic inflammatory disease caused by mixed bacterial infections is more common in older women, those with previous infections, and users of an IUD. It has a more acute and rapid onset, with fever and prostrating effect.

If PID becomes chronic, the symptoms of pelvic pain, pain during coitus, pelvic swelling, and tenderness may remain long after the original infection is gone. Furthermore, chronic PID will often lead to blockage of the fallopian tubes and sterility, as well as an increase in the chances of ectopic pregnancy (Chapter 6).

The diagnosis of PID is complicated by the difficulty in obtaining material from the endometrium and the fallopian tubes for a culture. However, a presumptive diagnosis is enough for a doctor to initiate treatment to relieve symptoms and to prevent sterility.

Treatment   The treatment of PID usually involves intravenous antibiotic therapy in a hospital. Treatment must begin as early as possible with more than one antibiotic, in order to provide a broad range of protection against the possible bacteria involved, even though they have not been identified.

Syphilis

Syphilis is caused by a *spirochete* (a corkscrew-shaped bacterium-like organism) called *treponema pallidum*. The term *syphilis* was introduced in 1530 by the Italian physician Girolamo Fracastoro, who wrote a poem in Latin about a shepherd boy named Syphilius (from the Greek for "crippled"), who was stricken with the disease as a punishment for having insulted Apollo. It is generally believed that syphilis was brought into Europe from the New World by Columbus' crew. The first syphilis epidemic spread through Europe shortly after his return in 1493; Columbus himself died of advanced syphilis in 1508.

Syphilis is usually transmitted through intimate sexual contact, including kissing, but the spirochete can penetrate through all mucosal surfaces and through minor abrasions in the skin. Conceivably, syphilis could be transmitted through skin-to-skin contact of a nonsexual nature, but it is highly unlikely.

Syphilis (also called *lues*) is one of the most serious of the sexually transmitted diseases. Before the advent of antibiotics, syphilis epidemics posed an enormous public health threat. After the discovery of penicillin, the prevalence of syphilis went down to 4 cases per

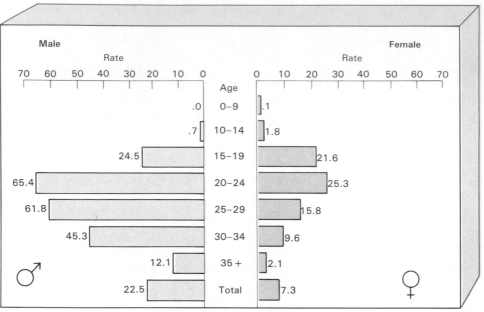

Figure 5.2   Primary and secondary syphilis: age-specific case rates (per 100,000 population) by sex, for the United States in 1982.

100,000 population in 1957, but it then went up again to 12 cases per 100,000 during the years 1965–1983 (Sparling, 1988). Currently, about 85,000 new cases occur yearly (Leary, 1988). It is more prevalent among younger age groups (Figure 5.2), especially among males in poor neighborhoods.

The rates of syphilis among gay men have been especially high, but these rates are declining currently as they adopt safer sex practices to protect themselves against AIDS.

Primary Stage Syphilis   The clinical course or progression of untreated syphilis is divided into three stages. The first stage is marked by a skin lesion known as a *chancre* (pronounced "shank-er") at the site where the spirochete has entered the body (Figures 5.3 and 5.4). The chancre is a hard round ulcer with raised edges that is usually painless. In the male it most commonly appears on the penis, the scrotum, or in the pubic area. In the female it is usually on the external genitals, but it may be within the vagina or on the cervix, so not readily visi-

ble. Chancres may also occur on the lips, the mouth, the rectum, the nipple, the hand, or anywhere the organism has entered the body.

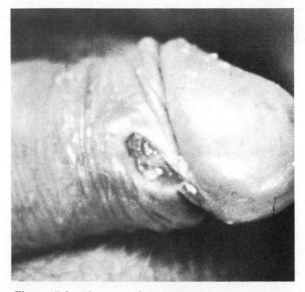

Figure 5.3   Chancre of the penis.

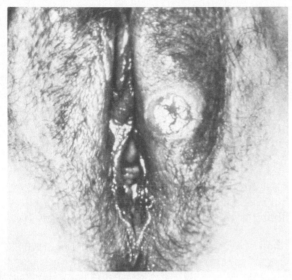

Figure 5.4   An unusually large chancre on the labia majora.

The chancre appears about two to four weeks after infection, and if not treated usually disappears in several weeks, giving the false impression of recovery.

The syphilitic chancre must be differentiated from a similar genital ulcer called *chancroid* ("soft chancre"), which is caused by the bacterium *Hemophilus ducreyi*. Chancroid is also transmitted sexually, but in contrast to syphilis, the skin lesion has soft edges and is painful. The definitive diagnosis must be made, however, by identifying the organism through the microscope or by cultures. Treatment with sulfa drugs is effective. This infection is rare in Western countries but common in Asia and tropical regions.

Various blood tests (such as the *VDRL*, or *fluorescent treponemal antibody-absorption test*) screen for syphilis by identifying antibodies produced against the spirochete. As in all other laboratory tests, there may be false positive results in the absence of the disease or false negative results even when the person is infected. The definitive diagnosis is made by identifying the spirochete through a special technique called *dark field microscopy*. The treatment of primary syphilis is with penicillin (or other antibiotics), which cures most cases promptly.

Secondary Stage Syphilis   When the primary stage is untreated, the secondary stage symptoms can appear anytime from several weeks to several months after the healing of the chancre. There is usually a generalized skin rash, which is transient and may or may not be accompanied by other symptoms, such as headache, fever, indigestion, sore throat, and muscle or joint pain. Many people do not associate these symptoms with the primary chancre as part of the same disease. At this stage of infection, the diagnosis is again made with blood tests and by attempts to isolate the organism from blood, genital secretions, or skin lesions.

During the first two stages of syphilis, the person is highly infectious. The spirochete may be shed in mucosal secretions (such as genital discharges and saliva), in blood, and in the skin lesions.

Following the secondary stage, there is a period called the *latency phase*, which may last from two years to many decades. During this period the person has no symptoms and is not infectious. However, spirochetes continue to cause internal damage—burrowing into blood vessels, bone, and the central nervous system.

Tertiary Stage Syphilis   Symptoms of the long-term infection constitute the third stage. Only about 50 percent of untreated cases actually reach the final or tertiary stage.

Tertiary syphilis may cause heart failure, ruptured blood vessels, loss of muscular control, disturbances in the sense of balance, blindness, deafness, and severe mental disturbances from brain damage. The cardiovascular system and the central nervous system are the main systems affected. Ultimately, the disease may be fatal, but treatment with penicillin even at the late stages may be beneficial, depending on the extent to which vital organs have already been damaged. Surgical or medical repair of the damage to the vital organs may also prevent or delay death. Because syphilis is usually treated early, few cases progress to this stage anymore.

Congenital Syphilis   Syphilis can be transmitted to the fetus through the placenta from the infected mother. Mandatory blood tests now identify untreated cases before the birth of a child. Nine out of ten pregnant women who have untreated syphilis will either miscarry, bear a stillborn child, or give birth to a living child with congenital syphilis. Congenital syphilis can cause a child to be mentally disturbed or retarded, to have facial and tooth malformations, or to have other birth defects. Treatment with penicillin during the first half of pregnancy prevents congenital syphilis (Murphy and Patamasucon, 1984).

Other STDs Caused by Bacteria

There are a number of other STDs caused by bacteria or bacteria-like microorganisms. They are relatively uncommon (less than 1000 cases a year) in the United States, but more prevalent in tropical and subtropical areas (including part of the southern United States). One of them is *chancroid*, which we discussed in connection with syphilis. Two others require brief mention.

Lymphogranuloma venereum (LGV) is manifested by enlarged and painful lymph glands in the groin, accompanied by fever, chills, and headache. It responds well to sulfa drugs. *Granuloma inguinale* is characterized by ulcerated, painless, progressively spreading skin lesions, usually around the genitals. It responds well to antibiotics (Sparling, 1988).

STDs CAUSED BY MISCELLANEOUS ORGANISMS

A number of STDs are caused by organisms that cut across various categories. But because of similarities in the way they are transmitted and the symptoms they cause they "fit" together.

Enteric Organisms

Organisms that live in the intestines are called *enteric*. Enteric organisms cause STDs through anal sex or fecal-oral contamination.

The rectum is affected by some of the conditions that involve the urethra and the vagina. Gonorrhea is one. In addition, certain infections that are more specific to sexual activity involve the anus, such as anal intercourse and anal-oral contact, either directly or through contaminated fingers. As a result, a number of diseases are transmitted during sexual activity that also commonly spread through nonsexual means, such as through food or water contaminated with fecal matter.

Causative Agents   The most important enteric organisms involved in STDs are the bacteria in the *shigella* and *salmonella* groups (Keusch, 1984); the *hepatitis* virus; and the protozoa that cause *amebiasis* and *giardiasis*. Other organisms, like *E. coli* bacteria, normally occur in the rectum; they may cause local infections when transferred to the urethra or the prostate.

Until the AIDS epidemic, these conditions were seen more frequently in gay men, who are more likely to engage in anal sex. Given the hazards of AIDS, most gay men now avoid these risks or take precautions.

Heterosexual couples who engage in anal sexual activities are no less vulnerable. Men who have coitus following anal intercourse without carefully washing the penis in between, or wearing and changing condoms, risk infecting the vagina and their own urethra with fecal organisms. Transmission of such microorganisms can also occur indirectly when hands touch the anus or the fecally contaminated penis and then the mouth.

There are other reasons to worry about anal intercourse. The rectum is vulnerable to injury during anal intercourse or during the insertion of objects for anal masturbation. Because of the "pull" of the anal sphincter, objects that are inserted in the anus can be inadvertently drawn into the rectum. Repeated stretching of the anal sphincter may also lead to fecal incontinence (Rowan and Gillette, 1978).

Symptoms   The symptoms of these enteric infections frequently include diarrhea, nausea, possibly vomiting, and sometimes fever. As these organisms multiply, they irritate the lining of the bowel. These symptoms are no different in cases that have been sexually transmitted. The symptoms of viral hepatitis will be described later, when we discuss virally caused STD. The long-term consequences of some of these enteric infections include weight loss, wasting, and possible damage to the intestines, but usually they are self-limiting.

The diagnosis of these conditions may involve taking a stool specimen, blood tests, and other procedures. Treatment is usually with antibiotics.

Parasitic Infections

The organisms we have discussed consist of single cells. Parasites are multicellular organisms, and some of them may be transmitted by sexual or close contact and infest the skin, the pubic hair, or other external regions.

Pubic Lice   *Pediculosis pubis* ("crabs") is an infestation of pubic hair by crab lice (*Phthirus pubis*). They spread usually through sex but occasionally through infected bedding, towels, or clothing (Billstein, 1984). The primary symptom is intense itching. Cream, lotion, or shampoos with benzene hexachloride (Kwell) eliminate both adult lice, which are about the size of a pinhead, and their eggs, which cling to the pubic hair. To avoid reinfection, clothing or bedding that comes in contact with the body must be decontaminated.

Scabies   *Scabies* is a contagious skin infection caused by the itch mite (*Sarcoptes scabiei*). It causes intense itching and may cause a red rash. It can be transmitted by close personal contact, sexual contact, or infected clothing or bedding. Scabies is commonly found in the genital areas, buttocks, and between the fingers.

The female itch mite burrows into the skin and lays eggs along the burrow. The larvae hatch within a few days. The body's reaction to these eggs and larvae is responsible for the itching, irritation, and redness that appear in small track patterns. The mites are more active at night, when they cause the most itching. The treatment is the same as for pubic lice (Orkin and Maibach, 1984).

VIRAL STDs

Viruses are the smallest microorganisms. Despite their simplicity, they are highly specialized. They can only reproduce within the cells of a host—human, animal, plant, or bacterium. Some viruses are so well adapted that they live in human cells without causing harm; others cause ailments ranging from the common cold to fatal illnesses like rabies. Among the STDs, they cause *genital warts*, *genital herpes*, *hepatitis*, and *AIDS* (which we shall discuss separately).

Once a person acquires the virus for an STD, it stays in the body. There may be occasional flare-ups of the condition. Even a person who stays asymptomatic continues to carry the virus and can pass it on under certain conditions.

The symptoms of viral infections usually subside by themselves, as in many cases of herpes, but they may become progressively more serious, as in many cases of AIDS. STDs caused by viruses are not cured by antibiotics. Other treatments may alleviate symptoms, but usually do not get rid of the infection. Because viruses mutate easily into different strains, it is difficult to develop effective vaccines against them.

Genital Herpes

Genital herpes is a skin lesion with painful blisters caused by the herpes simplex virus, usually of the type II (HSV II) variety (Corey, 1984). Type I (HSV I) of the same virus usually causes *oral herpes*—"cold sores" with lesions on the lips, mouth, or face. HSV I can also cause genital herpes and HSV II, oral herpes.

Although recognized for some time, genital herpes did not become highly prevalent in the United States until the 1960s. The number of cases seen by physicians increased almost ten-fold during the period from 1966 to 1984 (Figure 5.5). An estimated one in five adults in the United States (over 20 million persons) has had at least one episode of genital herpes. Each year another half million cases are added, making herpes a major focus of medical and public concern (Leary, 1988). A burst of attention in the popular media in the late 1970s generated much public fear and concern about this sexually transmitted disease.

Transmission of Herpes   Herpes is typically transmitted by contact with the infected areas of the sexual partner. Self-contamination is also possible if a person touches an infected lesion and then immediately touches another body surface or an eye. Genital herpes most often spreads through coitus, mouth-genital contact, or anal intercourse—any contact with the lesions or an infected site that is about to develop lesions. For this reason, all contact with an infected region must be avoided until the blisters have healed completely. When there are early signs of a recurrence—itching, burning, and tingling where the blisters are to appear—touching this region should also be avoided.

The chances of infection during an active period of disease are high. The risk of infection is much lower during asymptomatic periods (Judson, 1983), so herpes does not mean that a person will never again be reasonably safe as a sexual partner. Using condoms and spermicides and washing with soap and water after coitus further reduces the risk. However, when lesions are not restricted to the area covered, a condom cannot provide complete protection against transmission of the herpes virus (Stone et al., 1986).

It is conceivable that herpes may also be transmitted through accidental contamination. The virus has been shown to survive for short periods outside the body in warm mucosal secretions, which may contaminate a toilet seat or hot tub. However, there is no evidence of significant risk in practical terms.

Symptoms   At first the lesions of herpes are small, fluid-filled blisters surrounded by inflamed tissue. They usually appear 2 to 20 days after first exposure to the virus (Figure 5.6). Common sites of herpes infection in women include the inner surface of the vagina and the surface of the cervix, in addition to the external genital region. In men, the penis, pubic region, and scrotum are frequent sites of infection.

Herpes blisters cause painful burning and itching. When they break open, the area may become secondarily infected with bacteria, causing prolonged painful sores; otherwise, the blisters clear up spontaneously within a few weeks. Meanwhile, the virus moves along nerve fibers into nerve clusters, where the body's immune system cannot get to it. In the case of genital herpes, these nerve clusters are

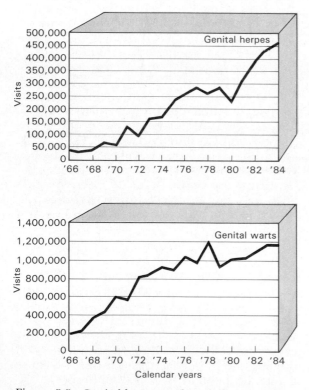

Figure 5.5   Genital herpes and genital warts: number of visits to private physicians' offices in the United States, 1966–1984.

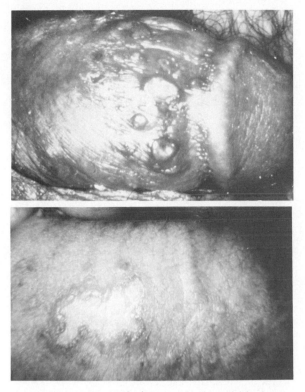

Figure 5.6   Herpes lesion on penis. Within hours or days, the blister-like eruptions (top) burst, leaving open sores (bottom).

in the sacral spinal cord. In the case of oral herpes, they are in facial ganglia deep in the cheek. There is no way at present of getting rid of the virus.

An infected person may have no further symptoms or may have recurrences as frequently as twice a month or as rarely as once a decade. The chance of the recurrence of herpes appears to be linked to a number of triggers. These circumstances lower the body's resistance, allowing the virus to cause symptoms again. Trauma, marked physical exertion, prolonged exposure to the sun, smoking, debilitating disease, menstruation, stress, and sexual activity itself (apart from reinfection) may all lead to recurrences of herpetic symptoms (Wickett, 1982). These triggers are also important in determining whether or not symptoms appear in the first place.

The long-term complications of herpes are rare but may be serious. Transferring the virus into the eyes after touching an infected area causes *herpes keratitis* and possible damage to the eye. Herpes on the lips may, in a small number of cases, lead to viral infection of the brain, or *viral encephalitis*. These complications are treated with the same drugs as other types of herpes.

Other long-term consequences of infection with the herpes virus may include an increased risk of cervical cancer. However, in most of the studies finding this risk, many of the patients were also infected with the human papilloma virus.

Occasionally a fetus may be infected before birth, and the virus may also be transmitted to babies during birth if the mother has an active lesion at the time of delivery. This condition affects about 100 to 200 babies per year in the United States. A pregnant woman who has an active outbreak at the time of delivery runs approximately a 50 percent risk that the newborn child will contract the disease. In babies infected with the disease, there is approximately a 50 percent risk of brain damage or other serious damage, and a high probability of death (Hatcher et al., 1988). To avoid this danger, women with active lesions deliver by cesarian section. Persons with active oral lesions should not fondle or kiss an infant.

Treatment   The treatment of herpes so far can only limit the symptoms, decreasing the pain and shortening the duration of the blisters. There is no cure because no drug can eliminate the virus from nervous tissue. Although all sorts of remedies have been tried, the only proven benefits have been achieved with *acyclovir*, which may be used as an ointment, taken orally, or in serious cases, administered intravenously. Acyclovir is thought to stop the herpes virus from multiplying. If used in the initial period the drug may alleviate the symptoms and shorten the duration of the attack. Washing with soap and water and the use of soothing agents also give some relief. Personal hygiene is important during active symptoms, to avoid reinfection by touching the sores and then touching other parts of the body.

The oral tablets of acyclovir, known as *zovirax* have been shown to reduce the length, severity, and frequency of outbreaks in people who take them before symptoms appear (Mertz, 1984). The long-term effect of taking this drug for many years is not known, so it should be taken for only six months to one year at a time, and not at all during pregnancy. Vaccines that could prevent acquiring the disease are actively being sought, but none are ready for marketing.

In comparison with other STDs, herpes is seldom a serious threat to health, yet emotional reactions to and fears of the disease may be severe. The stigma of having it may be more harmful than the disease itself. The threat of herpes infection was reportedly having widespread effects on patterns of sexual behavior among singles in the early 1980s, following strong media coverage (*Time*, 1981). Herpes has since been completely overshadowed by AIDS.

Genital Warts

Genital warts (*Condylomata acuminata*) are commonly, though not always, transmitted by sexual contact. Though this fact has been known since the time of ancient Greece, it was only in the 1930s that their cause was identified as the *human papilloma virus*, of which there are over 50 varieties. The prevalence of genital warts has been rapidly rising (Figure 5.5). An estimated one million new cases appear each year. Ten percent of the adults in the United States are estimated to be infected with the papilloma virus (Schmeck, 1987).

The transmission of genital warts is similar to the transmission of syphilis: a mucosal membrane is exposed to virus particles shed by an infected person. People may not realize they are infected with the virus and may show no visible genital warts, though they are still infectious.

Symptoms   Warts are growths on the skin that appear within three to eight months after the infection. Genital warts in women most often appear on the vulva or inside the vagina or on the cervix (Figure 5.7). In some pregnant women, previously acquired warts grow rapidly, causing annoyance, itching, irritation, and unsightly masses several centimeters in size. In males, they are usually seen on the surface of the penis or around the anus. Warts may last only a few weeks or may become permanent.

Another kind of genital wart caused by the same virus is not visible to the naked eye. These "flat warts" usually occur on the cervix or on the penis. They may become apparent under a special microscope after they are exposed to vinegar, which whitens infected cells.

The long-term consequences of flat genital warts include predisposition to cervical cancer and (rarely in this country) cancer of the penis and anus (Gal et al., 1987). This type of infection may be recognized on a Pap smear in women. Usually men are not screened for it unless they are the partners of infected women.

The diagnosis of visible warts is usually based on their appearance. It may be confirmed by examining a small section, or biopsy, under the microscope to confirm the presence of abnormal "giant cells" (*koilocytes*), which indicate infection with the human papilloma virus (Oriel, 1984).

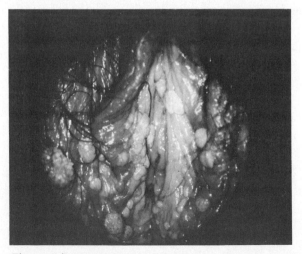

Figure 5.7   Genital warts on vulva.

Treatment Warts are treated by surgery, cauterization (burning), cryosurgery (freezing), or the application of various chemical compounds to kill the cells infected with the wart virus. Most forms of treatment require follow-up, especially the chemical forms, and all of them may be followed by a recurrence of warts in the same or adjacent areas.

Hepatitis

Hepatitis is an inflammatory illness of the liver. When caused by the *hepatitis B virus* (HBV), it is usually acquired through sexual contact or blood transfusions (Lemon, 1984). In the United States, 5–20 percent of the general population has evidence of past hepatitis B infections, though not everyone has been ill with it. In developing countries and in homosexual male populations, previous hepatitis B infection may exceed 80 percent.

The symptoms of hepatitis B infection take two to six months to appear. They include skin rash, muscle and joint pain, profound fatigue, loss of appetite, nausea and vomiting, headache, fever, dark urine, jaundice, and liver enlargement and tenderness. Although this picture is similar to other forms of hepatitis, a patient with typical symptoms and history is considered to have hepatitis B. The definitive diagnosis is by special blood tests.

There is no specific treatment for hepatitis B, but supportive and systematic care can improve rate of recovery and comfort. There is a vaccine highly effective at preventing the infection. However, once a person has already become ill, vaccination does not help. Vaccination is urged for high-risk groups—medical personnel, partners of known carriers, and people in the inner city, where the rates of HBV infection are high.

The potential complications of hepatitis B include persistence of the hepatitis in chronic form, cirrhosis of the liver, liver cancer, liver failure, and death. In rare cases, hepatitis may cause death within a few months. Some people with hepatitis remain, even after recovery, chronic carriers who may unknowingly transmit the virus to others through sexual contact or blood contact.

AIDS AND THE HUMAN IMMUNODEFICIENCY VIRUS

In June 1981, the Centers for Disease Control (CDC) reported an unusual outbreak of a rare disease in Los Angeles. Five gay men had developed a pneumonia caused by *Pneumocystis carinii*, a protozoan infection usually seen in individuals whose immune system is seriously deficient (Gottlieb et al., 1981). The month after, CDC revealed that another rare disease, this time a skin cancer called *Kaposi's sarcoma*, also associated with compromised immune systems, had been diagnosed in 26 gay men over the previous two and a half years.

From these isolated beginnings emerged a pattern of illness called the *Acquired Immunodeficiency Syndrome*, or *AIDS*. Along with a bewildering variety of related manifestations, AIDS has become the most formidable challenge to health that the world has faced during the past 50 years. Randy Shilts (1987) has reported how society reacted at first. It is a tale of initial neglect by the government (by the time President Reagan delivered his first speech about the epidemic, over 21,000 persons had already died of AIDS); scientists concealing crucial information from each other to garner credit; news media failing to focus on public policy issues; and gay leaders fearing that the truth about how AIDS is spread would compromise their hard-won liberties. Some find Shilts' critique to be unduly harsh (Reinhold, 1987). AIDS now receives high public attention, and the gay community is actively confronting the issues it raises.

The head of the World Health Organization has called it a "disaster of pandemic proportions" (Altman, 1986, p. 1), predicting that 100 million people will be infected by the AIDS virus by 1990, making it a global threat (Mahler, 1986). A National Academy of Sciences report predicts that by 1991, 270,000 people in the United States will have AIDS; 54,000 will die a year (more than deaths due to car accidents). Infection with the virus would have spread to 5 to 10 million people, of whom perhaps half will eventually die as a result (Morgenthau and Hager, 1987).

The disease has so far been identified in over 120 countries. The largest number of reported cases is in the United States. By March 1988, a total of 56,212 cases had been reported in the United States since the disease was identified; over 31,400 had died of AIDS. Estimates of the number of people who carry the virus range from 1 million to 1.5 million Americans (*Morbidity and Mortality Weekly Report, 1988*).

AIDS is not a disease of homosexuals alone nor of the United States alone. Both in prevalence and in capacity to deal with it, central Africa (Zaire, Rwanda, and Burundi) is the most threatened by the epidemic; its annual incidence is 550 to 1000 new cases per million adults. The cost of caring for ten AIDS patients in the United States (about $450,000) is greater than the entire yearly budget of a large hospital in Zaire, where nearly 25 percent of all adults and children admitted test positive for the AIDS virus (Quinn et al., 1986). Most cases of AIDS in central Africa are in the heterosexual population, and the prevalence of AIDS is about the same among men and women (Peterman and Curran, 1986). As we shall see, this is not the case in the United States, where over 90 percent of AIDS occurs among men.

How could such a devastating disease suddenly appear out of nowhere? The prevalence of the AIDS virus in central Africa suggests that the disease started on that continent, possibly in the African green monkey. Seventy percent of these animals are infected with the AIDS virus, although they do not suffer from its effects. It is suspected that during the past several decades the virus was transmitted to humans through monkey bites (this process has actually happened with another virus) (Essex, 1985).

It was at first assumed that the AIDS virus appeared in the United States sometime in the mid-1970s. It has now been established that a teenage boy (referred to as Robert R.) died in St. Louis in 1969 of what appears, in hindsight,

to have been AIDS, suggesting that the virus had already entered the country in the 1960s. Robert was sexually active, and it is suspected that he was gay. The fatal illness that baffled his doctors fits well with the symptoms of AIDS, and frozen sections from his autopsy have tested positive for the AIDS virus (Kolata, 1987).

Epidemics are nothing new. They decimated the population of medieval Europe time and again. More recently, in 1918, 20 million people (including 500,000 in the United States) died of the Spanish flu. Syphilis epidemics also have raged in this century.

Will the story of AIDS match that of syphilis? Like syphilis, AIDS is sexually transmitted, has varied effects, is highly virulent, and evokes panic and prejudice (Brandt, 1987). However, AIDS is different in alarming ways. It is transmitted not only sexually but through blood transfusions and between intravenous drug users. The most serious consequences of syphilis come late in the disease, when people are older; AIDS kills mostly the young. Syphilis can be cured; AIDS cannot. Most importantly, AIDS is caused by a virus, which makes the discovery of treatments and vaccines much more difficult.

One of the reasons that the AIDS epidemic caught health workers by surprise was that it did not fit the classic picture of an epidemic illness. AIDS is not highly contagious. It is not transmitted through casual contact even over long periods of time. By all counts, it should be restricted to isolated pockets of the population.

What led to its rapid spread were the changed social conditions in the 1970s. Extensive air travel exposed widely separated populations to each other. The dramatic rise in sexual permissiveness rapidly expanded the range of sexual contacts. The increase in intravenous drug use similarly set the stage for rapid infection (Kolata, 1987).

The Virus and the Immune System

The pattern of AIDS transmission pointed early to a virus as the causative agent. By

1984—only a few years after the disease had become known—the AIDS virus was identified in France and in the United States. It had taken several centuries for the causative agent of syphilis to be identified.

The AIDS virus was initially called LAV (*lymphadenopathy virus*) by some researchers, HTLV-3 (*Human T cell lymphotropic virus*) by others. It is now generally known as *HIV* or the *human immunodeficiency virus*. So far virtually all AIDS cases in the United States have been caused by the HIV-1 type of virus. A second type, HIV-2, has been isolated in cases of AIDS in West Africa and may become established in the United States as well. It seems to spread the same way but may cause a less severe form of AIDS.

HIV belongs to a group of viruses called *retroviruses*. A retrovirus contains a small number of genes made of RNA (ribonucleic acid) instead of DNA. These viruses are able to replicate themselves inside other cells, using the machinery of the host cell to multiply in large numbers. The newly formed viruses trickle out of the cell, eventually destroying it, and they enter other cells and repeat the process.

Through this process, HIV destroys the cells that form the key link in the *immune system* of the body. To understand how this happens we need to know how the immune system works.

The immune system has several defensive lines. *Macrophages* ("big eater") provide the first line of defense. These large cells are like scavengers that enter the circulation and engulf cells that have already become infected. At the same time they sound the alarm by activating white blood cells, called *helper T cells*. These in turn initiate the immune response. They alert other T cells to destroy infected cells, thus cutting off multiplication of the virus, and induce *B cells* to produce *antibodies* that help destroy the virus in circulation (Raven, 1986).

The AIDS virus breaks this chain. Invading helper T cells, HIV stops them from signaling T cells and B cells to go into action; moreover, HIV turns the helper T cells into minifactories, producing more viruses and killing the cells in the process. The virus also mul-

tiplies in macrophages, disrupting their activities and infecting yet other cells (Kolata, 1988a). The result is an unchecked spread of the virus and the collapse of the body's immune system, which leaves the person defenseless to other microorganisms, such as bacteria, fungi, and other viruses. It is the diseases caused by these "opportunistic" invaders that account for the symptoms of AIDS and eventual death.

Whether they are successful or not in combatting the invaders, the presence of antibodies against an organism is indirect but reliable evidence that the body has been infected with it. This fact is the basis of testing for the AIDS virus, which we will discuss later.

Transmission of AIDS

Viruses are tiny organisms that, unlike bacteria, cannot be seen under an ordinary optical microscope. If an electron-microscope photograph of an AIDS virus were to be enlarged to look as big as a fingernail, a human hair, by comparison, would appear to be 25 feet wide (Lertola, 1986).

Unlike the viruses that cause the common cold, which are airborne, the AIDS virus can only be transmitted to another person through body fluids. Ten body fluids may contain the virus: blood, semen, vaginal secretions, menstrual blood, breast milk (not cow's milk), tears, saliva, urine, cerebrospinal fluid, and alveolar fluid (in the lungs). However, not all these fluids actually transmit the virus from one person to another. The few that can transmit it are not all equally likely to do so. There is no documented case of infection from saliva or tears. Only blood and semen effectively transmit the virus; vaginal fluid and breast milk are less likely to do so (Friedland and Klein, 1987).

How will someone exposed to AIDS catch it? It depends on two factors. First, to *which* of the body fluids has the person been exposed? Second, *where* did the infected fluid enter the body? For example, infected semen is more likely to lead to AIDS if deposited in the rectum than in the vagina or the mouth. These two variables determine how risky a certain sexual behavior is with an infected person.

Anal Intercourse   The most common sexual form of transmission of AIDS in the United States has so far been anal intercourse. Especially at risk is the person who receives infected semen in the rectum ("receptive" anal intercourse). In a study of over 2000 gay men who tested negative for the HIV, 11 percent of those who engaged in receptive anal intercourse had acquired the virus a year later; as against 0.5 percent who only engaged in the insertive role in anal intercourse. All of those who avoided anal intercourse altogether still tested negative at the end of the year (Kingsley et al., 1987).

Because anal intercourse is most common among gay men, they have been the most at risk. The spread of AIDS among gay men has also been facilitated by the tendency of some of them to have numerous sexual partners. In the early days of the AIDS epidemic, gay bathhouses (most of which have now closed down) served as an "amplification system." A concentrated group of infected people quickly infected many others, who in turn passed the infection to an ever-expanding group (Shilts, 1987).

Shared Needles   The most common nonsexual form of transmission is the sharing of needles contaminated with blood containing the AIDS virus. Second only to gay and bisexual men, who account for about 70 percent of cases (78 percent of male cases), intravenous drug users account for about 17 percent of cases (15 percent of male and 53 percent of female cases). They are the most likely "bridge" for infection of the heterosexual population. These figures reflect the fact that there are many more gay men than IV drug users. However, in some cities IV drug users account for two-thirds of AIDS cases. Sharing needles is no less risky than anal intercourse.

Currently, as a result of greater awareness of AIDS, marked changes have occurred in the sexual behavior of gay men. The spread of the AIDS virus among gay and bisexual men has slowed dramatically in San Francisco and possibly in other urban areas with large homosexual populations (although half of the gay men in San Francisco are already infected) (Dowdle, 1987). However, AIDS continues to spread unabated among intravenous drug users, of whom there are 750,000 at risk (Des Jarlais, 1987). Unlike gay men, drug users are hard to reach, educate, and change. Attempts to make clean needles available to addicts have so far met with little headway; but arguments rage for and against it (Johnson and Joseph, 1987).

There has been a gradual concentration of cases in certain ethnic minorities that have higher prevalence of intravenous drug use and needle sharing; blacks and Hispanic people account for 17 percent of the adult population in the United States, but 39 percent of all AIDS cases. The Surgeon General reports that in the United States one in every four people with AIDS is black; nearly half of those under age 30 are either black or Hispanic; more than half of infants with AIDS are black, and one-quarter are Hispanic (*New York Times,* July 9, 1987). AIDS threatens to become one more massive problem for a segment of the population that is already heavily burdened.

Heterosexual Intercourse   Women and "straight" men also can catch AIDS. In the United States, 93 percent of adults with AIDS were male, 7 percent female—a total of 30,160 men against 2205 women—as of 1987 (Rubinstein, 1987). However, in central Africa the sex ratio is even.

The first studies showing AIDS among heterosexuals appeared in the early 1980s. These cases accounted for 1 percent of all AIDS cases in 1983, and 1.8 percent in 1987. Contrary to early fears, the disease did not appear to spread widely among the mainstream heterosexual population.

This reassuring picture was challenged by sex researchers Masters, Johnson, and Kolodny. Based on a study of 800 sexually active heterosexual adults, they found 6 percent of those with at least six sex partners during the preceding five years to be infected with HIV. From this and related findings, they concluded that the AIDS virus has established a beach-head in the heterosexual population, and its spread would begin to escalate at an alarming pace (Masters et al., 1988).

The conclusions and recommendations of these investigators were widely criticized by public health experts and the media (Eckholm, 1988). Their findings were at odds with most other studies conducted by epidemiologists using much larger samples. For example, screening of 25 million blood donations and over 3 million military personnel and recruits had shown rates of infection to be a fraction of 1 percent; these rates had not changed over two years (Boffey, 1988).

Hearst and Hulley (1988) have presented more detailed and less alarming data about the chances of heterosexual transmission of AIDS. The risks vary tremendously based on circumstances. In the most dangerous case, engaging in coitus without a condom with a partner infected with HIV, there is a 1-in-500 risk of infection. With over 500 acts of intercourse with such a partner, the risk of infection goes up to 2 in 3. At the opposite extreme, engaging in coitus using a condom with a partner who has tested negative for the virus, the risk of infection is 1 in 5 billion. In between these extremes, one time sex with someone who is not in a high-risk group but whose infectious status is untested carries a risk of 1 in 5 million (about the same as getting killed in a car accident while driving for 10 miles). These estimates may prove to be overly optimistic.

Although using condoms and limiting the number of partners continue to be important considerations, it appears that whom you have sex with is the more critical factor in who gets infected. This is consistent with the finding that in the United States nearly all of heterosexually transmitted infections (which account for 4 percent of all AIDS cases) have involved partners of drug users, bisexual men, or people from Africa or Haiti, where heterosexual spread is common. This population is largely concentrated among impoverished black and Hispanic communities in large cities. Until we know more about the heterosexual transmission of AIDS, guarded optimism combined with the exercise of caution provide the sensible approach.

The type of sexual activity among heterosexual couples also has a bearing on the risk of infection. Even in coitus, a woman is more likely to be infected by the man than the man by the woman: semen carries more virus and lingers in the vagina; the man is only exposed briefly to vaginal secretions, which carry less virus. (The same considerations make ejaculation in the mouth riskier than cunnilingus—oral stimulation of the vulva.) Anal intercourse carries a higher risk because of structural differences between the rectal and vaginal mucosa. The rectal wall will let the virus through more easily. Moreover, it is more likely to suffer slight tears during insertions of the penis (or a finger), allowing the virus to enter the bloodstream. The AIDS virus may also get attached to certain cells in the rectal mucosa and spread from this foothold.

Why is heterosexual coitus a far more common source of infection in central Africa than in the United States? One reason may be the high rates of other STDs that cause genital ulcers, such as chancroid and syphilis. Sores in the genital areas can be highly effective sites of viral shedding or attachment. Even in the United States, patients with genital ulcers seen at STD clinics are more likely to be infected with HIV than other patients, although the rate of such infection remains remarkably low (Quinn et al., 1988).

Blood Transfusion   Transmission through blood tranfusions has received much public attention because it is a potential threat to large numbers of medical patients, including children. Since no one chooses to need blood, getting AIDS through transfusions seems more "unjust" than acquiring the disease through high-risk activities in which a person chooses to participate.

A total of 500 transfusion-related cases of AIDS have occurred in the United States since HIV was isolated; meanwhile, over the past decade, 30 million patients have received 100 million units of blood. *Hemophiliacs* (of whom there are 20,000 in the United States) require weekly infusions of blood-clotting factors, so they are at greater risk; 1 percent of all AIDS cases have occurred in patients with hemophilia and 6 percent of children with AIDS have been hemophiliacs (Klein and Alter, 1987). With more stringent screening of donors and testing of blood, the likelihood of getting AIDS from a blood transfusion is now no more than 1 in 100,000 (Lipson and Engleman, 1985). Some people choose to store a supply of their own blood before a planned operation, to be perfectly safe.

Childbearing   Some 750 children had developed AIDS by the end of 1987. A small proportion of these children were infected through blood transfusions; most of the rest (70 percent) had gotten the disease from their mothers before birth. The majority of these children are born to mothers who have been infected by IV drug users. Four out of five are black or Hispanic (Eckholm, 1986).

A woman infected with the AIDS virus who wants to get pregnant faces a serious dilemma. Pregnancy itself, with its alterations of the immune system, may increase the chances of an infected woman developing AIDS. There is also a 50 percent chance of transmitting the virus to the child (Hatcher et al., 1988). Breastfeeding the infant possibly exposes the child to further risk. Nonetheless, pregnant women are not at present routinely tested for AIDS; whether to do so is part of a larger issue of testing, which we shall address later (Ledger, 1987).

Accidental Infection   Unbroken healthy skin is an effective barrier to the entry of infectious organisms, unlike the mucous membranes that line body openings (like the mouth, nose, anus, vagina, and urethra), which are more permeable. Accidental infections with the AIDS virus therefore involve contact of infected fluid with mucous membranes or cuts in the skin.

Fourteen health workers, by 1988, have been reported to have been infected with the AIDS virus on the job. Eight of them had suffered accidental punctures with infected needles. Two had chapped hands and were not

wearing gloves when exposed to infected blood. One was splashed with it in the face. Doctors, dentists, nurses, laboratory workers, and ambulance crews now wear gloves, masks, or goggles when they might be handling infected fluids. Special syringes and plastic lab containers also reduce the risk of accidental contamination (Pear, 1987).

How AIDS Is *Not* Transmitted   How safe is it to interact with individuals infected with the AIDS virus? The Surgeon General of the United States states that *you cannot get AIDS from casual contact*, such as shaking hands, hugging, social kissing, crying, coughing, or sneezing, or from swimming pools, hot tubs, or eating in restaurants (even if the cook or waiter has AIDS). You cannot get AIDS from toilet seats, doorknobs, telephones, office machinery, or household furniture—not even from shared bed linens, towels, cups, dishes, or eating utensils. Donating blood is absolutely safe. No child has ever gotten AIDS from another child in school. Dogs, cats, and other domestic animals are not a source of AIDS infection, nor are insects such as mosquitos (even though they can retain the virus in their bodies for a few days after ingesting infected blood) (Koop, 1986).

Symptoms

AIDS is not a simple disease. It is the most severe stage of infection with HIV. This virus makes the body vulnerable to many other infections, so there are many possible symptoms or illnesses that are associated with AIDS (Groopman, 1988). The course of infection with HIV has been classified by the Centers for Disease Control into four categories.

Category 1   Category 1 forms the pyramid base, comprising people who have been infected with the virus and have developed antibodies against it. Most people in this group either have no symptoms or show a flu-like syndrome, with fever, fatigue, and muscular aches and pains. These symptoms, if present, usually occur about two to six weeks after infection. Antibodies usually begin to appear in the blood in two months, but they may take six months or even a year to reach detectable levels. During this period, the person does not yet have AIDS, and may never get it, but he or she is infectious, able to transmit the virus to another person. In the absence of antibodies, blood tests are negative, so these individuals are undetected, silent carriers.

Category 2   Persons in category 2 still have no significant clinical symptoms, although the infection has now taken hold. Laboratory tests show the presence of antibodies and may also show small reductions in the level of T-4 cells—evidence that the immune system is being undermined, though the person is not aware of it.

Category 3   Persons in category 3 are unmistakably ill. The lymph nodes of the body swell up and can be felt like little lumps in the armpits, groin, neck, and elsewhere in the body constituting the *lymphadenopathy syndrome* (*LAS*). In addition, there may be persistent fever, night sweats, diarrhea, weight loss, fatigue, and uncommon infections such as yeast infections of the mouth (thrush) and of the vagina, or reactivation of the chicken pox virus, causing a painful skin condition called shingles.

These symptoms were formerly referred to as ARC (AIDS-related complex). Typically, they are not life-threatening unless diarrhea and weight loss are severe. There is now a serious threat of progressing to category 4 diseases.

Category 4   Only conditions in category 4 are called AIDS. It ushers in one fatal disease or another, most commonly *Pneumocystis carinii pneumonia* or *Kaposi's sarcoma*. The symptoms of pneumonia are concentrated in the chest, including cough and shortness of breath. The external lesions of Kaposi's sarcoma consist of painless blue or brown nodules of varying size.

A large number of other *opportunistic infections* and other cancers may further complicate the picture, causing meningitis, tuberculosis, toxoplasmosis, and so on. Because the virus invades the nervous system, AIDS may actually present itself initially as a neurological or psychiatric illness (Price and Forejt, 1986).

The four categories form an "iceberg" (Figure 5.8). Categories 1 and 2, including the bulk of infected people, are "below the water"—they show no major clinical symptoms. Categories 3 and 4 have clear symptoms and constitute the visible part of the iceberg.

Most people are thought to ultimately progress from category 1 and 2 to category 3 and 4. As of 1987, 20–30 percent of people initially infected with the HIV virus had developed AIDS within five years. Similarly, about 30 percent of patients with LAS (category 3) go on to develop AIDS in five years

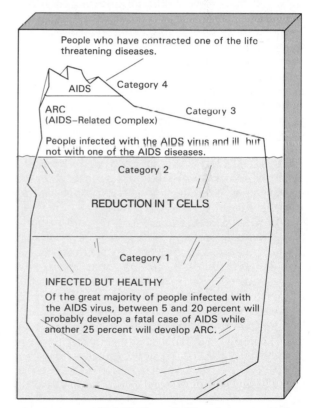

Figure 5.8   AIDS "iceberg." (Not drawn to scale; otherwise, category 1 would be about 20 times as large as category 4.)

(Kaplan et al., 1987). Current estimates suggest that at least 50 percent of all of those infected with the virus will go on to develop AIDS within ten years. Until more time has passed, it will be impossible to know what percentage of infected people develop AIDS over a lifetime.

Of patients diagnosed with AIDS (category 4), 50 percent die within 18 months of diagnosis; about 80 percent die within three years of diagnosis.

However, some people beat these odds. There are men infected with HIV who have developed no symptoms for almost ten years. A few patients have been known to live as long as six or more years with full-blown AIDS. It is as important to know why some people resist the disease as it is to know why others succumb to it (Altman, 1987). The long-term prognosis of AIDS may change rapidly as we find out more and as new drugs become available for its treatment.

Testing

How do we find out who carries the virus and who is ill with AIDS? The first question entails screening large groups of people; the second, diagnosing individuals.

A useful tool for both purposes is the *antibody test* (Box 5.3). As we discussed, the presence of antibodies in blood serum is indirect but reliable evidence of past exposure to the virus. Once the virus enters the body, it stays there indefinitely, so a *positive* test for antibodies also indicates current infection.

The antibody test is highly accurate, but it is not perfect. Antibodies to other, similar viruses can cause a *false positive* result. On the other hand, antibodies may not show up in an infected person, giving a *false negative* result; remember that it takes time for antibodies to develop. Even a true positive only reveals a potential for AIDS, if a person has no other symptoms. Tests cannot predict whether a virus carrier will develop AIDS in the future.

Counseling

The results of a positive antibody test must be communicated with sensitivity. The person needs help to place the issue in perspective and come to terms with its emotional and practical consequences.

Most people respond to having any life-threatening illness predictably: anxiety and fear mingle with anger, despair, bewilderment, and a sense of helplessness. Persons with AIDS carry some additional burdens. Guilt and remorse may affect those who have acquired the disease through sexual activity. Anger may be directed at those who were the source of infection. The person may feel "unclean" and dangerous to others, or fear rejection. The knowledge that the virus is there for life and represents a threat to life is likely to have a major impact on facing and planning for the future. It will affect all significant human relationships, especially intimate and sexual ones.

In helping people cope with the knowledge that they carry the AIDS virus, two points are particularly important to get across. First, having a positive antibody test does not mean having AIDS. Although there is at least one chance in three of developing AIDS over the next six to eight years, it is not possible to tell who will develop it. A positive test for HIV is not a death sentence. It does carry certain serious implications; facing them, a person can live fully, and if necessary, prepare to die with dignity.

Second, a positive test for HIV means that the person is potentially infectious to others by intimate sexual contact, by sharing needles, by childbearing, or by donating blood, semen, or body organs. No one should expose others to potential serious harm, at least not without their knowledge and consent. A carrier has an obligation to inform a sexual partner beforehand and take all possible precautions against infection through "safe sex" practices.[1]

Treatment

Despite intensive efforts, there is as yet no cure for AIDS or any way to eliminate the virus

[1]The Public Health Service offers information and counseling on these and other issues related to AIDS through hotlines (call 1-800-342-AIDS; 1-800-443-0366).

Box 5.3

AIDS BLOOD TESTING

Testing for infection with HIV is done on *serum,* the clear fluid of blood without cells. The first test, called *enzyme-linked immunosorbent assay (ELISA),* is sensitive enough to detect almost all infected persons, but it also will give a large number of false positives in a low-risk population such as blood donors.

If the ELISA test is negative, the person is considered not to have been exposed to the AIDS virus, provided enough time has elapsed for antibodies to develop. If the test result is positive, the test is repeated twice on the same serum sample. If one of the two repeat tests is positive, then the ELISA test is considered positive; if both repeat tests are negative, the ELISA test is considered negative. In other words, for an individual to be reported as positive by ELISA at least two out of three tests on the same serum specimen should have been positive (Saah, 1987).

A confirmed positive by ELISA is followed by a more refined test called the *Western Blot,* to determine whether the antibodies present are specific to HIV antigens. Other refined tests are currently being developed or in limited use. For instance, the RIPA relies on radioisotopes; the IFA, on infected cells from tissue cultures; antigen detection tests use monoclonal antibodies. Technical problems and expense make the wider use of these tests impractical at this time.

If the Western Blot test is also positive, the person is considered to be infected with HIV. If it is negative, the person is probably not infected with HIV. Nevertheless, that person's blood should not be used for transfusions, and he or she must behave

as if the test were positive for another six months. At that time if a repeat Western Blot test is negative, the individual can be considered not to carry the AIDS virus.

Double testing with ELISA and the Western Blot is expensive; it is only done if ELISA is repeatedly positive. Therefore the Western Blot test does not affect the false negative rate, but it does reduce the false positive rate. If the ELISA shows 0.25 percent of a low-risk group to test positive, the Western Blot will give positive results for only 0.1 percent, but they will be more specific to HIV and not some related virus (Saah, 1987).

Moreover, the significance of these tests depends on whether they are carried out in low-risk populations (such as blood donors, who are unlikely to have the virus) or high-risk populations (such as IV drug users, who are more likely to have the virus). In a high-risk population, where the prevalence of HIV infection may range from 30 percent to 70 percent, a positive ELISA is almost always confirmed by Western Blot, giving it a high prediction value. In a low-risk population, if 10 million people are tested by ELISA, 25,000 (0.25 percent) will be positive, whereas only 10,000 (0.1 percent) will be confirmed as positive by the Western Blot. This means 15,000 false positives by ELISA.

The use of ELISA for mass screening of low-risk populations will label many people as infected with HIV when in fact they are not. The social and psychological consequences of such a false verdict are considerable. Therefore, testing large segments of the population raises important social and political concerns.

---

from the body. However, two types of drugs are being tested: antiviral agents to keep the virus from multiplying, and immune system stimulants to restore the damaged immune system.

The most promising drug so far is the antiviral agent *azidothymidine* (AZT). When given to ARC patients (category 3), AZT significantly reduces the chances of developing AIDS

(Shilts, 1988). Given to patients suffering from *Pneumocystis carinii* pneumonia, it has shown a definite ability to lessen symptoms and prolong life. AZT interrupts the conversion of RNA to DNA by the virus, interfering with its multiplication. However useful it may prove to be, AZT is not a cure for AIDS.

Other antiviral agents under investigation are *ribavirin* and several other compounds. Immune system stimulants include *interleukin* and *interferon,* which also has antiviral properties. A combination of antiviral drugs and immune system stimulants may hold the most promise, especially if treatments start early, before the virus has caused severe damage.

Cost of AIDS

Caring for people with AIDS is going to be costly—how costly remains to be seen. Here again, we should avoid inertia and panic.

Assuming that caring for each AIDS patient will cost no more than $50,000 a year, by 1991 AIDS is estimated to cost $10.9 billion—a substantial sum, yet only 1.4 percent of the nation's annual health budget (Morganthau and Hager, 1987). In this broader context, AIDS will be about as costly as cancer.

Directly or indirectly, everyone will pay for the treatment of AIDS patients. The money will have to come either from higher health insurance premiums or from Medicaid (the federal subsidy for the indigent) through higher taxes.

Private insurance companies fear that the cost of an unchecked AIDS epidemic will destroy their business. They are accused of "dumping" AIDS patients or rejecting their insurance claims; life insurers in turn claim that the illness preexisted at the time of enrollment and that some applicants have taken out policies after learning they carry the virus. Here again, although the sums are large, they are modest in the overall context of health care costs: less than 1 percent of the total payouts in the care of one major insurer. However, the total health insurance costs of AIDS is expected to reach $10 billion a year by 1991 (*New York Times,* July 13, 1987).

Even more serious is the fate of those left out of the U.S. health care system. An estimated 35 million people have no medical insurance. Faced with catastrophic illness, they have to exhaust their personal assets before becoming eligible for public assistance. In addition to the personal cost, a substantial increase of AIDS cases in this population will place a tremendous strain on hospitals, especially on tax-supported big city hospitals. New hospitals will have to be established to care for AIDS victims or wider use will have to be made of alternative-care facilities such as hospices, nursing homes, and in-home care by visiting nurses. There are enormous problems to be faced with each of these alternatives. Unless there is a revamping of the existing health care system (such as by establishing a national health insurance), the care available to AIDS cases who are needy may be seriously hampered.

Prevention

The ideal way to deal with a disease is to prevent it. The best way to accomplish that with infectious diseases is either to avoid exposure or to protect the body by bolstering its defenses. *Vaccines* protect by stimulating antibody production *before* the body is invaded by the microorganism, rather than merely in response to it.

Vaccines   Substances called *antigens,* which are akin to a disease-causing organism but harmless to the body, stimulate the body to produce antibodies that will provide immunity against the organism itself. For example, the vaccinia virus (from which the word "vaccine" derives) causes an infectious disease in cattle, called cowpox. Inoculated into humans, it causes the immune system to produce antibodies that protect it against smallpox. So successful has been this vaccine that smallpox has been virtually eliminated from the world.

Other viruses have proven difficult to vaccinate against; the viruses that cause the common cold are one example. There are so many variants and so many shifts in their properties that they are like a multiple moving target. The same is true for HIV. Therefore, although some vaccines are under trial, there is no way of telling if and when an effective and safe vaccine will be available (Scarpinato and Calabrese, 1987).

Safe Sex   Currently, avoiding high-risk activities and following safe sex practices remain the only safeguards against AIDS.

If you are not an intravenous drug user who shares needles, sexual contact is virtually the only way that you are likely to get infected with AIDS. You are *absolutely safe* from the threat of AIDS if you fulfill one of the three following conditions: you *abstain* from sex altogether; you engage in sex of any kind only with a partner with whom you have had a strictly *monogamous* relationship (no other sexual partners for either of you) since 1977 (when AIDS appeared in the United States); you engage in sex with a new partner with whom you establish a monogamous relationship, provided that neither of you is already infected with the AIDS virus. This condition means both of you test negative for AIDS antibodies and have not engaged in sex for a year with anyone who might carry the virus. You must fully trust yourself and your partner to know that; if in doubt, you and your partner must wait for a year and be tested again. During this interval, you may engage in the safer sex practices described below but avoid all genital contact or exposure to each other's body fluids. That means restricting yourself to hugging, caressing, "dry" kissing, and mutual masturbation that does not involve direct contact with semen or vaginal fluids (Kaplan, 1987).

These are stringent rules. Not everyone is willing to follow them. As we shall discuss below, what you do in any activity in life is in part a function of what level of risk you are willing to take. The only way to avoid an airplane crash is not to fly; the only truly safe sex with regard to AIDS is what is described above.

The term "safe sex" is now commonly used for sexual practices that more properly should be called "safer sex"—they provide considerable safety but fall short of being altogether safe (Ulene, 1987). Admittedly, there is a problem of semantics here. Vehicles and machinery approved as "safe" are not entirely safe either; they merely are expected to meet legal requirements. You will have to decide for yourself what "safety" should mean for you.

Once you move out of the truly safe sex category, the chances of exposure to AIDS are determined by what you do sexually, how often, and most importantly, with whom. *Safer sex practices* include vaginal and anal intercourse with condoms; oral-penile contact (fellatio or "sucking") that stops short of ejaculation or uses a condom; mouth-vaginal contact (cunnilingus) with a dental dam (a piece of stretched thin rubber), which prevents the mixing of saliva and vaginal secretions; and dry kissing. *Unsafe practices,* starting with the most risky, are anal intercourse without a condom (especially in the receptive role); vaginal intercourse without a condom; oral-anal stimulation ("rimming"); unprotected fellatio; and mouth-vaginal contact (Gong, 1987; Ulene, 1987).

Condoms have received a good deal of favorable publicity as safeguards against transmitting or acquiring AIDS. Laboratory tests have actually confirmed that the AIDS virus will not cross the walls of an intact latex condom (but may get through "natural" condoms made of animal intestines). The virus also cannot survive long periods of contact with the chemicals in spermicides. As a result, condoms certainly help reduce the chances of infection, but they are far from foolproof. Condoms could not be more effective in preventing AIDS than they are in preventing pregnancy. As we shall discuss (Chapter 7), condoms have a 2–10 percent failure rate. In other words, 10 percent of women who rely exclusively on their partner's using a condom will get pregnant every year. The failure rate of condoms with respect to HIV must be at least as high. Spermicides may increase the level of protection condoms provide but it is unclear to what extent. The spermicide foam or cream must be placed in the vagina, not inside the condom, which would make it easier for the condom to slip off the penis during intercourse.

The most critical question is, with whom should you have sex? A few simple rules prevail. First, the greater the likelihood of your partner being infected, the greater the chance of your being infected. Intravenous drug users are hazardous partners for both sexes; bisexual men are far riskier partners than heterosexual men (all else being equal) for women. Men and women who have had some other STDs are more likely to have the AIDS virus. Even geographical location makes a difference in the probability of infection. Although no state or city is immune to AIDS, some areas have a higher incidence than others. In 1987, New York City and San Francisco each had over 1000 cases of AIDS per million population; Chicago had 96; much of the rest of the United States had 53. It is to be expected, therefore, that a sexually active person in a high-risk city is more likely to come in contact with the AIDS virus than someone in a low-risk city, even if the number of sexual partners they had were the same. (This risk applies only to those who are sexually active; simply living in New York or San Francisco does not carry a higher risk.) Box 5.4 summarizes the factors that determine a sex partner's risk for AIDS.

Second, the more partners you have and the more partners your partners have, the riskier the contact becomes (Goedert et al., 1987). Even after getting infected, repeated exposure to the virus seems to increase the chances of infection progressing to the development of symptoms. In a group of 1034 single men aged 25 to 34 in the San Francisco Bay area, 49 percent of the gay and bisexual men tested positive for the AIDS virus, but those with 50 or more partners were 71 percent positive. None of the heterosexual men had evidence of infection (Winkelstein et al., 1987).

Prostitutes of either sex present a high-risk group because they have multiple partners and many are drug users; at least some female prostitutes are now more likely to insist that the customer use condoms, which makes them less of a hazard. Some studies have actually failed to implicate female prostitutes as a major source of transmission (Rabkin et al., 1987). Much may depend on where the prostitute is from: 57 percent of prostitutes in Newark have tested AIDS positive, as against 1 percent in Atlanta or Colorado Springs and none in Las Vegas (Ulene, 1987).

Finally, how often you engage in an unsafe activity is of obvious significance. Even unprotected anal intercourse might not result in infection the first time, or even the tenth time, but the odds will catch up with you. The longer you persist in the behavior, the higher the risk of infection.

Risk-Taking Behavior   Driving motorcycles is much more dangerous than driving cars. Over a period of a year, 1 out of 1000 motorcycle users as against 1 out of 6000 drivers of cars will die in an accident (Hatcher et al., 1988). Similarly, various occupational and recreational activities differ widely in the level of risk they entail. How much chance we take, how often, and why defines the pattern of our risk-taking behavior. How dangerously we live can be a function of necessity (as in occupational hazards) or choice. Both are determined by complex psychological and social reasons.

Sexual behavior also entails a certain degree of risk. One in 50,000 women dies each year because of pelvic infections acquired through coitus. The risk of death in pregnancy is 1 in 10,000 per year (Chapter 6). Exposure to STDs, especially AIDS, greatly increases the risks in sexual activity.

It is up to every individual to make an informed choice about how much to risk in being sexually active in general and in a particular case. You must assume responsibility for protecting yourself, weighing the costs and the benefits. Consider, for instance, the use of condoms. Some people simply do not bother with them. Others consider the failure rate to be unacceptable, "when the price for failure is getting a disease that can kill you" (Ulene, 1987, p. 30). Abstinence from sex except under fully safe conditions is the only sensible choice for some, but too restrictive for others. At a minimum, you must take *calculated risks*, knowing the most likely consequences of your actions and being as clear as possible about your motives.

The Joy of Sex   There are two unfortunate reactions to a condition like AIDS. One is to ignore and deny it, endangering yourself and others. The other is to allow it to stifle the joy of sexuality. With every sexual partner a potentially deadly source of infection, every sexual encounter a tangle of latex barriers, the

Box 5.4

ESTIMATING A SEX PARTNER'S RISK FOR AIDS

	No Risk	Low Risk	High Risk
Number of sexual partners	None, or one with mutual monogamy since 1977	Few	Many
Sexual preference of partners		Heterosexuals or homosexual females	Homosexual or bisexual males
Use of barrier contraceptives with others		Always used condoms	Rarely or never used condoms
AIDS antibody test results	Negative and no sexual exposure for past six months	Negative and no sexual exposure for past three months	Positive or untested
Prior history of sexually transmitted disease	No	No	Yes
Use of drugs	No drug use by subject or partners	No IV drug use by subject or partners	Uses IV drugs, or sex partner uses them
Transfusions	None, or has been tested and found negative for AIDS antibodies	Transfused but not tested	Transfused many times or with large volume of blood; not tested
Places of residence		Low-incidence area for AIDS	High-incidence area for AIDS

From *Safe Sex in a Dangerous World* by Art Ulene, M.D. Copyright © 1987 by Feeling Fine Programs, Inc. Reprinted by permission of Random House, Inc.

fame after a while seems no longer worth the candle.

There is much to be said for abstinence or playing it absolutely safe, but if you are willing to take calculated chances, as you do in other realms of your life, it should be possible to engage in safer sexual activities without throwing caution to the winds. It will require the realization that there is far more to sexual intimacy than intercourse, or direct genital contact. By focusing on the erotic potential of the entire body—through physical closeness, caressing, hugging, and expressions of affection—our sexual interactions are enriched, even in situations that lead to intercourse. "Outercourse" rather than intercourse, "dry" rather than "wet" sex, and sensuality rather than sexuality will have to provide the security needed in the age of AIDS.

Similarly, we need to learn to inquire about our sexual partner's past without becoming an inquisitor, to be alert to signs of illness without being clinical, to be romantic while remaining realistic.

Handling the STDs will call for massive education, radical changes in risk-taking behavior, and the exercise of self-restraint. The personal and social challenges of AIDS in particular are immense, and each of us will have to deal with them.

REVIEW QUESTIONS

1. What cancers affect the female and male reproductive systems?
2. Which STDs cause an unusual vaginal discharge in women and a urethral discharge in men?
3. What STDs cause skin lesions in the genital area?
4. Rank the STDs by the threat they represent to personal health.
5. How is the AIDS virus transmitted?
6. What are the treatments for STDs?

THOUGHT QUESTIONS

1. How would you respond to the charge that the STDs are a just punishment for immoral sexual behavior?
2. If you were a college president, would you let the health service test students for AIDS voluntarily? How about making the test mandatory? Explain.
3. As a public health official, how would you deal with the possibility that the STDs may be more prevalent among the gay male population in your city?
4. How should society deal with an individual who has AIDS yet continues to engage in behavior that is likely to transmit the virus to others?

SUGGESTED READINGS

Federation of Feminist Women's Health Centers. (1981). *A new view of a woman's body*. Illustrated guide to female health care.

Holt, L. H., and Weber, M. (1982). *Woman care*. New York: Random House. Overview of issues related to women's health, written for a general audience.

Rowan, R., and Gillette, P. J. (1978). *The gay health guide* (1978). Boston: Little, Brown. Useful guidance for health problems affecting gay men.

Silber, S. S. (1981). *The male*. New York: Scribner's. Guide to male health care.

Ulene, A. (1987). *Safe sex in a dangerous world*. A thoughtful and concise overview of AIDS and coping with its threat.

Pregnancy and Childbirth

CHAPTER

6

And God blessed them, and God said unto them, "Be fruitful and multiply."

GENESIS 1:27–28

OUTLINE

The desire to generate a new life runs deep in human nature. Having children is not for everyone, but for the human species as a whole no event is more critical. Though sex serves other purposes than reproduction, no other aspect of sexual relations carries more important biological, psychological, and social consequences. No decision requires more serious thought than becoming a parent and accepting responsibility for a new life.

You do not have to become a parent. Many individuals are voluntarily and involuntarily childless. The desire for parenthood and the fact of parenthood do not always match. On the one hand about one in nine married couples in the United States is infertile. On the other hand, roughly half the pregnancies that do occur in the United States are unintended, and one-fourth are voluntarily aborted (Hatcher et al., 1988).

Motherhood entails much more than pregnancy, and there is more to fatherhood than impregnating a woman. Parenthood entails many important phychological and social considerations beyond the biology of reproduction. Furthermore, there are significant changes now in parental relationships and roles. The traditional ideal has been to have children in a stable marriage. Although most children are still born to couples who are married or in a stable relationship, a small but growing number of single women and men are choosing to raise children alone or sometimes within a gay relationship.

No one knows when in human history the momentous discovery was made connecting coitus with reproduction. The association between the two is by no means obvious, and until fairly recent times it remained unknown to people in isolated cultures. For instance, Trobriand Islanders in the South Pacific thought a woman conceived when a spirit embryo entered her body through her vagina or head; the Kiwai of New Guinea ascribed pregnancy to something the woman ate (Ford and Beach, 1951). Myths persist even after people become aware of the true nature of reproduction. However, herding people as early as biblical times had already made the connection between the mating of their flock and the birth of baby animals in the spring.

There are many compelling reasons why people want children, apart from the enjoyment of engendering them. Some of these reasons are socioeconomic, others more personal and psychological. Traditionally, children have worked to help support the family, and the cycle of generations has depended on parents looking after their children and then the children looking after their parents. Also, children are a contribution parents make to the clan, ethnic group, and nation; parenthood makes them feel like full-fledged adult members of society. Though in modern industrialized societies children are much less of an insurance against the future, there is still considerable social pressure on young adults to become parents, both from their own parents and from society at large (Pohlman, 1969; Fawcett, 1970).

Powerful inner psychological forces also motivate us to have children. The notion of a "parental instinct," though hard to define and even harder to substantiate, aptly conveys the deep and elemental urge to have a baby, even when the rewards are uncertain or the costs prohibitive. To love and to be loved by a child, to share that love with others, and to feel the sheer enjoyment that children provide are not matched by many other human experiences.

Some pregnancies are motivated by the wrong reasons: to cement an uncertain or faltering marriage; to entrap another into a permanent relationship; to assert manhood or womanhood; to enhance self-esteem; to take a shortcut to adulthood; to fill idle time, and other manipulative and self-serving reasons.

CONCEPTION

The fertilization of the egg by the sperm marks the beginning, or *conception,* of a new life.[1] For a fertile and sexually active woman trying to have a baby, it takes on the average six months

---

[1] The discussion of embryonal development is based on embryology tests, in particular Moore (1982) and Sadler (1985).

to get pregnant: a single act of intercourse at the time of ovulation has a 21 percent chance of resulting in pregnancy. Approximately 90 percent of women trying to get pregnant conceive within one year and another 5 percent within two years (Trussell and Vost, 1987; Hatcher et al, 1988). Some women get pregnant after the first act of coitus, others after more than a decade of trying.

The Journey of the Sperm

Sperm cells develop within the seminiferous tubules (Chapter 2) and travel through the testicular duct system to the epididymis, where they attain full maturity. Then they move through the vas deferens to the ejaculatory duct. During emission (Chapter 3) sperm are mixed with the secretions of the seminal vesicles and the prostate gland, which nourish them and help them move.[2] This mixture, called *semen*, is ejaculated through the urethra. Up to this point, the movement of sperm has been assisted by the contractions of the male tubal system; now the sperm must move on their own.

The spermatic fluid in a normal ejaculation has a volume of 2 to 3 milliliters (approximately a teaspoonful), containing 300 million to 500 million sperm. Most of the fluid in the ejaculate comes from the seminal vesicles and the prostate (whose secretions are also responsible for its odor). Semen is whitish and semigelatinous, but it gets more watery after repeated ejaculations within a short time. It coagulates on exposure to air.

In the vagina the sperm begin to make their way into the uterus, leaving most of the fluid of the ejaculate behind. If the woman is lying on her back, more of the sperm will reach the cervix and enter the uterus. Keeping the penis for a while in the vagina following orgasm also helps the chances of impregnation. If the secretions of the cervix and vagina are

strongly acidic, sperm are destroyed quickly; even in a mildly acidic environment the movement of sperm ceases fairly quickly.

Sperm swim in fluid by lashing their tails, moving at a rate of one to two centimeters (less than one inch) an hour. The notion that orgasmic contractions suck in sperm through the cervix has not been substantiated. Actually, the sperm have to get by the plug of cervical mucus that blocks the entrance to the endometrium during most of the menstrual cycle. Passing through the uterus, they get into the fallopian tubes and complete the final two inches or so of their journey, assisted by contractions of the tubes and swimming against the current generated by the hairlike cilia that line them (which propel the ovum in the opposite direction).

The journey of the sperm ends several hours after ejaculation when they reach their usual fertilization site, at the lower third of the fallopian tube (Figure 6.1). Of the several hundred million sperm that start on this journey only about 50 actually make contact with the egg, and then only one sperm eventually penetrates it.

The Migration of the Egg

At the time of ovulation, the mature Graafian follicle protrudes from the surface of the ovary. It is filled with fluid under pressure, and its wall has become thin. The egg has become detached within the follicle and is floating freely in its fluid. At ovulation, the follicle wall breaks through the ovary's surface, and the ovum is carried in a stream of fluid into the fallopian tube by sweeping movements of the tube's fringed end. The follicle walls that remain behind develop into the corpus luteum.

Once the egg has entered the fallopian tube, it begins a leisurely journey toward the uterus, taking about three days to move three to five inches. The egg, in contrast to the sperm, has no means of self-propulsion. It is swept along by the current generated by the cilia lining the tube. If the egg is not fertilized, it will be expelled with the subsequent menstrual flow (Kaiser, 1986).

[2]Prostatic secretions are rich in *acid phosphatase*. Its presence in the vagina is used as presumptive evidence that coitus has occurred in legally contested cases (such as rape).

Box 6.1

BOY OR GIRL

The sex of the baby is determined at the time of fertilization by the type of sperm that fertilizes the egg: if the sperm carries an X chromosome, the child will be a girl; if it carries a Y chromosome, the child will be a boy.

If you could pick the sex of your child ahead of time, would you do it? Such attempts are as old as recorded history. Aristotle recommended having intercourse in a North wind if boys were desired, in a South wind if girls were desired. By the Middle Ages the formula had gotten more complicated: to ensure a son the man had to drink lion's blood, then have intercourse under a full moon (Wallis, 1984).

Recent attempts have been based on the fact that Y- and X-bearing sperm vary slightly in physical characteristics. The X-bearing sperm are heavier, and hardier in an acid environment.

One formula is to engage in intercourse two to three days before ovulation and then to abstain, if a girl is desired. (Supposedly, because X-bearing sperm are heavier, they will travel more slowly and get to the ovum at the right time.) To conceive a boy, the couple must limit intercourse to the time of ovulation. (The lighter Y-bearing sperm move faster.) This timetable is combined with an acid douche, shallow penetration, and no orgasm for the woman to conceive a girl, or an alkaline douche, deep penetration, and orgasm for the woman to conceive a boy (Shettles, 1972; Guerrero, 1975). This approach has now been largely discredited.

A more promising approach is to let semen drift down a solution in a glass column. In a sticky albumen solution, the Y-sperms make it faster to the bottom than the X-sperms; in gelatinous powder, the X-sperms get ahead. In either case, the concentration of sperm at the bottom can be used in artificial insemination to enhance the chances of sex selection. The proponents of this method claim a 77 percent success rate. Some investigators have been able to replicate these results; others have not.

Some couples go so far as to determine the sex of the fetus by amniocentesis (Box 6.3), and then abort it if it is the "wrong" sex. The practice is said to be more prevalent in certain third world countries that place a high premium on male infants.

A study of 5981 married women in the United States indicates that if sex selection could be successfully practiced, the long-term effect on the ratio of male to female births (currently 105 male to 100 female) would be negligible. However, there was a strong preference for their first child to be male, so the short-term effect would be a preponderance of male births, followed several years later by a preponderance of female births—provided that most couples have more than one child. When asked about the desirability of sex preselection, 47 percent of the women in the sample were opposed, 39 percent were in favor, and 14 percent were indifferent (Westoff and Rindfuss, 1974).

Fertilization and Implantation

Fertilization is the fusion of sperm and ovum. The egg usually must be fertilized within 12 hours after ovulation; it ordinarily does not survive longer than 12 to 24 hours. Sperm also usually live for 24 hours, but some may remain capable of fertilizing the ovum for about three days. Frozen sperm are said to survive for up to ten years. (The use of such sperm for artificial insemination is discussed later.) For a woman to get pregnant, intercourse must usu-

ally take place within a day before or after ovulation. Rhythm methods of contraception rely on avoiding this fertile period, but the exact timing of ovulation is not easy to determine (Chapter 7).

Sperm that reach the egg are held to its surface by minute projections (microvilli) from the follicular cells (corona radiata); this is why in electron microscope photographs, sperm appear over the egg surface. The fertilizing sperm must undergo critical changes before it can penetrate the ovum's protective layers. In this process of *capacitation* the sperm sheds the protein coating from its head, releasing enzymes that digest a path through to the ovum (the *acrosome reaction*). Normally, once a sperm has made its way into the ovum, no other sperm can get through. An immediate inhibitory reaction in the egg blocks the wall. Occasionally more than one sperm will penetrate the egg, but the result of such abnormal fertilization will abort sooner or later (Moore, 1982).

The head of the sperm that has successfully penetrated the ovum detaches from the rest of its body. The nucleus of the sperm and that of the egg merge, intermingling their sets of 23 chromosomes and restoring the full complement of 46 chromosomes typical of human cells. This new combination of genes from the parents determines the genetic makeup of the new organism, including its gender (Box 6.1).

The process of fertilization takes about 24 hours. The fertilized egg, which is called a *zygote*, now continues to move toward the uterus. After some 30 hours, the zygote divides into two cells; these two become four, the four become eight, and so on. Though there is no significant change in volume during these first few days, the zygote becomes a round mass of numerous cells called a *morula* ("mulberry") by the time it reaches the uterus in three days (Figure 6.1).

The cells of the morula arrange themselves around the outside of the sphere, leaving a fluid-filled cavity in the center. This structure, called a *blastocyst*, floats about in the uterine cavity. Sometime between the fifth and seventh days after ovulation it attaches itself to the uterine lining. This *implantation* is the real start of pregnancy. Though the normal site of implantation is in the uterus (usually in the back wall), occasionally it is outside it. Such a pregnancy leads to serious complications as we shall discuss further on.

Two ova that are fertilized simultaneously give rise to fraternal or *dizygotic twins*. When a single fertilized egg subdivides before implantation, identical or *monozygotic twins* develop. Genetically, fraternal twins are like ordinary siblings. Identical twins have the same genetic makeup; they are always of the same sex and look alike. Twins occur in 1 out of 90 births;

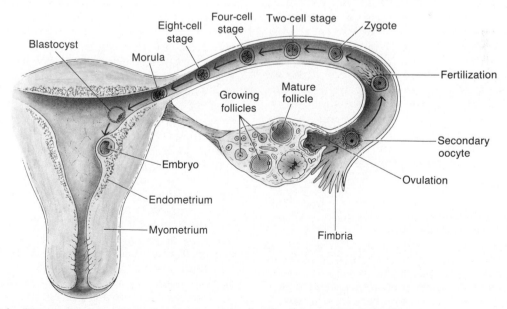

Figure 6.1   Diagram summarizing the ovarian cycle, fertilization, and implantation.

two out of three sets of twins are fraternal. Triplets occur about once in 5000 births; quadruplets, once in 500,000 births. Births of more than four children are extremely rare. The first quintuplets known to survive were the Dionne sisters, born in Canada on May 28, 1934.

PREGNANCY

It is common practice to divide the nine months of pregnancy into three-month periods or *trimesters*.[3] During this period the growing organism is called an *embryo* (Greek, to "swell") for the first eight weeks and a *fetus* (Latin, "offspring") thereafter. The average length of pregnancy is 266 days. Some pregnancies are shorter or longer.[4] Babies born before the 36th week (252 days) are considered *premature*, but fetuses are *viable*, or able to survive, when they are 28 weeks and older.

The First Trimester

The Embryo Develops   At the time the blastocyst is implanted, the uterine endometrium is at the peak of the secretory phase of the menstrual cycle (Chapter 4). The blastocyst burrows in as its enzymes digest the outer surface of the uterine lining, reaching the blood vessels and nutrients below. By the 10th to 12th day after ovulation, the blastocyst is firmly implanted in the uterine wall; but the woman does not yet know that she is pregnant, for her menstrual period, which she is going to miss, is not yet due for several more days.

This time is critical for the embryo's sur-

[3]The discussion of pregnancy is based on obstetrical texts, including Pernoll and Benson (1987), Danforth and Scott (1986), and Pritchard et al., (1985).

[4]An authenticated upper limit of pregnancy is 349 days (Haynes, 1982). The possible length of pregnancy assumes legal importance in establishing the legitimacy of a child when the presumed father has been away for more than ten months. In the United States the longest pregnancy upheld by the courts as legitimate lasted 355 days.

vival. The outer layer of the embryo, consisting of *trophoblast cells* (which will become part of the placenta), starts producing *chorionic gonadotropin*. This hormone stimulates the corpus luteum to maintain its output of estrogens and progestins. If the level of these hormones drops, the uterine lining will slough off and the embryo will be lost without the woman even realizing that she was pregnant (Yen, 1986).

In the early stages of development, a layer of cells forms across the center of the hollow blastocyst. From this *embryonic disk* will eventually grow all of the parts of the embryo. The surrounding cells will develop into the placenta and the fetal membranes. The membranes will form a sac filled with *amniotic fluid* ("bag of waters"), within which the fetus will float, cushioned and protected until birth.

The embryonic disk elongates into an oval shape by the end of the second week after fertilization. The embryo is still barely visible to the naked eye at this stage. Figure 6.2 shows the actual sizes of the embryo during the first seven weeks of life.

During the third week, the embryonic disk differentiates into its three distinctive layers: *endoderm*, *mesoderm*, and *ectoderm*. From the inner, endodermal layer will develop the internal organs; from the middle, mesodermal layer will arise muscles, skeleton, and blood; and from the outer, ectodermal layer, the brain and nerves, the skin, and other tissues.

The head end of the embryo develops faster than the rest of the body during these early stages. By the end of the third week the beginnings of eyes and ears are visible and the brain and other portions of the central nervous system are beginning to form. By the end of the first month, the embryo has a primitive heart and the beginnings of a digestive system. In the fifth week, precursors of arms and legs become visible; finger and toe forms appear between the sixth and the eighth weeks. Bones are beginning to ossify (harden with calcium), and the intestines are forming. By the seventh week the gonads are present, but cannot yet be clearly distinguished as male or female (Chapter 2).

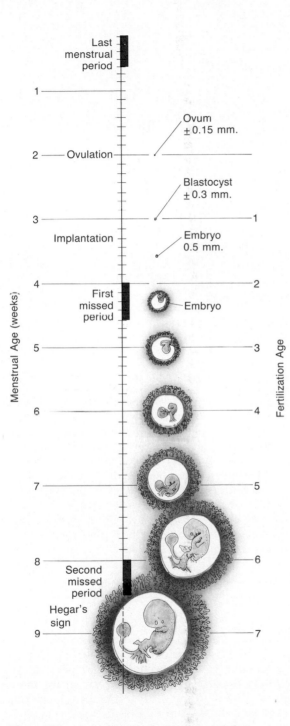

Figure 6.2   Actual sizes of embryos and their membranes during the first seven weeks of life.

By the end of eight weeks, when the embryonal period ends, the rudiments of all essential structures are present, and by four months the major organs are fairly well formed. The clearly defined external features now make the tiny fetus look human. From here on fetal development is mainly a matter of growth and differentiation. The way a simple cluster of cells develops into a complex organism has been a source of wonder, speculation, and study throughout history (Box 6.2).

The Placenta Develops   The embryo's lifeline to the mother is the *placenta*. The placenta develops from both fetal and maternal tissues, eventually growing to a sizeable organ weighing about one pound (450 grams). It sustains the life of the fetus by conveying to it nutrients and oxygen and carrying away its waste products (Knuppel and Godlin, 1987). This transport takes place through the blood vessels of the *umbilical cord*, which connects the fetal circulatory system to the placenta. There nutrients and waste products are carried across and seep through the walls of an extensive capillary network. The fetal blood comes very close to the maternal blood without direct contact or intermingling.

The placental membrane that separates the maternal and fetal blood also allows the passage of hormones, electrolytes, and antibodies. It acts as a barrier to some but not all substances that may be harmful to the fetus. Most drugs pass the placental barrier. So do many infectious agents, which is how infants are born with congenital syphilis and AIDS (Chapter 5).

The placenta also functions as an endocrine gland, first producing chorionic gonadotropin to keep the corpus luteum active and then itself taking over the production of estrogens and progestins.

Signs and Symptoms   How does a woman suspect she is pregnant? There are both objective signs and subjective symptoms (Taylor and Pernoll, 1987). Most common in early pregnancy are feelings of fatigue and drowsiness, but some women experience a sense of height-

Box 6.2

THEORIES OF EMBRYONAL DEVELOPMENT

The discovery that coitus results in pregnancy, momentous as it was, did not resolve the mystery of how a complex being develops from formless elements. People speculated about this process until the modern science of embryology was able to elucidate its underlying mechanisms.

Aristotle believed that the human embryo originated from the union of semen and menstrual blood. This same notion has also been held in preliterate societies. For instance, the Venda tribe of East Africa believed that "red elements" like muscle and blood were derived from the mother's menses (which, they explained, ceased during pregnancy because the menstrual blood was being absorbed by the developing fetus); the "white elements"—like skin, bone, and nerves—developed from the father's semen (Meyer, 1939).

Following the invention of the microscope in the 17th century, sperm and ovum could be seen and their role in reproduction surmised, yet neither looked remotely "human." How could a child develop from them? Two schools of thought emerged to explain the mystery. The *ovists* claimed that a minuscule but fully formed baby was contained in the egg, and that the sperm functioned only to activate its growth. The *homunculists* held the opposite view, that the preformed baby resided in the head of the sperm but did not begin to develop until it arrived in the fertile uterine environment. Looking through their crude early microscopes, some homunculists claimed to have actually seen a homunculus ("little man") inside the sperm (see the figure).

Ovists and homunculists were both *preformationists*—they presumed that a fully formed minia-

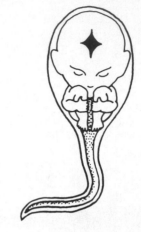

A homunculus as drawn by Niklaas Hartsoeker in 1694.

ture being simply grew bigger during development. This idea meant that all generations of humans, past and future, were stacked inside each other, like so many Russian dolls-within-dolls. Curiously, modern genetics provides some vindication for this fantastic notion: our genetic endowment is transmitted in a continuous line from one generation to another.

Preformationist theories died hard. Aristotle had considered the possibility of differentiation rather than simple growth. However, it was not until the 19th century that the theory of *epigenesis*, which embodies this doctrine, became established. In this theory, simple components develop into more complex parts in a continuous sequence of growth and differentiation (Arey, 1974).

---

ened energy and well-being. Physiological factors affect how a woman feels. The mood of the woman who knows she is pregnant also depends on circumstances. If the pregnancy is genuinely wanted, there is a sense of satisfaction and anticipation. If for any reason the pregnancy feels wrong, the experience may be colored by worry, anger, and depression.

The sign most commonly associated with pregnancy is a missed menstrual period. Although pregnancy usually stops menstruation, so can many other causes, including vigorous exercise, illness, and emotion. Women younger than 20 and older than 40 are more likely to skip a period. Conversely, the presence of a vaginal bloody discharge does not rule out pregnancy. About 20 percent of women have *spotting*, a short period of slight bleeding connected with implantation. Such bleeding is usually harmless but may also be an early sign of miscarriage. This bleeding may also be mistaken for a menstrual period.

Other physical signs of early pregnancy are swelling and tenderness of the breasts, frequent urination, irregular bowel movements, and increased vaginal secretion. A particularly bothersome symptom in many women during the first six to eight weeks of pregnancy is *morning sickness*. Queasy sensations upon awakening are accompanied by an aversion to food, or to the odors of certain foods. In some cases there is nausea, with or without vomiting. Some women experience these symptoms only in the evening. About one in four pregnant women experience no morning sickness, whereas in about one in 200 cases vomiting is so severe that the woman must be hospitalized. Treatment keeps this condition, known as *hyperemesis gravidarum*, from causing serious consequences, including malnutrition.

A doctor can find more objective evidence of early pregnancy. The vagina shows a purplish coloration (*Chadwick's sign*), as does the cervix. By the sixth week of pregnancy (a month after missing a period) a doctor can feel a soft and compressible area between the cervix and body of the uterus (*Hegar's sign*).

Occasionally, a woman develops some of the symptoms of pregnancy without being pregnant ("false pregnancy" or *pseudocyesis*). These women, who are usually intensely desirous of having a child, stop menstruating and develop symptoms of morning sickness, breast tenderness, a sense of fullness in the pelvis, and the sensation of fetal movements in the abdomen (caused by contractions of the abdominal muscles). The absence of objective signs and negative pregnancy tests reveal the true condition.

Pregnancy Tests   The signs and symptoms above do not prove pregnancy. More definitive confirmation comes from laboratory tests for *human chorionic gonadotropin* (*hCG*) in blood or urine—a substance normally found only in pregnant women.

In earlier versions of these tests, urine or serum was injected into female mice or rabbits: the presence of hCG made the animals ovulate (or, injected into male frogs, made them ejaculate). At present, hCG is detected by immunologic tests that are simpler, less expensive, and 90–95 percent accurate.

The test consists of mixing a woman's blood or urine sample with specific chemicals; if hCG is present, even in small amounts, it can be detected within minutes. The test for hCG using urine is now the most common test for determining pregnancy. It can detect hCG as early as 7 to 12 days after conception. The blood test for hCG is used less frequently, in part because it is more expensive, but it can detect hCG within 6 to 8 days after fertilization.

The most sensitive test for hCG is *radioimmunoassay*, which is almost 100 percent accurate and can detect hCG within the first week of pregnancy. This method determines the concentration of substances in blood plasma through the use of radioactive antibodies.

Another version of the hCG test is available in pharmacies in the form of kits for home use. If used correctly, it can detect hCG in the urine of a pregnant woman as early as nine days after a missed period. Recent studies show that only in 3 percent of cases, home tests wrongly indicated pregnancy, but in 20 percent of tests giving negative results, the women were actually pregnant. When women with negative results performed a second test eight days later, test accuracy increased to 91 percent (McQuarrie and Flanagan, 1978).

All pregnancy tests can give *false negative* results when the woman is pregnant and *false positive* results when she is not. Absolute confirmation of pregnancy can be established only by one of three means: hearing the fetal heartbeat, seeing the fetal skeleton, or observing fetal movements. Until recent developments in technology, none of these signs of pregnancy could be verified until well into the second trimester; now they can be verified earlier.

Using a conventional stethoscope, a physician can hear the fetal heartbeat by the fifth month. (The fetal heart rate is 120 to 140 beats per minute, so it sounds different from the mother's heartbeat, which is usually 70 to 80 per minute.) A fetal *pulse detector* can detect the fetal heartbeat as early as 9 weeks, and quite reliably after 12 weeks.

Photographic evidence is obtained through *ultrasound*. Variations in the echo from an ultrasonic pulse reflect off the fetal skeleton. They are converted to a photographic image of the fetus in action. The image is far more distinct than a conventional X-ray; it is also much safer, because it does not involve radiation. The fetal heartbeat can also be seen on the ultrasound screen. Neither ultrasound nor X-ray is used routinely, but each can verify suspected complications, such as fetal head size larger than the pelvic opening, gross malformation, and multiple fetuses. Further information about the fetus can be obtained through newer techniques of *amniocentesis* and *chorionic villi sampling* (Box 6.3).

The mother begins to feel fetal movements in the abdomen usually by the end of the fourth month ("quickening"). They feel like the fluttering of a bird in the hand. The movements not only confirm pregnancy but indicate that the fetus is alive.

Expected Date of Delivery   Once pregnancy has been confirmed, the next question usually is, "When is the baby due?" The expected delivery date or *expected date of confinement (EDC)* can be calculated by the following formula: add one week to the first day of the last menstrual period, subtract three months, then add one year. For instance, if the last menstrual period began on January 8, 1988, adding one week (to January 15), subtracting three months (to October 15), and adding one year gives an expected delivery date of October 15, 1988. In fact, only about 4 percent of births occur on the day predicted by this formula; but 60 percent occur within five days of it.

The Second Trimester

The Fetus Grows   Internal organ systems mature during the second trimester, and there is a substantial increase in size. By the end of the sixth month the fetus weighs about two pounds and is some 14 inches long. Among the changes during this period is the development of a temporary fine coat of soft hair (*lanugo*) on the body.

By the end of this period the facial and bodily features are well formed. The fetus moves its arms and legs and alternates between periods of wakefulness and sleep. Though the uterus is a sheltered environment, a loud noise or a rapid change in the position of the mother can disturb the tranquility of the womb and provoke a vigorous reaction. Changes in outside temperature are not perceived; the intrauterine temperature stays slightly above the temperature of the rest of the mother's body.

The Mother in the Second Trimester   For many women, the second trimester is the happiest time of pregnancy. The nausea, lethargy, and other troublesome symptoms of the earlier period have usually subsided, and the concerns about delivery are still in the future.

The pregnancy now feels secure, and the expanding abdomen and bustline (which may necessitate maternity clothes) make the pregnancy "public." The experience is also much more real because late in the fourth month the mother can feel the fetal movements more strongly. The "kicking" of the fetus becomes outwardly visible on the abdomen. The mother is now directly aware of the new life growing within her, and begins to relate to it. Other family members may do the same. Being more comfortable than earlier, and not yet as burdened as she will be later, more women are able to keep active during this time, at work, at home, or in sports and leisure; there is also heightened sexual interest.

Nevertheless, the second trimester has its own possible discomforts. Pressure from the expanding uterus may lead to indigestion and constipation. Varicose veins and hemorrhoids may appear or get worse. Fluid retention

Box 6.3

AMNIOCENTESIS AND CHORIONIC VILLI TESTING

Under certain circumstances it is important to know about the fetus as early as possible. For instance, if there is a likelihood of genetic abnormalities, because of family history or the age of the mother, an early determination gives the parents either reassurance or the option of aborting the fetus. In other cases, it may be possible to treat the problem while intrauterine development is still in progress.

Amniocentesis is the analysis of fetal cells floating in amniotic fluid in order to obtain genetic information about the fetus. First, ultrasound scans the womb for the position of the fetus. Next, a long needle is inserted through the abdominal wall into the uterus, and a small amount of amniotic fluid is drawn without harming the fetus (see the figure). Now fetal cells are recovered from the fluid and analyzed. The condition and number of the chromosomes in these cells (as well as the presence of XX or XY sex chromosomes, showing the sex of the child) can be determined accurately.

Amniocentesis is an accurate and relatively safe procedure for both mother and fetus. It is usually performed after the 14th week. Among the abnormalities that can be detected by amniocentesis are Down's syndrome (mongolism), neural tube closure defects (open spine), Tay-Sachs disease, cystic fibrosis, and Rh incompatibility. The analysis of amniotic fluid can reveal a wealth of information on such chromosomal, metabolic, and blood conditions. When abnormalities are diagnosed, parents can choose to have an abortion—although by the time the chromosome analysis is complete (the 20th week to 24th week of pregnancy), abortion is no longer a simple, risk-free operation (Chapter 7). Parents who choose to continue the pregnancy have time to prepare for the extra challenge.

Because a woman in her forties is 100 times more likely than a woman in her twenties to have ova with abnormal chromosomes, amniocentesis is recommended for women who become pregnant after the age of 35, and who have previously had a child with an abnormality that can be detected by analysis of fetal cells. It is also recommended for couples with family histories of certain hereditary

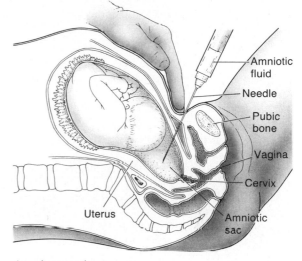

Amniocentesis.

problems. About 3 percent of newborns in the United States suffer an obvious defect, so procedures like amniocentesis could be of wide benefit.

The newest test, *chorionic villi sampling (CVS)*, uses a small sample of tissue from the surface of the chorion, the outermost layer of the membranes enclosing the fetus. Though it may not be as accurate as amniocentesis or detect as many abnormalities, it can be done as early as the eighth week (but usually in the ninth to eleventh week) and the results are obtained in a few days (it takes two weeks for amniocentesis). This timing allows the mother to have a first trimester abortion with fewer complications. The risks with CVS are not significantly higher than in amniocentesis, but there is a 1–2 percent chance of causing a miscarriage. This risk is especially significant for women over 35, who may have more trouble conceiving again (Brozan, 1985).

Doctors can now draw fetal blood samples directly from the umbilical vessels. Using ultrasound to "view" the fetus, a thin needle is passed through the mother's abdomen into the umbilical cord. This allows for a greater range of abnormalities to be detected early and treated by drugs or transfusions (Kolata, 1988b).

causes swelling of the feet and ankles (*edema*), and the woman may begin to gain excessive amounts of weight.

The Third Trimester

The Fetus Matures   The last phase of development is mainly a period of further maturation. The fetus is becoming well enough formed to survive on its own; by the middle of the last trimester the prematurely born baby (six weeks early) has a 70–80 percent chance of survival.

Early in the trimester the fetus is in an upright or *breech* position. It keeps shifting around in its increasingly cramped quarters, and in 97 percent of cases assumes a *head-down* position by the time it reaches full term.

During the ninth month the fetus gains more than 2 pounds (0.9 kg), and essential organs like the lungs become ready for life in the outside world. At full term the average baby weighs 7.5 pounds (3.4 kg) and is 20 inches long (50.8 cm), but there is a great variation in birth weights, ranging usually from 5 to 9 pounds. (The largest baby known to have survived weighed 15.5 pounds at birth.) Ninety-nine percent of full-term babies born alive in the United States now survive, a figure that could be improved even further if all expectant mothers and newborn babies received proper care.

The Mother in the Third Trimester   The relative comfort and tranquility of the second trimester gradually give way to the special discomforts of the third. Most of them have to do with the increased activity and size of the fetus, which displaces and presses upon maternal organs (Figure 6.3). What were occasional fetal movements now turn to periods of seemingly perpetual kicking, tossing, and turning, which may keep the mother awake at night.

The woman's weight, if not controlled up to this point, may become a problem. The optimal weight gain during pregnancy is 24 to 27 pounds (11 to 13 kg). The average infant at

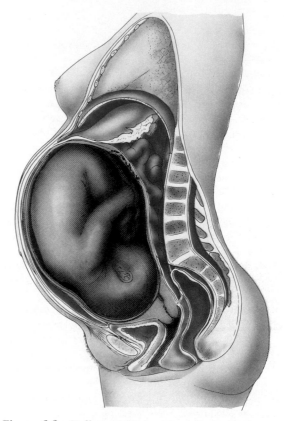

Figure 6.3   Full-term fetus.

nine months weighs about 7.5 pounds (3.4 kg). The rest of the weight gain is accounted for as follows: the placenta, about 1 pound (0.4 kg); amniotic fluid, 2 pounds (0.9 kg); enlargement of the uterus, 2 pounds (0.9 kg); enlargement of the breasts, about 1.5 pounds (0.7 kg); and the retained fluid and fat accumulated by the mother, the balance of 10 pounds (4.5 kg). Excessive weight gain is associated with a higher incidence of medical complications during pregnancy, such as strain on the heart and high blood pressure.

By the ninth month of pregnancy a woman is usually impatient. She speculates about the sex of the child and worries whether or not the baby is going to be all right. She may also feel some anxiety about the process of delivery. However, the properly prepared woman in good health also counts the days with pleasure and anticipation before she finally meets her baby face to face.

Psychological Aspects

Almost any parent could tell you that pregnancy brings out strong feelings. There has been too little research on these vital aspects of pregnancy: the fluctuating emotional states, the formulation of new identities and roles for the mother and father, and the shifts in relationships.

Pregnancy and childbirth are among the most significant events of many women's lives. Childbearing has major consequences for a woman's health, intimate relationships, career, and sense of fulfullment. Few experiences have as profound an effect on women's lives as bearing children. Traditionally, having children has been expected of women in most societies, but modern women in the industrialized world exercise a great deal more choice. Whichever way a woman decides, her choice will affect her life.

It is hardly surprising that pregnancy and childbirth are emotionally rich and psychologically challenging experiences. How satisfying or distressing they will be depends to a large extent on the circumstances. To have a child with the right man, at the right time, is obviously a very different experience than when one or more key requirements are lacking. Some of the special problems of unwanted pregnancies will be discussed in connection with abortion (Chapter 7) and adolescent development. We shall focus here on pregnancies that occur under ordinary, if not always optimal, conditions.

The first set of psychological concerns has to do with the physical experience. As we have described, the female body undergoes a remarkable series of changes over a nine-month period, some of which entail considerable discomfort. This expectation and fear of the unknown cause anxiety: Will I get sick? Will I be able to keep active? Will childbirth be painful? Such questions naturally occur to many women. Closely related are concerns about the baby's health and safety: Will my child be normal? An expectant mother can hardly avoid wondering, especially if there is the slightest indication of something wrong.

Another set of questions, just as important, involves the relationship with the baby's father—typically the husband: Will our marriage change? Will we be able to afford this child? Will he still find me attractive? Some women worry that childbearing will make them less romantic, less sexy, more domestic and humdrum; other women expect the child to strengthen the marital bond, to deepen the commitment of the husband, or to complete the relationship.

Finally, women wonder about motherhood itself: Will I be a good mother? Will I know how to take care of a child? These considerations may stretch all the way to the child's adolescence. Questions about their own future tie in for working women: Will I be able to take off enough time after the baby's birth? Will I earn enough to provide for the baby? Will I be able to balance mothering with a career? How much responsibility will my husband assume?

Sexual Interest and Activity

Pregnancy changes a woman in body and mind. No wonder it affects her sexual interest and behavior. There is much variation in its effect among women and in the different phases of pregnancy. In general, sex in pregnancy is quite safe for the mother and the fetus. Nonetheless, people often worry that it will affect the mother's well-being and the baby's safety.

Orgasm makes the uterus contract. It does have an effect on fetal heart rate, and mothers report an increase in fetal movements, but the significance of these effects is unclear (Chayen et al., 1986). Some cultures prohibit coitus during part or all of pregnancy; others permit sex all the way to the time of delivery (Ford and Beach, 1951). Due precaution or abstention are necessary in certain circumstances, such as when there is danger of miscarriage (Herbst, 1981). Otherwise, how a couple's sex life changes during pregnancy is usually a matter of personal choice.

Some studies show a steady decline in the frequency of coitus throughout the pregnancy (Pepe et al., 1987). Other studies reveal a different pattern. In the first trimester, women show a highly variable response. There may be a heightening of sexual feelings, no change, or a decline in sexual interest, often associated with morning sickness, fatigue, fear of disrupting the pregnancy, or high progesterone levels. In the second trimester, some 80 percent of women report an increase in sexual desire and responsiveness (Masters and Johnson, 1966; Tolor and DiGrazia, 1976). In the last trimester there is a consistent drop in the frequency of coitus linked with physical awkwardness and discomfort, a woman's feeling of being less attractive (a perception not always shared by the husband), and fear of hurting the baby (Calhoun et al., 1981).

There are two health concerns in this area: the chance that orgasm might cause miscarriage or premature labor, and the risk of infection. The risk from orgasm is realistic for women with a history of miscarriage or who are showing early signs of it (like vaginal bleeding). There seems to be no other correlation between experiencing orgasm and giving birth prematurely (Perkins, 1979). The risk of infection becomes a serious consideration if the membranes are ruptured or if the man is infected with a sexually transmitted disease (which may be asymptomatic).

Some of these problems related to sex during pregnancy are easily resolved. The woman-above position or rear entry coitus is less awkward in later months and less taxing on the woman. Use of condoms will help prevent infection, although it will not protect against herpes, which can be fatal for the baby. Noncoital sexual activities like masturbation or mouth-genital stimulation provide alternatives to coitus, but orgasm still can stimulate miscarriage. The practice of blowing air forcefully into the vagina (which some people do during cunnilingus) is dangerous in pregnancy, because it may introduce air bubbles into the woman's bloodstream, causing an air embolism that may be fatal (Sadock and Sadock, 1976).

Sex during pregnancy involves factors besides health. Just because a woman is pregnant, her sexual needs and those of her partner should not be ignored. Of the 79 husbands of pregnant women interviewed by Masters and Johnson (1966), 12 had turned to extramarital sex. Discovery of such an occurrence is likely to be particularly upsetting for a pregnant wife.

Sex is but one aspect of a couple's relationship significantly modified by pregnancy. In many other ways, a woman's pregnancy also affects the expectant father.

The Father's Experience

The prospect of becoming a father is an enormously satisfying and joyful experience for many men. Much depends, of course, on the couple's relationship and the social and economic circumstances under which the pregnancy is taking place. To have a child in a loving and compatible relationship can be a tremendous satisfaction to a man, as it is to a woman, but an unwanted pregnancy in a tenuous or doomed relationship leaves the man feeling helpless. Some men distance themselves from the woman ("It's her problem"); others are racked by guilt and despair ("How could I have done this?") or anger ("How could she do this to me?").

Despite the difference in reactions, the prospect of fatherhood is likely to affect men in some common ways. This is particularly true when a man becomes a father for the first time, which is what we shall focus on here.

At a psychological level, a man may see parenthood as a confirmation of his manhood, just as a woman may see it as an affirmation of her womanhood. His capacity to impregnate is public evidence of his potency, but beyond that it confers on him, just as it does on a woman, an added measure of being *generative*, which is the quintessential mark of adulthood. Hence, a man feels a deep pride and a strengthened self-esteem.

Having a child irrevocably changes a man's self-image, no less than it does that of a woman. His entire perspective on life is likely to be revised in the light of his expected responsibilities as a parent. All major life decisions must now take into account their impact on his children.

These positive reactions also have their negative counterparts. Most often, there is a mix of contradictory feelings. Expectant parenthood is never a wholly positive or wholly negative prospect.

Traditionally, marriage and parenthood have been supposed to provide a woman with security and fulfillment. For men they have supposedly meant getting "tied down." Friends joke that the man is giving up his carefree independence.

Currently, both men and women balance the anticipated joy of parenthood against the loss of freedom and social and financial responsibilities looming ahead. They both struggle with the same doubt: "Will I be a good parent?"

At the relational level, significant shifts take place. Some husbands identify with their pregnant wives so closely that they too suffer from morning sickness and other signs of early pregnancy. In one study, 23 percent of husbands in the United States showed such psychological manifestations (Lipkin and Lamb, 1982). In some preliterate societies, men actually went through "labor" in concert with their wives (*couvade*) in an effort to distract evil spirits away from the mother and baby (Davenport, 1977).

A man needs to adjust to the physical and psychological changes in the expectant mother. Those men who put great stock in the figure of their wife may be alarmed to see it change; other men are fascinated by these changes and even find them an added source of attraction.

Pregnancy makes a woman turn inward. She has a new being within her to relate to—a feeling that a man can never experience, hence never fully comprehend. Her preoccupation with the pregnancy and developing infant may draw her apart from the husband; yet other couples develop a greater intimacy than ever before. At this time a woman may also become closer to her mother or female friends, which is likely to make the man feel excluded.

The combination of the wife's physical changes, the novelty of the experience, and not knowing what to expect, makes the first trimester a particularly trying time for the man. The second trimester provides a breather—the woman feels better and the man is less anxious about her. The signs of pregnancy are now visible enough for him to be able to share more directly the mother's experience. For instance, by feeling the movements of the fetus through her abdomen, he too can now relate to this new "person" as yet unseen. At the doctor's office, he can hear his child's heart beating. He can begin to attend prenatal classes with his wife, and talk there with other "expectant fathers."

In the third trimester, the man confronts a new set of adjustments. As the time of birth approaches, the reality of what is happening becomes more tangible. The woman is no longer as physically active a companion as he is used to for housekeeping chores, work, play, or sex. More than ever before he must confront the question of whether by becoming a mother, she will be less of a wife, lover, companion, and friend. Jealousy and envy may begin to creep into his expectation of what life is going to be like. On the other hand, there is the exhilaration of meeting his child face-to-face and the expectation that the relationship will revert to what it was or even get better; but it will never be quite the same.

CHILDBIRTH

In the past it was believed that babies struggled out of the uterus at birth. Actually the baby plays no active part in the birth process; it is passively propelled out of the birth canal. Most societies have devised various ways of assisting women in childbirth. In preliterate cultures, sitting, squatting, or kneeling postures for the mother were usually considered optimal. Women were also "helped" by being suspended, sat upon, tossed about in blankets (to shake the baby loose), and having smoke blown into their vagina.

Traditionally, women have been assisted in childbirth by other women, usually an experienced *midwife* (from "with wife"). In Europe, men were barred from attending women at childbirth until the 16th century: in 1552, a Hamburg physician called Wertt was burned at the stake for posing as a woman to attend a delivery. As late as the 18th century most physicians considered it beneath their dignity to care for pregnant women, and notions of modesty precluded their assisting at labor (Speert, 1982).

With the advent of modern obstetrics early in the 19th century, the delivery of children was taken over by physicians and became almost exclusively a medical practice in the industrialized world. Recently, objections have been raised to this practice on the grounds that the hospital setting turns childbirth into an impersonal, expensive, and needlessly medicalized procedure (Arms, 1975). As a result, a number of alternative methods of *natural childbirth* have been advocated. Some women now choose to deliver at home, or in *birthing centers* that combine medical facilities for any emergency with a home atmosphere. Meanwhile, hospitals have modified their routines to counter some of the criticisms.

Midwifery is making a comeback (Figure 6.4). Nurse-midwives are trained to assist throughout the pregnancy, delivery, and postnatal care, and to counsel about sexuality. Most of them work in cooperation with obstetricians.

Labor

As the end of pregnancy nears, the expectant mother experiences sporadic contractions in her uterus—referred to as *false labor*. Three or four weeks before delivery the fetus "drops" to a lower position in the abdomen. The next major step is the softening and dilation of the cervix. Then, just before true labor begins, a small, slightly bloody discharge (*bloody show*) appears in the vagina, when the plug of mucus that blocked the cervix comes out. *Labor* follows this event usually within a few hours but sometimes after several days.

Labor consists of regular and rhythmic uterine contractions, which dilate the cervix and culminate in the delivery of the baby, the placenta, and the fetal membranes. It is the fetus rather than the mother who actually triggers labor (Daly and Wilson, 1978). A number of hormonal factors are known to be involved (Russell, 1987). The fetus' adrenal gland produces hormones that make the placenta and uterus increase the secretion of *prostaglandins*, which in turn stimulate the muscles of the uterus to contract. *Oxytocin*, a hormone produced by the mother's posterior pituitary gland, is also released in the late stages of labor; it stimulates the more powerful contractions required to expel the fetus. The effects of oxytocin depend on the presence of estrogen, and the uterus becomes much more responsive to the action of oxytocin late in pregnancy.

Labor is divided into three stages. The *first stage* is the longest, extending from the onset of regular contractions until the cervix is fully dilated to about 4 inches (10 centimeters) in diameter. This stage lasts about 15 hours in first pregnancies, and about eight hours in subsequent ones. (Deliveries after the first child are generally easier in all respects.) Uterine contractions begin at intervals as far apart as 15 to 20 minutes, occurring more frequently and with greater intensity and regularity over time. The fluid-filled amniotic sac surrounding the baby cushions it from the effects of these early contractions.

The *second stage* of labor starts when the cervix is fully dilated and ends with the delivery of the baby (Figure 6.5). It may last for a matter of minutes, or up to several hours in particularly difficult births. At some point during the first two stages, the fetal membranes rupture, and amniotic fluid gushes out. In 10 percent of the cases, premature rupture of the membranes initiates labor. In other instances

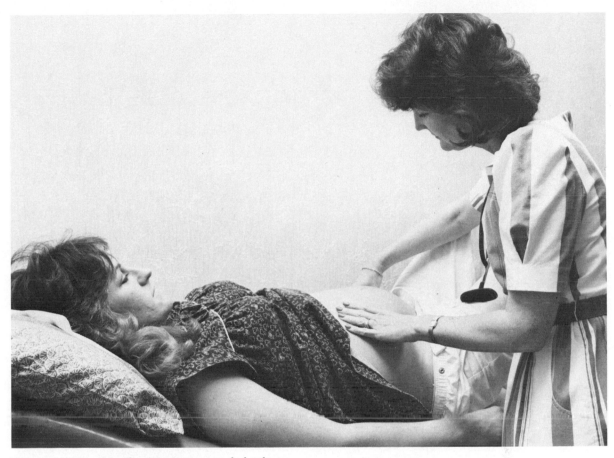

Figure 6.4   A midwife giving a prenatal checkup.

the physician breaks the membranes deliberately to speed labor. After the baby is born and starts breathing, the umbilical cord is cut, severing the last physical link to the mother.

In the *third stage* of labor the placenta separates from the uterine wall and is discharged with the fetal membranes as the *afterbirth*. The uterus contracts markedly, and there is some bleeding, usually limited. The third stage of labor lasts about an hour. During this time the mother and baby are carefully examined for signs of trauma and other possible problems. If all is well, the parents can use this time to touch and bond with their child.

Methods of Childbirth

Chances are that your grandmother was born at home. Your mother was probably born at the hospital, as were you. Where do you think your child will be born? Most women in the United States today give birth in a hospital, but some women prefer to deliver at home, with a physician or a nurse-midwife attending.

Home Delivery   The advocates of home delivery argue that hospital settings, with their impersonal and forbidding atmosphere, place undue stress on the mother and turn what should be a natural and joyous family occasion

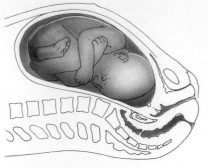

1. Head floats.

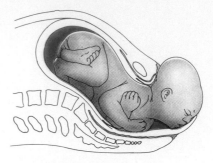

5. Neck bends back fully.

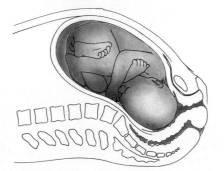

2. Head enters canal, neck bends down.

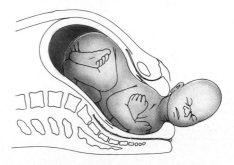

6. Head emerges, baby turns sideways.

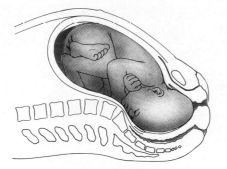

3. Baby descends and head turns.

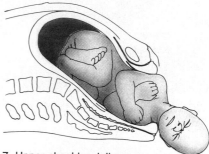

7. Upper shoulder delivers.

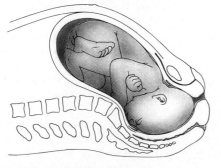

4. Neck begins to bend back.

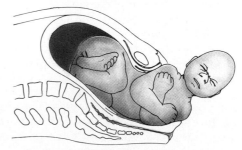

8. Lower shoulder delivers.

Figure 6.5   The process of birth.

into a costly surgical procedure. A safe choice exists between delivery in a hospital or at home only when the pregnancy has progressed normally and there are no anticipated complications. Childbirth can lead quickly to acute emergencies that endanger the life of mother and child, which can be best handled in a hospital. Only about half of the complications of delivery, such as a fetal head too large for the mother's pelvis, can be predicted before labor begins.

However, the majority of deliveries are quite normal, so many women, in fact, do have a choice. A couple considering home delivery should learn all the possible risks and the probable benefits. They should not let themselves be either pressured by convention or swayed by fads.

Hospital Delivery   The main advantage of a hospital delivery is the security it provides against unforeseen complications. It also relieves members of the family or friends of the responsibility of caring for the new mother and baby right away. A woman may also prefer the privacy of the hospital setting.

The experience of giving birth is basically the same wherever it takes place; yet a number of significant differences depend on the setting. In the hospital, the woman who is to give birth will be usually placed in a *labor room* where she is prepared for labor (having her pubic area shaved and/or cleansed, her bowels voided, and so on) and monitored; the actual birth takes place in the *delivery room* (which is like an operating room). Currently many hospitals allow labor and delivery to occur in the same *birthing room*, which is set up to convey a warm, soothing atmosphere.

While labor progresses, the baby's father, or a relative or friend, is encouraged to stay with the laboring mother. The comforting presence of a trusted companion makes the experience easier for the mother, allows the father or friend to share in it, and may actually reduce the complications of childbirth associated with stress and anxiety.

Hospital delivery usually involves procedures to forestall future problems. An IV needle is started, so no time will be lost if the woman winds up needing drugs, surgery, or glucose. Especially with first deliveries, an *episiotomy*—a cut in the perineum between vagina and anus—is made to avoid a jagged tear when the head passes. These precautions may well pay off, but sometimes prove needless.

Giving birth can entail considerable discomfort, even though it is a natural process and not an illness (Figure 6.6). After general anesthesia was introduced in the 19th century, physicians sought to use it to make childbirth painless. This practice was resisted for a while, because it seemed to be "unnatural" and nullified God's judgment on Eve ("I will greatly multiply your pain in childbearing; in pain you shall bring forth children"—Genesis 3:16). The practice only became popular after Queen Victoria delivered her eighth child under chloroform anesthesia in 1853.

General anesthesia is now used much less than in earlier decades. It entails considerable risk to the mother, slows down labor, and may interfere with the newborn's respiration. It also deprives the mother of assisting and witnessing the birth of the baby.

Hospitals sometimes give drugs routinely. A woman and her doctor should agree ahead of time on how to handle discomfort. A mild *analgesic* can lessen pain and anxiety; an *anesthetic* can block it. A *caudal* or *spinal block* is popular today. Temporarily the mother loses sensation and power to move below the waist. This method leaves her awake but unable to push. Some women prefer full awareness and participation. They ask the doctor to withhold all anesthesia.

Hospital deliveries can allow for the scheduled *induction of labor* through infusion of oxytocin when the mother is near the end of her term or is overdue. This procedure carries some risk but is necessary in select cases. The same is true for deliveries by *cesarian section*, a major surgical procedure in which the fetus is delivered by cutting through the abdominal wall into the uterus. (Although the procedure is named after Julius Caesar, it is highly unlikely he was in fact delivered in this fashion. The Roman practice was to remove

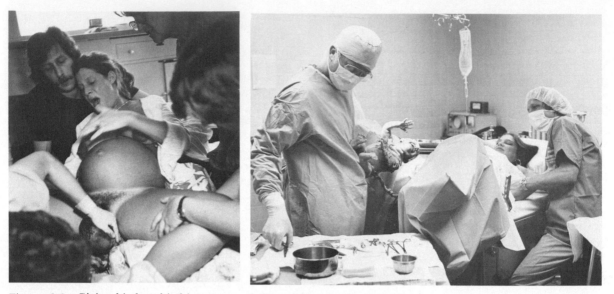

Figure 6.6   Giving birth at birthing center (left) and in a hospital (right).

the entire uterus to save the child if the mother was dying. Caesar's mother survived his birth.) The first authenticated cesarian delivery on a living patient was performed in 1610 (Speert, 1982). Almost one out of four births in the United States currently takes place through cesarian section (906,000 deliveries in 1986). This practice has quadrupled over the past two decades, and critics say it is overused. Cesarian sections can increase the risk of complications for the mother and substantially increase the cost of having a baby. The mortality rate is higher than the rate of maternal deaths in vaginal deliveries. The fact that a woman was delivered by surgery once does not preclude her giving birth vaginally on subsequent occasions.

However, in 12–16 percent of cases, there are compelling reasons for this procedure. Among its legitimate indications are discrepancies between the size of the baby's head and the mother's pelvis (making passage difficult or impossible); complications endangering the fetus (such as abnormal positioning or premature separation of the placenta); venereal disease in the birth canal (Chapter 5); and other conditions interfering with normal childbirth.

Prepared Childbirth   *Natural childbirth* is a set of attitudes and practices that seek to free the woman giving birth from unnecessary pain, anxiety, and medical intervention. To some, the term simply means giving birth at home, with the help of a midwife, without drugs and doctors. To others, it signifies one of several methods to facilitate birth through "natural" means rather than through the use of anesthesia, instrumentation, and so on.

The term "natural childbirth" was coined by the English physician Grantly Dick-Read in 1932 in his book *Childbirth without Fear*. Dick-Read postulated that the pain of childbirth is primarily related to muscular tension brought on by fear. He sought to educate women about the birth process in order to break the cycle of fear, tension, and pain.

The *Lamaze* form of natural childbirth originated in Russia, but it was popularized by the French physician Bernard Lamaze in 1970. It entails teaching women how to relax their muscles, concentrate, and breathe properly during labor, coupled with various massage techniques (Lamaze, 1970). An expectant mother and her partner (usually the baby's father or a friend) attend prenatal classes for six to ten weeks before the date of birth. These classes are designed to inform prospective parents about each step of labor, to answer questions, and to dispel anxiety. The women learn a variety of exercises that increase muscle control (Figure 6.7). The tension of abdominal and perineal muscles is thought to make it harder for the baby to emerge. By learning to relax these muscles, women can allow the baby to pass through the birth canal more comfortably. By learning to contract them, they can help to push the baby out. Other techniques, such as massaging the abdomen or concentrating on visual targets, further distract attention from the contractions.

The woman's partner plays an important part. This "coach" not only provides comfort and encouragement but helps her to breathe properly and to relax, using all the techniques they have been taught together. Coaching is hard work—but the teamwork is rewarding. Fathers need not pace the corridors, nervous and useless, as used to be the case.

Another method of childbirth that has become popular in recent years was expounded by the French physician Frederick Leboyer in *Birth without Violence* (1975). This method is mainly concerned with protecting the infant from the trauma of childbirth. It is a slow, quiet method of delivery, often done in hospitals. Everything is aimed at protecting the infant's delicate senses from shock. Birth takes place in a quiet and warm room where the lights are kept low, and unnecessary noises are avoided. Following birth, the baby is gently settled onto the mother's abdomen to adjust to its new environment for a few minutes while still attached by the umbilical cord. Only after breathing starts spontaneously is the cord cut and the baby gently lowered into a warm bath. (Leaving the cord uncut causes some of the baby's blood to drain into the placenta, which may weaken the baby.) This gentle entry into the world is a far cry from the usual image of the kicking and screaming infant, dangled by the feet and spanked to induce breathing. However, whether Leboyer's method results in

Figure 6.7   Couples attending a Lamaze childbirth class.

more relaxed and healthy children in the long term remains to be substantiated (Nelson et al., 1980).

Natural childbirth approaches have already had an impact on hospital practices. Increasingly, hospitals have modified their regulations to allow husbands or partners to be present at delivery and to allow *rooming-in*—having newborn babies live in the same room with their mothers rather than in a separate nursery.

Early Parent–Child Interaction

A related area of recent interest is the interaction between parent and infant soon after birth. Just as there are *critical periods* during which mother and offspring become *bonded* among animals, similar processes are assumed to operate among humans. Klaus and Kennell (1976) hypothesize that "the entire range of problems from mild maternal anxiety to child abuse may result largely from separation and other unusual circumstances which occur in the early newborn period."

To enhance early attachment, the newborn is shown to the parents right after birth and then placed next to the mother face-to-face to facilitate touching and eye contact. Sometimes the baby is put to the mother's breast right on the delivery table.

Cuddling, fondling, cooing, and other forms of tender and affectionate contact are encouraged from the start. The baby actively participates in this process by responding to its parents. In a few weeks, the *smiling response* further reinforces the parents' affectionate attention, encouraging more smiles (Scharfman, 1977). Such interactions help the child's psychological development. However, there is no convincing evidence that they work on the same bonding principles that operate among animals and their infants.

THE POSTPARTUM PERIOD

The *postpartum period* (or the *puerperium*) is the period of six to eight weeks that follows deliv-

ery and ends with the resumption of ovulatory menstrual cycles. It is an important time for the mother, as well as the baby and the father, from physiological, psychological, and practical perspectives (Nory, 1987).

Physiological Changes

We tend to take for granted the adaptation of the female body to the burden of pregnancy, losing sight of the tremendous changes it entails. In the postpartum period, tissues, organs, and physiological systems revert to their previous states—a considerable task.

After delivery, the uterus will shrink markedly (*involution*) and gradually regain its prepregnancy size; its weight drops from about two pounds to two ounces over a six-week period. The cervix, which is stretched and flabby after delivery, regains its tone and tightness within a week. For several weeks there is a uterine discharge, like menstrual flow, called *lochia,* which gradually turns from reddish-brown to yellowish-white.

The woman's body is meanwhile undergoing tremendous hormonal changes, as the high levels of progestin and estrogen that maintained the pregnancy decline. These hormonal changes affect not only the onset of milk production and uterine involution, but also the woman's emotional state.

Emotional Reactions

The relief of having given birth and the pleasure of having the baby make the first few days after delivery an exhilarating time. A woman usually leaves the hospital two or three days after an uncomplicated delivery. The first week at home may be quite taxing to the new mother, who is trying to cope with the many needs of the baby (especially around-the-clock feedings). Fatigue may be a major complaint at this point, and a general feeling of "let-down" follows the earlier euphoria.

About two-thirds of women experience transient episodes of sadness and crying sometime during the first ten days after delivery, a phenomenon known as *postpartum depression* ("baby blues"). In one to two cases out of 1000 births, the disturbance is severe enough to lead to profound depression with hallucinations, and suicidal and infanticidal impulses, a condition that requires urgent psychiatric attention (Yalom et al., 1968).

These problems can seem puzzling, because they occur when the mother is expected to be especially happy over the arrival of the baby. However, in the emotional upheaval of the postpartum period, a woman may have considerable mixed feelings toward the infant and may fear that she could hurt it. Doubts about her competence as a mother, fatigue, feelings of being rejected or neglected by the father, and biochemical changes add to the upset (Simons, 1985). When the baby is unwanted, abnormal, or to be given up for adoption, additional concerns inevitably compound the depression. On the other hand, under more usual circumstances the woman needs only a little cheering up. The postpartum period then is a time of great joy and deep satisfaction, as both parents revel in the presence of the fledgling human being that they have brought into the world.

Nursing

An infant animal sucking milk at the breast is *suckling*; a human infant is *nursing* or breastfeeding (Figure 6.8). Immediately following delivery the mother's breasts contain not milk but *colostrum*. This thin fluid has more protein, less fat, and as much sugar as normal milk. It is rich in antibodies that may help provide immunity to the infant.

Human milk production (*lactation*) begins two to three days after childbirth. Two pituitary hormones are involved: *prolactin* (from the anterior pituitary) stimulates the mammary glands to make milk, and oxytocin (from the posterior pituitary) causes the milk to flow from the breast to the nipple when the baby is nursed. After weaning, when a woman does not breastfeed, the breast is no longer stimulated by the nursing infant, and lactation ceases.

Nursing has been the universal method of

Figure 6.8   Breastfeeding.

feeding infants during most of human history and remains so in many parts of the world. Its decline elsewhere is mainly due to urbanization and industrialization. For mothers working outside the home, regular breastfeeding is not practical, unless the child is kept close by or the mother "pumps" out her milk for it. Mothers who work in the home have been influenced by trends or by manufacturers of milk substitutes to rely on alternative methods of feeding their babies. Women have been further discouraged from nursing by concerns that it might affect the shape of their breasts, although nursing has no permanent effect on breast size.

Breastfeeding is becoming popular again in the United States. It can be emotionally satisfying and pleasurable. It is not unusual for women to be sexually aroused, or in some cases to reach orgasm, while nursing. Some enjoy the experience; others react with anxiety and guilt. Commonly milk will ooze out of the nipples during orgasm, which may startle some people. The larger breasts of the lactating woman may be erotic to some men and disconcerting to others.

For the baby, the mother's milk is unquestionably superior to cow's milk or commercial formulas. Human milk contains the ideal mixture of nutrients and antibodies that protect the infant from certain infectious diseases; it is free of bacteria; it is always at the right temperature; and it costs nothing. On the other hand, breastfeeding is impractical for some women. Some are sick or absent; others do not produce sufficient milk. Drugs the mother must take are usually secreted in the milk. A few infants have a milk allergy. For such reasons a mother may decide not to nurse her infant, and she need not feel guilty about it.

Ovulation and Menstruation

If a woman does not nurse her child, the menstrual cycle will usually resume within a few months after delivery, though in some cases it may take as long as 18 months. Lactation inhibits ovulation in women whose infants nurse around the clock, but it is not a reliable method of contraception, even though it has been widely used in many parts of the world (Chapter 7).

It should be realized that ovulation can occur before the first postpartum menstrual cycle; consequently, a woman can become pregnant without having had a menstrual period after the birth of her baby. The first few periods after pregnancy are usually somewhat irregular in length and flow, but they do become regularized in time. Also, women who have had painful periods may suffer less discomfort after they have had a child.

Sex in the Postpartum Period

There is considerable variation in sexual activity after delivery. Fatigue, physical discomfort,

sexual interest, and the obstetrician's advice play an important part in determining when a woman resumes sexual relations after childbirth. Women who nurse their babies reportedly have a higher sexual interest than those who do not.

Doctors used to advise women to refrain from intercourse "six weeks before, six weeks after." This rule does not hold any more (Easterling and Herbert, 1982). There is no medical reason why a healthy woman cannot have vaginal intercourse as soon as the episiotomy scars or other lacerations of the perineum have healed and the flow of lochia has ended, which usually takes about three weeks. The only medical risk at this time is the possibility of infection, but couples who use condoms or practice sexual activities other than vaginal intercourse need not be hampered by this concern.

The impact of the baby on a couple's relationship becomes even stronger after birth. Particularly with the first baby, the exclusive relationship of the couple must now make room for a third person, also an intimate part of the family, whose helplessness must take precedence over the personal needs of the adults. A child can boost the affection linking a couple and enhance their sexual attraction to each other. It can also be an intruder and a competitor for the affections, time, and energy of the parent. All these considerations can operate at once, making it necessary for a couple to go through a period of adjustment.

Eventually, most women recover their previous levels of sexual interest and orgasmic responsiveness, but it is also not uncommon to experience shifts in this regard. Some women reach orgasm more easily after becoming a mother. Others develop dysfunctions due to pain or the loss of tone in circumvaginal muscles (discussed in Chapter 8).

PRENATAL CARE

As soon as she gets pregnant, a woman should have a doctor. Though pregnancy is a normal physiological process, not a disease, it is a period of increased medical risk. Before modern obstetrics, countless women died from complications of childbirth. Pregnancy nowadays carries minimal risk for most women who are under proper care.

In most cases, pregnancy should not seriously interfere with normal work, social activities, and sexual relations. Some adjustments may need to be made, but there is no reason why a healthy pregnant woman should be treated as if she were ill or handicapped. However, for the pregnancy to progress normally, appropriate *prenatal* ("before birth") *care* is necessary to safeguard the health of the mother and the baby (Taylor and Pernoll, 1987). Teenage mothers (below 15) and older mothers (over 35) are especially vulnerable to complications of pregnancy and need to be watched with particular care for signs of complications.

Nutrition and Exercise

During pregnancy, a woman does not need to "eat for two," as conventional wisdom claims, but she does need about 200 calories above her normal daily intake (a grand total of 40,000 calories throughout pregnancy) (Moghissi and Evans, 1982). Her diet should be rich in proteins, supplemented by vitamins and minerals. Calcium, in particular, is necessary for the growth of the fetal skeleton, iron and niacin for the prevention of anemia.

Pregnancy is no time to indulge in dietary fads. Poor nutrition may endanger the mother's health; it will also seriously compromise the development of the fetus, leading to premature or low-weight babies who are subject to higher death rates and more vulnerable to brain damage and mental retardation. The pregnant woman therefore walks a thin line with regard to food intake. On the one hand she must make sure she gets adequate nutrition; on the other she must watch for excessive weight gain, which carries its own penalties.

A similar situation exists with regard to physical activity. Ample exercise is as essential as adequate rest and sleep. Being pregnant is an added source of stress for the mother's body, and allowance must be made for it. The fatigue and lethargy that mark certain periods of pregnancy are adaptive reactions to induce the mother to rest more. How much to exert oneself is a matter of individual judgment. Some women are in better physical shape than others and can do more. Common sense would exclude hazardous sports, but activities like swimming and walking are beneficial to women with normal pregnancies. Running is a more controversial form of exercise; vigorous exertion may rob the uterus and fetus of oxygenated blood flow, and the repeated bouncing may also be harmful.

Smoking, Alcohol, and Drugs

The fetus is vulnerable to all harmful substances that cross the placental barrier. Every chemical introduced into the mother's body must be considered with that risk in mind.

Cigarette smoking during pregnancy is harmful above and beyond the risks it ordinarily entails. Fetuses carried by women who smoke while pregnant have an increased risk of prematurity, lower birth weight, and death in the perinatal period (before and soon after birth) (Baird and Wilcox, 1985; Neiberg et al., 1985). Children born to mothers who are heavy smokers carry almost twice the risk of the children of nonsmokers of developing impulsive behavioral disturbances in childhood. They also are more likely to have lower IQs and poorer motor skills (Dunn et al., 1977). A woman who is a smoker but quits no later than mid-pregnancy does not risk similar effects to the fetus.

Heavy drinking by a pregnant woman can cause serious abnormalities in the child (*fetal alcohol syndrome*), including birth defects and mental retardation (Haynes, 1982). Even moderate drinking (one or two drinks a day) or a single "binge" may be enough to cause fetal damage.

The active ingredients in marijuana cross the placenta and have been shown to cause fetal damage in some animals (Harbison and Mantilla-Plata, 1972). Though the influence of smoking marijuana in pregnant women remains unclear, it would be safer to avoid its use during pregnancy and nursing.

The use of narcotics by the mother exposes the fetus to especially high risks, including addiction, which necessitates a gradual withdrawal after birth. Cocaine-addicted babies have recently increased in number in hospitals in the United States, with long-term consequences for growth and behavior.

In principle, *any* chemical substance or drug can be harmful to the fetus. Not only *teratogens* (substances that cause birth defects) like Thalidomide but commonly used items like aspirin and Valium pose risks. Similarly, some steroid hormones can have serious effects on fetal development that are manifested years later. A pregnant woman should take as little medication as possible, and only with her doctor's approval.

COMPLICATIONS OF PREGNANCY

Until the middle of the 19th century, *puerperal fever* or childbed fever caused the death of one woman in ten giving birth in hospitals. In 1847, after Ignaz Semmelweiss established the infectious origin of this illness, physicians began to disinfect their hands before delivery, and the mortality rate dropped dramatically. Currently, the maternal death rate in the United States is 9.4 maternal deaths per 100,000 births, or about one death in 10,000 births (as against 70 per 10,000 births in Bangladesh) (Thompson, 1982). Uncontrolled bleeding following delivery is currently the most common cause of maternal death in the United States.

Difficulties in Carrying to Term

The most serious problem in the first trimester is *miscarriage* or *spontaneous abortion,* which accounts for the termination of 10–20 percent of pregnancies (three out of four occur before the sixth week). In about 60 percent of these cases a defect in the fetus is the probable cause (Scott, 1986). In the rest the miscarriage is due to some condition in the mother, such as illness, malnutrition, or trauma.

Another complication of pregnancy is *tox-emia* (or *eclampsia*). The cause of toxemia is unknown; presumably a toxin produced by the body causes high blood pressure, headaches, protein in the urine, and the retention of fluids, causing swelling in feet, ankles, and other tissues. Toxemia is a condition that occurs only in pregnant women, usually in the last trimester, affecting some 6 percent of all pregnancies, but as high as 20 percent of women who are not under prenatal care (Mabie and Sibai, 1987).

As mentioned before, the blastocyst may become implanted in a site other than the uterus (97 percent in the fallopian tube), resulting in *ectopic pregnancy.* Its reported incidence varies from 1 in 80 to 1 in 200 of live births (25,000 to 30,000 cases a year in the United States (Droegemueller and Bressler, 1980). It is usually caused when the progress of the blastocyst down the fallopian tube is blocked, often by pelvic inflammatory disease (Chapter 5). It is more common among older women. Though most ectopic pregnancies abort early, if they advance far enough the tube will rupture, with dangerous bleeding in the mother and death of the fetus.

During the third trimester, *premature birth* is the most serious concern (Pernoll and Benson, 1987). Because the date of conception is not always accurately estimated and because the age and weight of the fetus are highly correlated, prematurity used to be defined by weight rather than age. However, modern diagnostic methods, including ultrasound, allow doctors to determine the developmental level of the fetus and the expected birth date more accurately. Also, other reasons for babies weighing less than normal have been discovered, as important as prematurity. An infant who weighs less than 5 pounds 8 ounces (2.5 kg) at birth is considered to be of *low birth weight.* The mortality rate among premature or small infants is directly related to size: the smaller the infant, the poorer are its chances for survival.

The lower limit of a successful pregnancy is one that results in an infant's survival. Normally, fetuses are assumed to be *viable*—able to survive outside of the uterus—28 weeks after the last menstrual period of the mother. Fetuses typically weigh about 1000 grams at this stage. Both weight and length of gestation are important in determining viability. If delivered at the end of six months, the fetus may live for a few hours to a few days. With heroic efforts 5–10 percent of babies weighing no less than two pounds (900 grams) may live. A few infants born at less than 24 weeks of gestation have survived, as have a few others weighing less than 600 grams, but there is no reliable case of survival of any infant born at less than 24 weeks and weighing less than 600 grams (Tietze, 1983).

Fetuses can now be legally aborted up to 24 weeks. It is possible that some of them could survive with extraordinary efforts. The availability of a highly specialized neonatal intensive care unit is one of the critical determinants of whether or not a baby will be able to survive a premature birth. Continuing advances in medical technology will probably make it possible to sustain life for increasing numbers of such very prematurely born infants, which further complicates the ongoing controversy on abortion (Chapter 7).

An estimated 7 percent of births in the United States are premature. Low birth weight may be associated with various maternal illnesses (such as high blood pressure, heart disease, and syphilis), heavy cigarette smoking, and multiple pregnancies; half the time the cause of prematurity or low birth weight is unknown.

Not all women are equally vulnerable to the complications of pregnancy. A woman's general health, her age, and the quality of prenatal care are all significant factors. Pregnancies before age 17 run a higher risk of toxemia, low birth weight, and infant mortality. These problems are often compounded by ignorance, negligence, and poor care.

Women who wish to become pregnant after age 35 face special problems. To start with, they are less likely to succeed, because of reduced fertility. If they do get pregnant, they run a higher risk of miscarriage and of having children with certain birth defects. However, with proper care many women who are in their thirties or even older are capable of bearing perfectly normal children.

Birth Defects

The *infant mortality rate* is the number of infants who die within the first year of life; the current rate in the United States is 14.0 per 1000 live births. The *neonatal mortality rate* is the number of live-born infants who die within 28 days after birth, currently 9.8 per 1000.

A variety of maternal ailments harm the fetus, but developmental problems also arise independent of the mother. Abnormalities of structure or function present at birth are called *congenital malformations* or *birth defects*. They affect 3 percent of all live births. In 70 percent of cases there is no identifiable cause; about 20 percent are due to genetic factors inherited from the parents; 10 percent are due to chemicals (often drugs and alcohol used by the mother), radiation, infections, and other causes (Oakley, 1978).

One chromosomal disorder is *Down's syndrome*, which results when an egg has an extra chromosome in the nucleus (more likely in women over age 35). The condition results in severe mental retardation and defective internal organs. It affects one in 800 births, but the incidence rises sharply with maternal age to one in 300 births at age 35, and 1 in 40 births at age 45. For men older than 55, age is similarly linked to a higher incidence of this defect in the offspring.

Among infectious conditions, particularly damaging are certain viruses. German measles (*rubella*), for instance, if contracted during pregnancy in the first trimester causes serious defects in hearing, vision, and the heart, and mental retardation in up to 50 percent of cases.

Congenital disorders other than birth defects include sexually transmitted diseases, such as syphilis and AIDS, which the fetus contracts from the mother (Chapter 5). Another serious condition, whose symptoms become manifest soon after birth, is *Rh incompatibility*. The *Rh factor* is present (positive) or absent (negative) in human blood. When the mother is Rh-negative and the baby Rh-positive, the

Box 6.4

CHILDBEARING AND AGE

Only in humans, among our near relatives, does the end of female reproductive life precede the tidal wave of aging by many years. Evolutionary theory has it that the amount of energy put into begetting offspring is directly related to mortality; that is, life cycles seem genetically disposed to last about as long as it takes to reproduce. Thus, once monkeys and apes can no longer produce babies in the wild, they soon die. Not so for humans. Women come to an end of their reproductive capacity and they and their spouses live on for another 20, 30, or 40 years.

But consider the world in which we evolved—say, the world of 50,000 years ago. Life expectancy at birth was around 30 years, with the average skewed by high mortality in infancy. If you look now at life expectancy *after* childhood, in hunter-gatherer tribes that still exist much as they did in the Stone Age—the !Kung San, or Bushmen, of the Kalahari Desert in southern Africa, for instance—old age, as we understand it, is far from a sure thing. According to studies done by anthropologists Nancy Howell and Richard Lee, the average life expectancy, if you've made it to 15, is 55.

But interestingly, the average age at which a !Kung woman has her last child is about 39. The numbers may tell an evolutionary story: one that explains that humans are not so different from other animals after all. It's just that the human mother needs to stay around to care for her offspring, and humans have an extraordinarily long childhood. There needs to be enough energy to enable an animal to complete its reproductive cycle—to perpetuate itself. The last child, born, say, when the mother was 39, needs care to grow to an age at which its own reproduction could begin, at 16 or so, to ensure continuity of the lineage. The mother could then die at 55 with a certain, as it were, evolutionary peace of mind.

Such theories help explain why the human reproductive clock may have been *designed* to run out about when it does. They don't explain how the clockwork slows down. But new technology and research are beginning to provide that explanation.

. . . An analysis published last year in *Science* by Jane Menken and James Trussel of Princeton and Ulla Larsen of Lunds University in Sweden is among those that have confirmed declines in fertility with age: childlessness rose from around 5 percent in a group whose members married between ages 20 to 24, to around 9 percent in the 25–29 age group. For those marrying in the early 30s it was over 15 percent, in the late 30s, more than 25 percent. For marriages beginning between 40 and 44, it was over 60 percent.

The scarring of the ovary, though dramatic, is only one mechanism of reproductive aging. . . . New research also implicates the womb itself. The uterus, it turns out, loses its hospitality. The environment it creates for implantation and for the maintenance of pregnancy begins to be less suitable. It, too, depends on hormones that help prepare the uterine lining and enable the embryo to function. As the hormones decline, the uterus ages, loses half its weight from age 30 to 50, and begins to dry out. Collagen and elastin—two crucial proteins that make it durable and flexible, as they do skin and connective tissue throughout the body—decline markedly.

. . . Controversy continues about how to advise women. Certainly there is little risk in waiting until the early 30s to have a baby. In the late 30s the risk of involuntary childlessness becomes substantial, and in the early 40s great. Yet motherhood is possible for many women even until age 50 . . . And despite the generally unfavorable odds, there have been many successful in vitro attempts for individual women in their late 30s and early 40s.

Life holds risks, and the intelligent young woman can theoretically try to assess the loss she would feel if she ended up infertile, add in the likelihood of childlessness if she waits to a given age, and weigh the sum against the personal advantages of waiting.

Of course, real life is not that simple. Careers have a logic of their own. And, because most women are not willing to try this alone, the right man must come along. There is the possibility of adoption, though this itself is not emotionally painless. It can result in as much parental satisfaction as comes to biological parents.

How to guess the future? The medical frontier is continuously moving forward. Artificial insemination, in vitro, surrogacy—who knows what's next? Surely one can count on some future chemical magic that will enhance implantation and maintain pregnancy. Yet, neither that hope nor the consola-

tion of evolutionary understanding can erase the discomfort that arises from an arbitrarily waning force of life.

mother's body produces antibodies against the Rh factor. They destroy the red blood cells of the fetus, causing anemia, jaundice, and sometimes death. The development of Rh incompatibility can be prevented in Rh-negative pregnant women by medications that neutralize antibody formation. The condition is treated in the newborn by special transfusions (Durfee, 1987).

INFERTILITY

In our preoccupation with new life, let us not lose sight of the opposite—the inability to reproduce. *Infertility* means failure of a woman to conceive or of a man to impregnate after trying for a period of time (usually one year). *Sterility* is permanent infertility (Marshall, 1987).

The world may have more people than it needs, but for an individual couple the failure to have children can be deeply frustrating. As we said earlier, about one in ten marriages in the United States is childless after attempts to conceive for a year or longer. Another one in ten couples would like to have more children than they do but cannot (Menning, 1977). The incidence of infertility among married women aged 20 to 24 (the most fertile age group) jumped 177 percent between 1965 and 1982 (Wallis, 1984). Much of this increase has been due to the increase in pelvic inflammatory disease (Chapter 5).

Causes
The most important general factor that governs fertility is age. Men and women become fertile after puberty. Women cease to be fertile

following the menopause, whereas the fertility of men is gradually reduced (but not completely lost) as they grow older. Even during the reproductively active years, age influences fertility (Box 6.4).

Infertility in a couple is due to the male partner in about 40 percent of cases and to the female partner in another 40 percent; in 20 percent both partners contribute to the problem. The most frequent cause of male infertility is a low sperm count; when the ejaculate contains fewer than 20 million sperm per milliliter, impregnation becomes highly unlikely. Additional causes are defects in a large proportion of sperm and blockage somewhere in the tubal system. These problems may be due to developmental abnormalities in the testes (including undescended testes), infections (including sexually transmitted diseases), exposure to radiation or chemicals, severe malnutrition, and general debilitating conditions. Steroid hormones will also inhibit spermatogenesis by suppressing gonadotropin production.

The most common causes of female infertility are the failure to ovulate and blockage of fallopian tubes. A large variety of causes may account for these problems, especially defects of reproductive organs, hormonal disorders, diseases of the ovary, severe malnutrition, chronic ailments, drug addiction, and scarring of the fallopian tubes by pelvic infections. The problem may even be psychosomatic—the anxiousness to get pregnant prevents it (Moghissi and Evans, 1982).

Psychological Impact
The inability to have a child can be quite distressing, and the desire to become a parent

may become almost an obsession. Traditionally, women have been thought to be especially desirous of having children as a fulfillment of their female "destiny."

Motherhood continues to be important for many women, but infertility also confronts many men with a difficult situation. In addition to the desire to continue their lineage, a man's fertility deeply involves his self-esteem, self-image, and sense of adequacy. To be sterile makes him feel that he has a damaged body and is biologically defective (Schreiner-Engel, 1987).

Men who react most negatively to their infertility treat themselves initially as if they were their only child, getting involved in activities like body building, health foods, and "macho sexuality." They are unlikely to remain married or adopt a child. Other men, who have a more positive approach, get involved in child-rearing activities (like leading youth groups) and remain married. Nonetheless, they are less eager than their wives to adopt a child.

Infertility also affects a man's sexual life. He may feel less attractive sexually. With initial reactions of disappointment, anger, and grief, sexual interest goes down. Adjustment to infertility is helped by sharing the problem in support groups.[5]

Treatment

Diagnosing infertility involves both the female and male. Once the problem has been pinpointed, it can often be corrected. Sometimes the woman takes a fertility drug like *clomiphene,* which induces the pituitary to produce LH and FSH, or, failing that, *human menopausal gonadotropin* (HMG), which acts directly on the ovary. In a high proportion of cases these drugs induce ovulation and make pregnancy possible. In fact, the ovaries are often overstimulated, resulting in multiple pregnancies.

Also, microsurgery may succeed in opening blocked tubes.

Among couples who seek medical help for infertility, 40 percent subsequently conceive; in another 40 percent the cause of infertility is found but cannot be cured; in the remaining 20 percent no cause is detected, yet pregnancy does not occur (Moghissi and Evans, 1982). Many couples who are not able to conceive will give up the idea of having children or will adopt a child, but some now go to greater lengths to conceive, by resorting to artificial insemination and surrogate mothers.

Artificial insemination is the placement of semen in the vagina by means other than sexual intercourse. When the source of the semen is the husband, this approach raises no unusual problems; it is simply a means of pooling sperm from several ejaculations (obtained through masturbation) to overcome a low sperm count. When the semen comes from a donor because the husband is sterile, then many concerns arise—psychological, social, and legal. There are no serious technical problems; 75 percent of women get pregnant if artificially inseminated with fresh donor sperm.

The donor is carefully selected for blood type, freedom from sexually transmitted diseases, physical appearance, general health, and genetic background. Though some couples select the donor personally, his identity is usually kept secret; he never sees the child, nor even knows if the artificial insemination worked.[6]

Artificial insemination can be done with fresh or frozen semen. Frozen sperm is safer, because it can be screened for the AIDS virus or other microorganisms before it is used, but it has a lower rate of success. Frozen sperm is kept in *sperm banks* as "fertility insurance" by some men who are to undergo vasectomy (Chapter 7).

When it is the wife who is sterile, the male counterpart of artificial insemination is to have

---

[5]A national organization called Resolve provides support networks for infertile individuals. Further information may be obtained by calling (617) 643-2424.

[6]For women who "qualify" (by having high IQ, professional achievements, and good health), Germinal Choice in California will provide frozen sperm from Nobel laureates.

another woman get impregnated artificially with the semen of the husband. The *surrogate mother* carries the baby to birth and then relinquishes it to the couple, as stipulated by a contract. The comparison with artificial insemination is, however, not quite apt; for the surrogate mother to give up her child is a much more wrenching experience than for the sperm donor to suspect that he may be the biological father of a child somewhere. The legality of surrogate contracts has come under question and the practice raises a host of social concerns.

In 1978, a more astounding alternative became reality with the birth of Louise Brown, after a successful *in vitro fertilization (IVF)*. Over the next ten years, 1000 "test-tube babies" were born at the clinic where doctors Patrick Steptoe and Robert Edwards had developed the technique.

The procedure is to induce ovulation in the mother (who usually has defective fallopian tubes) through drug stimulation, which makes multiple eggs mature simultaneously. These mature ova are "harvested" about six at a time through a small incision in the follicles protruding from the ovarian surface. The eggs are extracted and fertilized with the father's sperm in an incubated laboratory dish containing appropriate nutrients. After the fertilized egg has reached the blastocyst stage, it is transferred to the mother's uterus, which has meanwhile been primed with hormones. From then on the embryo develops as in any ordinary pregnancy, making it possible for a woman to conceive who otherwise could not have done so.

Because the chances of success are not better than 20 percent, several fertilized eggs are implanted at once. As a result, multiple pregnancies are common, including triplets, unless some of the embryos are selectively aborted. The cost of finally achieving pregnancy through IVF may run into the six figures.

Finally, in cases where a woman cannot ovulate, it is now possible for her to become pregnant by *embryo transfer*. In this method a donor woman is artificially impregnated with the sperm of the infertile woman's husband. The developing embryo is then flushed out of the donor's uterus and implanted in the infertile woman's uterus. Sometimes the embryo is frozen and implantation is attempted later. A number of normal babies have already been delivered by this method.

These technological developments and yet others to come are scientifically dazzling, but they have unleashed social, ethical, and legal problems of unprecedented complexity (Francoeur, 1985). Some see these procedures as a ray of hope for thousands of sterile women who wish to experience motherhood; for others they herald an Orwellian world of assembly-line produced humanity. For some, surrogate motherhood reduces women to commercial producers of infants and denigrates and exploits them. A counterview argues that a woman should have control over her body and use it for reproduction as she chooses, with the proper safeguards for all concerned. These issues are of great importance and quite controversial.

REVIEW QUESTIONS

1. Trace the journey of the sperm from its origin in the testes till it fertilizes the egg; then list the stages the fertilized egg goes through until implanted in the uterus.

2. How do pregnancy tests work?

3. Describe the main features of a woman's physical and emotional experience of pregnancy through childbirth.

4. What does prenatal care entail?

5. What are the causes and remedies of infertility?

THOUGHT QUESTIONS

1. How can you determine on physiological grounds when a developing organism is a "human being"?

2. To how long a period of maternity leave should a working woman be entitled at full pay? When should she use her leave time?

3. What should the role of the father be during a woman's pregnancy, at the time of birth, and in the postpartum period?

4. What type of artificial insemination would you prohibit, if any?

SUGGESTED READINGS

Ashford, J. (1983). *The whole birth catalogue*. A broad survey of childbirth options. Trumansburg, N.Y.: Crossing Press.

Ingelman-Sandberg, A., Wirsen, C., and Nilsson, L. (1980). *A child is born* (2nd ed.). New York: Delacorte. Superb photographs of embryonal development with a concise and nontechnical text on pregnancy and childbirth.

Pritchard, J.A., MacDonald, P.C., and Gant, N.F. (1985). *Williams obstetrics* (17th ed.). New York: Appleton-Century-Crofts. A standard obstetrical text. Comprehensive and technical.

Witt, R.L., and Michael, J.M. (1982). *Mom, I'm Pregnant*. New York: Stein and Day. Clearly written book addressed to teenage girls who think they are, or are, pregnant.

Contraception and Abortion

Birth control affects nearly everybody—people either have used it, will use it or, at the very least, are against it.

CARL DJERASSI,
The Politics of Contraception

OUTLINE

Will you have children, how many, and when? If you live in the United States, you probably take for granted your right and your ability to control that. Nevertheless, despite many cheap, effective contraceptive options, millions of women still have unwanted pregnancies. Worldwide, the availability and willingness to use effective contraception is one of the most critical issues. The growth of the world's population is potentially catastrophic.

There is nothing new about trying to have sex without pregnancy (Box 7.1). What is new today is that we have highly effective and reasonably safe ways to do it, and that we can offer them to large populations as a matter of public policy (Noonan, 1967; Djerassi, 1981).

Contraception ("against conception") and *birth control* (a term coined by Margaret Sanger in 1914) both mean attempts at avoiding reproduction and attaining *fertility control*. The spacing of children through birth control is called *family planning* (Tatum, 1987). Although the result is the same, there is a valid distinction between preventing pregnancy through contraception and terminating it through abortion, as there is between abortion and infanticide (killing a newborn or letting it die). Infanticide has been practiced in some cultures to survive harsh conditions (among Eskimos), to ease population pressure (in Polynesia), and to eliminate abnormal offspring or the issue of illicit relations (in ancient Greece). The ritual of first-born sacrifice has also been known since biblical times as a means of offering to the deity one's most precious possession (Genesis 22). Infanticide is no longer permitted in any contemporary society, although it is still practiced illegally.

This chapter looks at the biological and behavioral aspects of contraception and abortion. These issues have also important legal and moral aspects.

PATTERNS OF CONTRACEPTIVE USE

There are many reasons for not wanting children, or not wanting them at a particular time. Some reasons are personal: many couples wish to plan a family, others to avoid parenthood altogether. Apart from individual concerns, the bearing of children has many consequences at the national and global levels. Personal and societal interests do not always coincide, so we will discuss them separately.

Reasons for Using Contraceptives

One major reason for avoiding parenthood is the couple's relationship. Although women occasionally choose to become single parents, most people wait till they are married, or have a stable relationship, before having children. Another is the general tendency to limit family size, especially among the middle classes. Unlike earlier times, when infant mortality rates were high, parents now can be fairly certain that their children will grow to maturity; there is no need to have many children to ensure that at least a few will survive to adulthood. In the industrialized world, fewer than 2 out of 100 babies die before the age of one; in the developing countries the infant mortality rate is 10 out of 100 babies (Camp and Spiedel, 1987).

Moreover, whereas in the past children were an added financial asset to the family through their labor, rearing children is now expensive. The estimated cost of raising one child to age 22, in a family of average income (excluding the cost of residential college education), is $215,000 in 1982 dollars (Ory et al., 1983). Therefore, not many parents can afford to have large families without straining their financial and personal resources. Other reasons for avoiding or postponing pregnancy include the health of the mother and career aspirations of both parents.

Childlessness used to mean the renunciation of sexual relationships or the inability to bear children. Sexually active individuals now have the choice not only to postpone childbearing but to avoid it altogether. Some people think the advantages of childlessness—in income, freedom, career, or health—outweigh the benefits of parenthood. To express the positive side of their choice, these couples are called *childfree* or said to follow a *childfree lifestyle* (Cooper et al., 1978).

Box 7.1

EARLY CONTRACEPTION

People have been attempting to prevent unwanted pregnancies for millenia.* With the exception of hormones, most of the contraceptive methods have their ancient precursors, which were based on the right ideas but were subject to the technological limitations of the time.

Probably the oldest and most common contraceptive practice has been withdrawal before ejaculation, attested to in the Old Testament (Gen. 38:8–11). The oldest known medical recipes to prevent conception are contained in the Egyptian Petri papyrus of 1850 B.C. Ancient Egyptians used vaginal pastes containing crocodile dung, honey, and sodium carbonate both as a barrier and a spermicidal agent. They also relied on vaginal douches containing mixtures of wine, garlic, and fennel.

In ancient Greece and Rome, there was much concern over contraception. People relied on absorbent materials, root and herb potions, devices to block off the uterus, and more permanent means of sterilization. Soranus, a Roman physician in the second century A.D., gives a clear account of the fertility control practices of the period, distinguishing between contraception and abortion-inducing agents (recommending reliance on the former). Through Islamic physicians, this classical knowledge passed on to medieval Europe and formed the basis of contraceptive practice up to the end of the 17th century.

Much of early contraception relied on barrier

*For a comprehensive history of contraception, see Himes (1970), Suitters (1967), and Draper (1976). Dawson et al. (1980) discuss fertility control in the United States before modern contraceptives were developed.

methods. Many materials were used in various cultures to provide obstacles to the passage of sperm (Cooper, 1928). Some Native American groups blocked the cervix with soft clay; the Japanese inserted paper; the French, balls of silk. Various types of sponge and cotton balls were also used, sometimes with medicated mineral oil or mild acid ointments to serve as spermicides. Pessaries made of gold, silver, or rubber have also been used in Europe to plug the cervix. They were inserted by a physician at the end of one menstrual period and removed at the onset of the next period. Intrauterine pessaries were extended into the uterus to decrease the likelihood of their slipping out of place.

Along with these rational, even if not very effective, methods, countless magical means have been resorted to, especially in preliterate societies, to avoid pregnancy (Ford and Beach, 1951; Gregersen, 1983). Among some North African tribes, for example, water that had been used for washing a dead person was secretly given to a woman to drink to make her infertile. In another group, a woman would eat bread into which had been ground a honeycomb containing a few dead bees. (The magical association between death and sterility was presumably the basis for these practices.)

Other contraceptive attempts were more far-fetched. Moroccan men would turn a special ring on a finger from one side to another following coitus. Papuan women of New Guinea who did not want a child tied a rope tightly around their waist during coitus. They also washed carefully following it, which probably helped a little and kept the two practices associated together.

Some couples arrive at the decision not to have children through a series of postponements; others decide early and enter marriage with that understanding (Veevers, 1974). The proportion of women who express their intention early to remain childless is quite small (about 6 percent). This group, especially its male members, has not yet been well studied.

It has been suggested that independence and achievement motivation among women make the choice of childlessness more likely. Such individuals must contend both with their own mixed feelings and with the pressures of our *pronatalist* society, which views parenthood as a normative part of adult life (Houseknecht, 1978).

Box 7.2

POPULATION CONTROL

The alarm over the dire consequences of population explosion was sounded by the English clergyman and economist Thomas R. Malthus (1766–1834). Malthus argued in *An Essay on the Principle of Population* (1798) that population grows exponentially by doubling, while the means of subsistence expands by simple increments. To avoid the inevitable prospects of famine and war that would result from this discrepancy, he advocated abstinence and late marriage. His successors shifted the emphasis to contraception, starting the *population control movement*, which attained great visibility in the 1960s (Ehrlich and Ehrlich, 1968).

From human beginnings in prehistory, it took until 1830 for the population of the world to reach 1 billion. During the next century this figure doubled to 2 billion, and in our century it doubled again to 4 billion in less than half that time and has already grown to 5 billion (see the figure).

This measure of *doubling time* provides a dramatic picture of population growth and also shows it to be a highly regional pattern. For instance, the Third World is growing 2 percent per year and will double its population in 34 years, assuming its present rate of growth. Meanwhile the population of the more developed world (including the United States) is growing at 0.6 percent per year and will double in 122 years. At a further extreme, Pakistan, growing at 2.8 percent per year, will double its population in 25 years, whereas Italy will require 693 years to do the same at its current growth rate of 0.1 percent per year (Camp and Speidel, 1987).

When births equal deaths in a population, we have *zero population growth* (ZPG), which the advocates of population control call ideal. Not everyone shares this view. Some Third World countries (and ethnic groups within the United States) perceive population control as an attempt to curtail their growth and their power (some call it "genocide"). Some economists in the United States advocate the faster growth of the population so as to generate a wider base of tax revenue and enrich the populace in other ways.

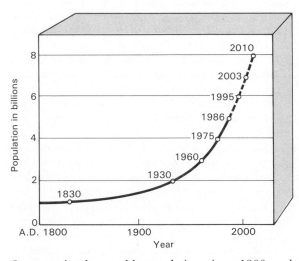

Increase in the world population since 1800 and projections until 2010.

A few people limit family size primarily out of concern for the world's expanding population. This matter is of great importance for social agencies and governments, particularly in developing countries with explosive rates of population growth (Box 7.2).

Prevalence of Contraceptive Use

We can group contraceptive devices by their modes of action. There are hormonal methods (like the birth control pill); intrauterine devices (IUDs); barrier methods (condom, cervical cap, diaphragm, vaginal sponge); spermicides (foam, jelly, cream); withdrawal; and methods that rely on periodic abstinence. In addition, there are various agents that induce abortion (abortifacients) and permanent ways of achieving infertility through sterilization.

Which of these methods of birth control could you use? It depends on where you live in the world. For example, in Nigeria—the

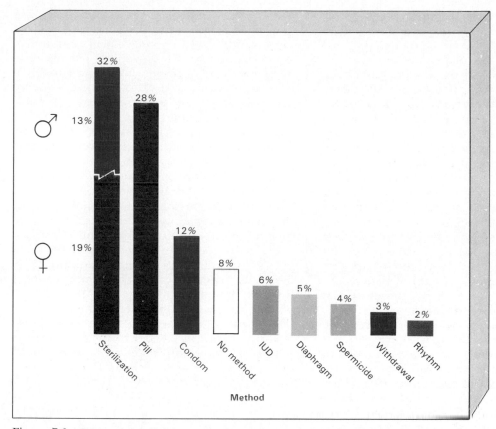

Figure 7.1   Percentage of women aged 15–44 and their partners using various contraceptive methods and no method (although not intending pregnancy) in the United States, 1982.

eighth most populous nation in the world—only 5.9 percent of all women have access to artificial birth control. About half the world population now has good to excellent access, 31 percent has fair access, and 19 percent have poor to very poor access. The level of industrial development generally correlates with availability of contraceptives, but not always. For instance, the USSR provides only fair access among developed countries; despite liberal laws on reproductive rights, contraceptives are difficult to obtain, and couples rely heavily on abortion. Relative costs and other factors (such as public education and dominant religion) are also different in other countries (Hatcher et al., 1988).

In the United States, because of free public birth control clinics and government funding for family planning services, everyone in most areas has access to safe, effective methods of birth control. As a result, the majority of the married female population aged 15 to 44 uses some contraceptive method. As shown in Figure 7.1, sterilization is the most common method in the United States, accounting for 19 percent of females and 13 percent of males. The Pill is next, with 28 percent, but it ranks as the single most frequently used device by women. It is also by far the most common method among the population aged 20 to 24 years, and to a lesser extent among those aged 15 to 19 and 25 to 29 years. Condoms and withdrawal are the only exclusively male methods; sterilization can be performed on either sex; the rest are all female methods.

At present, the main contraceptive responsibility clearly falls on women, burdening them with what many people think should be a shared concern. There are physiological reasons why it has been more feasible so far to develop female methods of contraception, especially hormonal methods (Djerassi, 1981). Technological breakthroughs and changes in social attitudes may alter this picture, but because it is only women who can get pregnant, contraception is likely to remain a matter of more concern to them.

Under ideal conditions, the only women who would get pregnant would be those who wished to do so. Instead, less than half (45 percent) of all pregnancies in the United States are intended; 40 percent occur before the woman wants them, or are mistimed (often involving unwed teenagers); and 15 percent are unwanted altogether. These figures do not mean that over half of all fertile women do not use contraceptives; rather, they reflect repeated nonuse by a smaller proportion. How many of these unplanned pregnancies result in unwanted children? Forty-six percent of unintended pregnancies are aborted, but 41 percent result in births (and 13 percent are miscarried). Some women are happy with an unplanned pregnancy after the fact, but others who do not get abortions are "stuck." Getting an abortion may not be an easy matter for them. Unfortunately, as with population problems on a global scale, these women are often the ones who can least afford the burden.

Especially serious is the situation for teenagers, who account for an estimated 1 million pregnancies a year—or one teenage pregnancy ever 30 seconds. Of these teenagers, 30,000 are younger than 15 (Hatcher et al., 1988). Half of teenage women aged 15 to 19 report having engaged in premarital coitus. Of these women, 27 percent never use contraception and 62 percent get pregnant. For the sexually active teenage population as a whole (including contraception users and nonusers) the pregnancy rate is 16 percent per year, as against 11 percent for all women of childbearing age. Recent trends show a somewhat greater use of contraception by young women. For instance, the percentage of those who used any method of contracepton at first intercourse went up from 38 percent in 1976 to 49 percent in 1979 (Zelnik and Kantner, 1980). This change has been mainly due to an increased reliance on condoms and on the relatively ineffective method of withdrawal.

In addition to age, contraceptive use correlates with marital status, socioeconomic class, ethnicity, religious affiliation, and other factors (Ory et al., 1983). Younger, unmarried women rely more often on the Pill, less on IUDs, and a great deal on their partners using condoms or withdrawal. These trends are quite fluid, especially among well-educated younger women. For example, on one University of California campus, between 1974 and 1978, the percentage of women using contraceptives who were on the Pill fell from 89 percent to 63 percent and diaphragm usage went from 6 percent to 33 percent, while IUD use remained steady at 8 percent (Harvey, 1980).

Reasons for *Not* Using Contraceptives

There are two populations that do not use contraceptives for obvious reasons: those who want to get pregnant and those who cannot get pregnant. In the latter group, some are sterile, and others do not engage in coitus. Apart from postmenopausal women, 10 percent of couples of childbearing age are involuntarily childless (Chapter 6). One out of five women aged 15 to 44 is not sexually active; half of teenage women aged 15 to 19 do not currently engage in coitus (Zelnik and Kantner, 1980).

Some 30 million women, or about 28 percent of the childbearing-age female population who are sexually active, do not use contraceptives (Hatcher et al., 1988). Close to 20 percent of them are teenagers; another 10 percent, women in their twenties (Forrest and Henshaw, 1983). It is with this group that we shall be concerned here.

Why is contraception not used? Religion is not the answer. The opposition of the Catholic church has attracted much public discussion and debate, but a relatively small proportion of women shun contraception because of it. Among married Catholic women in the United States, two out of three now use a contraceptive device other than rhythm; among those married between 1970 and 1975, the proportion of Catholic couples using contraception was 90.5 percent (Westoff and Jones, 1977). These Catholics use various birth control methods to the same degree as non-Catholics, except for sterilization, which they use less frequently.

What stops people from using birth control is not conscience but ignorance, lack of access to contraceptives, and various psychological and social considerations. There is a great deal of sexual misinformation, particularly among teenagers, even sexually experienced ones. Many think they are "too young" to get pregnant. Although there is in fact a period of relative adolescent infertility before the ovarian cycle becomes well established, it does not offer contraceptive security (Chapter 4). Others assume that it takes repeated intercourse to lead to conception, or that if girls do not reach orgasm they will be protected (Evans et al., 1976). Some have the wrong idea about the safe period of the month; in one study of sexually active teenage girls, 70 percent of whom had taken sex education courses, only one-third knew correctly the period of highest risk (Zelnik, 1979). Some women think they must be sterile because coitus in the past did not result in pregnancy. Teenage boys, on their part, have a tendency to view contraception as a "woman's problem" and to assume that if she says nothing, she must be "safe." Some couples forego contraception for nothing more serious than because it is "too messy" or "too much trouble."

The element of guilt may also play a role. To use contraception clearly implies the intent to engage in sex. Some teenagers believe that it is not morally right to have sexual intercourse outside marriage or a committed love relationship. To be fitted with a contraceptive device or carry condoms on a date, then, proves an immoral intention. To "solve" this dilemma, they engage in sex without contraception and by implication without forethought. On the spur of the moment, the sexual indulgence feels less like a premeditated act.

Similar considerations apply to those who find the expectation of engaging in sex to be unromantic. They think sex should be spontaneous, sweeping the couple off their feet, with caution thrown to the winds. The man with a condom at the ready comes across in this perspective as "exploitative" and interested only in his sexual pleasure; the woman who is always ready is perceived as "promiscuous."

The unpredictable association between sex and pregnancy itself generates a false sense of security. In fact, there is only a 2–4 percent chance that any single act of coitus will lead to pregnancy, but the odds go up to 21 percent on the day of ovulation, and to 90 percent over the period of a year. The more often one has sex, and the more often one has sex in high risk times, the greater is the chance of getting pregnant; as in any gamble, the odds sooner or later catch up.

The willingness to take a chance is a universal human characteristic. We do it all the time, either "just this once" or regularly. Risk-taking behavior in sexual activity is no exception. The relative risks entailed in sexual intercourse and its reproductive consequences are listed in Table 7.1 along with several nonsexual behaviors for comparison.

Some people get pregnant even when it seems to others that they should not, because they want to. Among one group of sexually active teenagers, 7 percent said they wanted to get pregnant and another 9 percent said they would not mind if they did. Many more may have similar motivations without quite being conscious of it. Even under the most inauspicious circumstances, having a baby carries enormous psychological significance. It may represent a teenager's desperate chance to be

Table 7.1   Voluntary Risks

RISK	CHANCE OF DEATH IN A YEAR (U.S.)
Smoking	1 in 200
Motorcycling	1 in 1000
Automobile driving	1 in 6000
Power boating	1 in 6000
Rock climbing	1 in 7500
Playing football	1 in 25,000
Canoeing	1 in 100,000
Using tampons (toxic shock)	1 in 350,000
Having sexual intercourse (pelvic infection)	1 in 50,000
Preventing Pregnancy	
Oral contraception— nonsmoker	1 in 63,000
Oral contraception— smoker	1 in 16,000
Using IUDs	1 in 100,000
Using barrier methods	None
Using fertility awareness	None
Undergoing sterilization	
Laparoscopic tubal ligation	1 in 20,000
Hysterectomy	1 in 1600
Vasectomy	None
Deciding about Pregnancy	
Continuing pregnancy	1 in 10,000
Terminating pregnancy	
Illegal abortion	1 in 3000
Legal abortion	
Before 9 weeks	1 in 400,000
9–12 weeks	1 in 100,000
13–16 weeks	1 in 25,000
After 16 weeks	1 in 10,000

From Hatcher et al., *Contraceptive Technology 1988–89.* New York: Irvington, 1988.

another, to force someone into marriage, to embarrass families, to punish others, and to assert oneself. Some teenagers reportedly get pregnant so as to be entitled to receive welfare payments, which then will allow them to leave home.

Additional obstacles to contraceptive usage are the practical problems of availability, cost, and access to contraceptive information and devices. Pharmacy shelves are full of condoms, but a person must still go to the counter to pay for them, which may be embarrassing, particularly for the young. Planned Parenthood clinics will give out advice and contraceptives, but clients first must be willing to accept their own sexual intentions in going to the clinic, which also may not be an easy matter. Where society makes it mandatory that teenagers' parents be notified of their receiving contraceptive help, you can imagine the reluctance to seek it.

Concern over the side effects of contraceptive agents is a final stumbling block. It may get so exaggerated that people fear to use them. However, the risks in contraception are minimal. Table 7.1 puts them in perspective.

Taking Responsibility

We have looked at the reasons why people use and avoid contraception. How will you act in this respect?

The first step in making a decision is coming to terms with your sexuality both in general and in a particular relationship, on a specific occasion. The question is whether or not you will use birth control, not just in principle, but in the concrete context of a given sexual act. There are many cogent psychological and moral reasons to engage or not to engage in sex, but there are never any good reasons, only bad ones, for getting pregnant unless the circumstances are right. Effective contraception starts with that conviction.

Of equal importance is knowing your vulnerability. A woman needs to monitor her high-risk periods around the time of ovulation and be aware of those times in her life when

someone, to have someone to love, and to receive a measure of affection and attention herself. For some young men, impregnating a woman is the ultimate sign of "machismo." Pregnancy is also a powerful tool to hold on to she may use contraception less reliably. These times tend to cluster around periods of transition; the time of becoming fertile during adolescence, and infertile during the menopause; at the start of a new sexual relationship; early in marriage; and following childbirth or abortion. Women are more apt to forget contraception in periods of stress, such as during or following separation, divorce, breakups, and lovers' quarrels. Moving away from home, joining a new circle of friends, and experimenting with a different lifestyle are other situations where sexual activity may take place before you are consistent about contraception (Miller, 1973a). Under any circumstances, the responsibility is still yours.

CONTRACEPTIVE METHODS

Whether you are a woman or a man, the ideal contraceptive would be socially acceptable, usable by either sex, fail-safe, free of side effects, esthetically inoffensive, and readily and cheaply available. No such contraceptive now exists or is likely to in the foreseeable future (Djerassi, 1981).

What we have instead is a set of alternatives with a varying mix of assets and liabilities. Each method must be evaluated not just in the abstract, but with regard to the needs of a particular individual at a given time. No contraceptive method, other than abstinence from intercourse, is 100 percent foolproof in preventing pregnancy, but some methods in combination come close to it.

The only way to avoid pregnancy with absolute certainty is not to have sexual intercourse or to become sterilized. Those who are not willing to do either must be prepared to take certain contraceptive risks, just as they take risks every time they step into a car or cross the street. The sensible course is to take calculated risks—to know what benefits are likely at what probable cost.

The effectiveness of contraceptive methods is measured in terms of failure rates: the number of married (therefore regularly sexually active) women out of a hundred who get pregnant when a given method is used over a year. A further distinction is made between the best or the *lowest reported rates* of failure and the *typical failure rate*, or the failure rate of users who have not been given the extra training, selection, and attention subjects get in the research studies in which the lowest rates are established (Trussell and Kost, 1987). Human error is responsible for the discrepancy between the two rates.

Table 7.2 (page 192) summarizes the first-year failure rates for all the methods to be discussed. For example, if a woman is using a diaphragm the typical failure rate is about 18 percent. This figure means 18 out of 100 sexually active "typical" women relying on the diaphragm are likely to get pregnant during their first year using it.

Abstinence

To refrain from engaging in sexual intercourse by choice, or abstinence, is the most certain and safest method of avoiding pregnancy. If sexual intercourse only served reproductive needs, abstinence would be the most sensible contraceptive method. However, sexual relations serve other needs, and abstinence may prevent their fulfillment, although abstinence from coitus certainly does not preclude other forms of sexual expression or orgasm.

Abstinence remains an important means of avoiding pregnancy. Until the advent of effective contraception it was actually the only safe and certain way. In the 19th century, abstaining to limit family size was the key principle of the birth control movement. Recently there has been a resurgence of the idea. Teenagers are urged to "just say no" to avoid pregnancy and sexually transmitted diseases.

However desirable abstinence is in principle, it is not a reliable method for many people in practice. There are always going to be those who engage in coitus no matter what the consequences, especially the young. There will always be the temptation to make love "just this once." For these people other methods of contraception are more realistic.

Bear in mind that getting pregnant poses a considerably higher threat to health than using a contraceptive, particularly before age 15 (Chapter 6). No less serious is the daunting prospect of bringing an unwanted child into the world.

Hormonal Methods

Hormones are the basic physiological way of dealing with birth control, because they turn off the reproductive process at its source. They are the most effective of all reversible contraceptive methods today. Properly used, the most effective hormonal methods provide over 99 percent protection. On the negative side, hormonal intervention intrudes the most on the body, and its side effects are the most serious. On the positive side, hormones have additional health benefits, which other methods do not.

Hormones can be taken as pills, by injection, and through other means like implantation under the skin. So far oral use has proven by far the most practical method.

Development of the Pill   Oral contraception became commercially available only in the early 1960s, but its use spread so rapidly and widely that it has come to be known simply as "the Pill."[1] Some 60 million women worldwide and 10 million in the United States now rely on the Pill, making it the most extensively used reversible contraceptive measure in the world. In the United States it accounts for 40 percent of contraceptive use by never-married women and 13 percent of married women (Hatcher et al., 1988). Pills are the most popular of all methods among younger women and teenagers; among women in this group who are sexually active, about half rely on it (Forrest and Henshaw, 1983).

The development of the Pill was a natural consequence of our modern understanding of the ovarian cycle. In the 1930s, it was established that progesterone inhibited ovulation and could prevent pregnancy. Simultaneously, the estrogens were chemically isolated and used to treat certain menstrual disorders. The first oral contraceptive chemical substance (*norethindrone*) was synthesized in 1951 at the Syntex laboratories in Mexico City, followed by wide-scale clinical trials in Puerto Rico in 1956. The first Pill (Enovid) became commercially available on the market in the 1960s.[2] Over the next two decades, oral contraceptives were studied more extensively than any other drug, because they are used by millions of healthy women for a critically important reason (Djerassi, 1981).

Types of Birth Control Pills   The Pill has the major advantage of freeing the couple from having to take contraceptive measures just before, during, or right after sexual intercourse; sexual activity is not interrupted. Also, more than any other contraceptive device, the Pill puts a woman in full and effective charge of her own reproductive process.

Birth control pills contain various combinations of *synthetic steroids* that imitate estrogens and progestins. They inhibit the secretion of LH and FSH by the anterior pituitary, preventing ovulation (Chapter 4). They also influence the structure of the uterine endometrium, making implantation difficult, and cause thickening and increased acidity of the cervical mucus, making it a more effective barrier to sperm (Gilman et al., 1985).

Oral contraceptives come in several forms (Figure 7.2). One of the most common is the *combination Pill*, which consists mainly of progesterone and a small amount of estrogen. The combination Pill is taken once a day for 21 days. The woman then stops for seven days, during which withdrawal bleeding simulates the normal menstrual flow; then she resumes taking the Pill.

To help the user stay on schedule, manufacturers sometimes add seven inactive pills (or vitamin tablets) colored differently to be

[1]It was Aldous Huxley who first referred to "the Pill" in *Brave New World Revisited* (1958).

[2]For a history of the Pill, see Pincus (1965). For an account of its development, see Djerassi (1981).

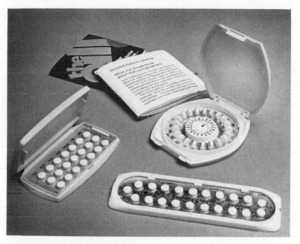

Figure 7.2   Various brands of oral contraceptives.

taken on the off days; thus the woman simply takes a pill of some variety every day. If she misses taking a birth control pill for a day, she takes two pills the next day; if two are missed, then both are added to the pill of the third day. If the Pill is missed for three days or more, the method may not be reliable for the rest of that cycle. Some other form of contraception must be used until the woman menstruates and then gets back on track by starting her next set of pills.

The combination Pill has been the most effective form of reversible contraception in the United States. Under the best conditions no more than 1 in 1000 (0.1 percent) of sexually active women studied got pregnant over a year. Some studies have shown that women who used the method correctly had no pregnancies at all. However, the more typical failure rate is 3 percent.

Hormones change daily in the natural menstrual cycle. To imitate this sequence better, a *sequential pill* was developed that gave estrogen for 15 days followed by a combination of estrogen and progesterone for the next 5. Though these pills were effective, they had a higher risk of serious side effects and were withdrawn from the market. They were replaced with the safer *diphasic pill* and *triphasic*

pill. These pills give lower daily doses in the early part of the cycle, reducing the total dose per cycle to less than most other Pills supply.

The biphasic Pill provides a small amount of progestin and a higher level of estrogen for the first ten days, followed by a much higher level of progestin and the same level of estrogen for days 11 to 21. The triphasic formulation uses three combinations of progestins and estrogens in varying doses over successive phases of the cycle. These new formulations appear as effective as the more standard forms of the Pill (*Population Reports*, 1982). They have become the most widely prescribed oral contraceptives in the United States.

There have been many attempts to minimize the undesirable side effects of the Pill while retaining its effectiveness. Manufacturers have used progressively smaller doses of the hormones, especially estrogen, which is responsible for most of the side effects. As a result, the estrogen content of most Pills is now a fraction of the original quantity.

The *Minipill* contains only a small amount of progestin and no estrogen. It is taken every day, with no break for menstrual bleeding. Though it does have fewer side effects, it may lead to more irregular bleeding. The Minipill is also somewhat less effective in suppressing ovulation than the combination Pill. It has a lowest observed failure rate of 1 percent, compared with 0.1 percent for the combination Pill, but the same typical failure rate of 3 percent.

Side Effects   All drugs may cause undesirable side effects or secondary reactions. A wide variety of such side effects complicates the use of birth control pills. We shall restrict ourselves here to the more common and the more serious undesirable effects (Hatcher et al., 1988).

The most common possible minor side effect of the Pill is nausea. It is usually mild and generally disappears after a week or two of starting the Pill. Other symptoms may include breast tenderness, constipation, skin rashes (such as brown spots on the face), weight gain, vaginal discharge, and headaches; these effects may occur only during initial cycles, or each month. All of them are similar to the symptoms of early pregnancy, which are also caused by the increased levels of the natural estrogens and progestins.

The most serious risks posed by birth control pills relate to the cardiovascular system. There is increased chance of heart attack, stroke, hypertension, and formation of blood clots (Wahl et al., 1983). These clots (emboli) cause only local discomfort if lodged, for instance, in a leg vein, but if they travel to the brain, the heart, or the lungs, the result could be serious or fatal. Older women and smokers are particularly susceptible to these risks; for instance, women in their thirties who smoke and use the Pill are four times more likely to suffer a heart attack then nonsmokers.

What has caused the greatest public alarm is the fear that birth control pills may cause cancer of the uterus or the breast. Research has linked estrogen with uterine and breast cancer in several animal species; yet no such effect has been substantiated in women despite a great deal of research (Murad and Haynes, 1985b). As time passes, it becomes less and less likely that any significant risk of cancer remains undetected (*Population Reports*, 1982).

Similarly, there is no evidence that birth control pills cause birth defects, unless they are taken by the mother when she is already pregnant (which is why a woman should only start taking the Pill right at the end of a menstrual period). Nor is there any evidence that oral contraceptives have a long-term effect on subsequent fertility, even though there may be a delay in returning to normal fertility levels for several months after they are discontinued. The use of the Pill has no effect on the onset of the menopause. Its influence on sexual interest follows no consistent pattern, because there are so many other factors (Chapter 4).

Some side effects are actually positive. The risks of getting cancer of the endometrium and cancer of the ovary are reduced by about 50 percent among users of the Pill. Women on the Pill are also protected to some extent from various other ailments, including some forms of pelvic inflammations, rheumatoid arthritis (the incidence of which the Pill reduces by half), anemia, ovarian cysts, menstrual irregularities, acne, and premenstrual tension (Droegemueller and Bressler, 1980; Altman, 1982; Hatcher et al., 1988). However, hormonal methods offer no protection against STDs, as condoms and spermicides do.

Nonetheless, a woman should think carefully and get medical advice before she decides to use the Pill. The risks should neither be minimized nor needlessly exaggerated. For many women who are young and who do not smoke, oral contraceptives are the best choice. For other women they are too risky: older women, smokers, and those suffering from cardiovascular disease, liver disease, diabetes, cancer of the breast, cancer of the reproductive organs, and some other conditions should avoid them.[3]

The Post-Coital Pill   A pill that works after coitus (*the morning-after pill*) has the great advantage of altering a woman's hormones only when coitus has taken place. It acts by interfering with implantation of the fertilized ovum, instead of preventing ovulation. Because the woman is not sure that she is pregnant (and she may not be), post-coital methods avoid the psychological turmoil of choosing abortion. To emphasize this fact, proponents call its effects *contragestion* instead of abortion. Opponents call it "chemical warfare" against the unborn (Murphy, 1986).

An early version of such a pill, not yet approved by the FDA for this purpose but legally available from physicians for other uses, contains a potent estrogen, *diethylstilbestrol* (DES), which prevents pregnancy if given to a woman twice a day for five days beginning within 72 hours (preferably within 24 hours) of unprotected intercourse. However, the estrogen content of the DES pill is 500 times the level of estrogen in ordinary birth control pills, so it causes severe nausea and vomiting. DES has also been linked with birth defects and increased risk of cancer in female children many years later if the drug is taken when the woman is already pregnant (Herbst, 1981).

---

[3]For a comprehensive list of contraindications to pill use, see Hatcher et al. (1988).

Hence, if the post-coital pill fails to interrupt the pregnancy, the risk to the offspring is an additional reason to consider abortion.

A more recent, safer, and less distressing version of the post-coital pill contains another form of estrogen (*ethinyl estradiol*), singly or in combination with a progestin (*norgestrel*). The first pill is taken as soon as possible after coitus, and a second dose in 12 hours (or sometime within 72 hours after coitus). Because the hormone content of these pills is not so high (only four times that of medium-dose oral contraceptives), nausea and vomiting are not as severe. Nevertheless, the dosage of estrogen in this newer form of post-coital pill is high enough to work (Yupze, 1982).

In a group of women exposed to single acts of unprotected coitus followed by the above regimen, only seven pregnancies occurred, instead of the 30 to 34 that would have been expected with no such intervention (*Population Reports*, 1982).

This post-coital pill is not yet available as a regular form of contraception. The number of times a person could safely use the method is probably limited, but studies are under way to explore broader uses.

French investigators have recently come up with a substance called *RU 486*. A single dose taken within ten days of a missed period has effectively prevented or interrupted pregnancy in 85 percent of cases without major side effects (Couzinet, 1986). RU 486 works by blocking progesterone receptors, canceling the progestins' essential role in sustaining pregnancy. As a result, the uterine lining sloughs off and the embryo is lost in a menstrual period.

RU 486 remains in the body for only 48 hours and does not impair fertility (Couzinet, 1986), but its long-term effects have not yet been fully studied, and the drug is not yet approved for use in this country. In many developing countries, where only one woman in five practices contraception, such a pill could stem the population growth (Sullivan, 1986). However, the newest (or most expensive) method developed in the West is not necessarily the best answer to the contraceptive needs of the developing world.

Most college health services and private physicians will prescribe post-coital contraceptives in cases of an "accident" or a rape, but federally funded clinics are not allowed to do so.

The Intrauterine Device

The *intrauterine device (IUD)* is a plastic object about the size of a small paper clip, which is inserted in the uterus to provide contraceptive protection. The IUD goes back to the early 1930s, when Grafenberg in Germany and Ota in Japan experimented with metal rings, but its use became widespread only in the 1960s. Some 25 different forms of the IUD have been used in the past. At present there is only one type of IUD available in the United States (Figure 7.3). Lawsuits from women who developed pelvic inflammatory disease and became infertile caused pharmaceutical companies to withdraw the other products from the market (Hatcher et al., 1988). About 88,000 cases of infertility were probably related to IUD use, especially to the badly designed Dalkon Shield. Among women with tubal infertility, the percentage who had IUDs was over twice the percentage of those who had no IUDs.

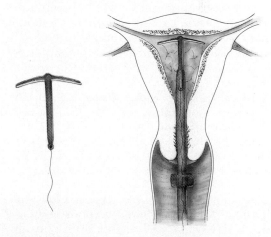

Figure 7.3   (Left) The Progestasert T, the only IUD currently in use in the United States. (Right) The IUD inserted into the uterus.

The IUD is a close second to the Pill in effectiveness and prevalence of usage. Some 50 million women throughout the world rely on it, including 35 million in China, where one type of IUD is called the "Flower of Canton" (Djerassi, 1981; Hatcher, 1988). It accounts for 5 percent of contraception by married women in the United States but less than 1 percent for those never married (Hatcher et al., 1988).

How the IUD works is unclear. The most widely accepted explanation is that it causes cellular and biochemical reactions in the uterine endometrium that interfere with implantation. Should implantation occur, it may also dislodge the blastocyst, in which case it would be more accurately viewed as inducing early abortion rather than providing contraception.

Early IUDs were made from various metals; the IUDs currently in use are made of a flexible plastic that can be squeezed into the inserter tube and then released in the uterus, where it returns to its usual "T" shape. A nylon thread or "tail" attached to the lower end of the IUD trails out of the cervix into the vagina, enabling the woman to check that the device is in place. Small amounts of barium are incorporated in the IUD to make it visible on X rays for a more definitive check. The IUD is taken out by special forceps.

The IUD has a lowest observed failure rate of 0.5 percent and a typical failure rate of 6 percent (including those who discontinue using it without starting another method) (Hatcher et al., 1988). The failure rate tends to decline after the first year, because long-term users are women who tolerate it better. To improve effectiveness, IUDs rely on added chemical measures. Some available abroad have copper filament wrappings that slowly dissolve in the uterus. The Progestasert T releases progesterone at a slow rate. It does not inhibit ovulation, but it changes the lining of the uterus enough to interfere with implantation.

The Pill's effectiveness depends on the woman who takes it, but the responsibility to make the IUD work well rests with the health professional who inserts it. Proper placement greatly influences the effectiveness of the device. Once the IUD is in place, the woman need not do anything except occasionally check the thread to make sure the device is still in place. Unless there are complications, some IUDs can be left in place for up to four years, but the progesterone-releasing type must be replaced every year.

IUDs have also been experimented with as post-coital contraceptives. When inserted within five days of unprotected coitus, the copper-containing IUDs have been shown to prevent pregnancy. Further research should be done to confirm this finding and to look for complications.

The IUD is just a mechanical device, not an ingested substance, so it does not cause systemic side effects. It does cause some local reactions that are undesirable. The two most common side effects are irregular bleeding and pelvic pain.[4] Mild bleeding or "spotting" may occur at various times in the menstrual cycle, and the menstrual periods of women with IUDs tend to be heavier than usual. Pelvic pain is caused by uterine cramps. It affects some 10–20 percent of users but tends to disappear after several months. In 10 percent of women, pain or bleeding is severe enough to necessitate removal of the IUD. Side effects are said to be less of a problem with IUDs containing progesterone. In 5–15 percent of users the IUD is expelled spontaneously (which is why women need to check the thread) (Sparks et al., 1981).

Less common, but more serious complications include perforation or piercing through the uterus, pelvic infection, and problems related to pregnancy (should it occur despite the IUD). The danger of uterine perforation is about 1 in 1000 insertions; it requires emergency surgery. IUD use among women with more than one sexual partner results in a five- to ten-fold increase in the risk of developing pelvic inflammatory disease (PID). The IUD tail acts as a conduit for bacteria from the

---

[4]For a more detailed discussion of the side effects of IUDs, see Osser et al. (1980). Contraindications of IUD use are listed in full in Hatcher et al. (1988).

vagina to the uterus, interfering with the protective mucus plug. Therefore, women with acute or recurrent pelvic infections are advised not to use IUDs.

Should a pregnancy occur while an IUD is in place there is a three-fold increase in the risk of spontaneous abortion, a ten-fold increase in the risk of ectopic pregnancy (5 percent of all IUD pregnancies are ectopic), and increased risk of infection during the pregnancy (Mishell, 1979; Droegemueller and Bressler, 1980). There is no danger that the IUD will cause cancer or birth defects.

Despite the possibility of these complications, the IUD is an effective and fairly safe device. Generally it does not interfere with sexual activity. Some women experience pain during orgasm when the contractions of the uterus press on the IUD. Though most men do not even feel the string, a few complain that it irritates the penis.

Barrier Methods

The principle of barrier methods of contraception is simple; they mechanically prevent sperm from reaching the ovum. Either the penis is covered with a *condom*, or the cervix is blocked by a receptacle—a *diaphragm, cervical cap,* or *contraceptive sponge*—filled with spermicide. Before the advent of the Pill and the IUD, barrier methods were the most reliable and most frequently used contraceptives. They remain in wide use today (Connell, 1979; Hatcher et al., 1988).

Barrier contraceptives are free from side effects and highly effective when used correctly with spermicides. Their main disadvantage is the likelihood of incorrect use; hence their higher typical failure rates, 18 percent for the diaphragm and 12 percent for the condom.

These devices may not be available when unexpectedly needed. Furthermore, some people dislike them for interrupting sexual activity and consider them unpleasant to use. It is possible to overcome these reactions. Couples may even learn to make placing a contraceptive part of lovemaking. For instance, putting on a condom can become part of the sex play preceding coitus.

The Diaphragm   Blocking the cervical opening for contraceptive purposes has been practiced for a long time (Box 7.1). The modern *diaphragm* was invented in 1882 in Germany. Like the modern condom, it was made possible by advances in the manufacture of rubber. Alone, the diaphragm is highly unreliable, so it is always used with a spermicidal cream or gel. It acts mainly by holding the spermicidal substance against the cervix. It is now used by around 4 million women throughout the world. In the United States it accounts for about 4 percent of all methods among married women, and a somewhat lower proportion for those never married. It is most popular with women aged 20 to 34 and among those from higher socioeconomic backgrounds (Forrest and Henshaw, 1983). The diaphragm has a lowest observed failure rate of 2 percent but a typical failure rate of 18 percent.[5]

Diaphragms today consist of a thin rubber dome attached to a flexible, rubber-covered metal ring. They are inserted in the vagina in a way that will cover the cervix and prevent passage of sperm into the cervical canal (Figures 7.4 and 7.5). The inner surface of the diaphragm is coated with a layer of contraceptive jelly or cream before insertion; otherwise its effectiveness is greatly reduced. After it has been used, the diaphragm is washed, dried, and stored for reuse. Diaphragms wear out over time and must be replaced. *Disposable diaphragms* are under study but not approved for general use.

Diaphragms come in various sizes (most are around 3 inches in diameter) and must be individually fitted by a physician or another health professional. A woman must be refitted following pregnancy, major changes in body weight, and other circumstances that may alter the size and shape of her vaginal canal. Once they have been fitted and instructed, most

[5]For the use of the diaphragm among college women see Hagen and Beach (1980).

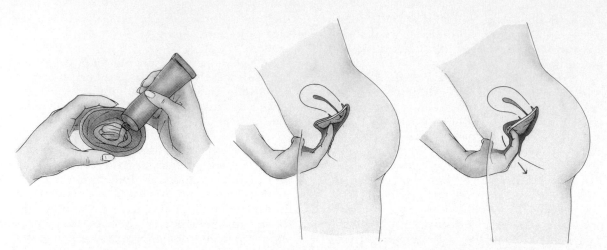

Figure 7.4   Diaphragm. (Left to right) Applying spermicide; checking placement; removing diaphragm.

women have no problem using the diaphragm, but because it involves the insertion of fingers into the vagina when putting it in and removing it, some women may feel uneasy about it.

The diaphragm must be inserted no earlier than six hours before intercourse in order for the spermicide to retain its effectiveness. If a woman has intercourse more than two hours after insertion of the diaphragm, she should leave it in place but add more spermicidal jelly or cream into the vagina. Similarly, more cream or jelly should be placed in the vagina (not the diaphragm) if intercourse is repeated. The diaphragm should be removed 6 to 8 hours after the last act of intercourse; though it can be left in place for as long as 12 hours. Keeping it in place for more than 12 hours is thought to increase the risk of toxic shock syndrome (Chapter 5).

Unless a woman knows in advance that she is going to have intercourse, she has to stop and insert the diaphragm in the midst of sexplay or risk pregnancy. This procedure is distracting to some, but others incorporate it into their foreplay, making a virtue of necessity. When the diaphragm has been properly inserted, neither partner is aware of its presence during intercourse. They also are likely to be unaware if the diaphragm becomes dislodged.

This problem is more likely if it fits loosely, or if the woman moves vigorously during coitus.

The Cervical Cap   Though it works on the same general principle as the diaphragm, the *cervical cap* fits snugly around the cervical tip by suction. This suction may cause some "erosion" or damage of the mucosal surface of the cervix. It should be used only by women with normal Pap smears. As effective as the diaphragm, the cap must be individually fitted, used with spermicide, and may be kept in as long as 48 hours (as against 12 hours for the diaphragm).

One of the most popular forms of contraception in the 19th century, the cervical cap was almost completely displaced by the newer methods. Although it remained in use in Europe, it was not approved by the Food and Drug Administration for use in the United States until 1988 (Figure 7.5).

The Contraceptive Sponge   A contraceptive device approved in 1983 by the Food and Drug Administration is the *vaginal sponge* (Figure 7.5). Made of polyurethane, the circular, highly absorbent sponge has a diameter of 2 inches and is permeated with a spermicide.

Inserted into the upper vagina, the sponge blocks the cervical opening and traps

Figure 7.5   Barrier contraceptives and spermicides for women. From left to right: contraceptive jelly, cervical cap, diaphragm, vaginal contraceptive film, "female condom," contraceptive sponge. The vaginal contraceptive film, like the vaginal suppository, dissolves in the vagina, releasing spermicide. (It is being considered for approval by the FDA.)

and kills sperm. The sponge stays effective for 24 hours (hence the trade name, Today), irrespective of the number of acts of coitus. Keeping it in place longer increases the risk of toxic shock syndrome (otherwise the risk is only 1 in 2 million uses).

The typical failure rate of the sponge is about 18 percent, comparable to that of the diaphragm. The sponge, for unclear reasons, is more effective in women who have never had a baby. So far, few side effects have been reported, mainly allergic reactions to the spermicide, vaginal dryness or irritation, and difficulty removing the sponge. There are anecdotal reports of cervical erosion and recent recommendations against using a sponge during menstruation.

The sponge does not require a prescription, does not need to be specially fitted, and is disposable. It appears to be most popular with women who have intercourse infrequently, who do not require contraceptive protection on a regular basis. Remember that its 18-percent failure rate refers to its use over a year of sexual activity; it does not mean that if 100 women use it for a single act of coitus, 18 will get pregnant.

The Condom   The *condom* is the only male contraceptive device in widespread use that is acceptably reliable (Figure 7.6). The rubber condom has been in use since the middle of the 19th century. Also known as a *prophylactic,* or *rubber*, its earlier prototypes were made of linen. Casanova referred to the condom as "the English vestment which puts one's mind at rest" (Himes, 1970).

Manufacturers in the United States and Japan supply over a billion condoms a year worldwide to roughly 20 million men. Some 10 percent of couples in the United States rely on condoms for contraception. They are used by 21 percent of sexually active teenage men (Forrest and Henshaw, 1983). The use of condoms is much more prevalent in Japan, where it is the preferred contraceptive method for three out of four married couples.

The modern condom is a cylindrical sheath usually made of thin latex rubber with a ring of harder rubber at the open end. Often a "nipple" at the closed end acts as a receptacle for the ejaculate; otherwise half an inch should be left loose at the end for that purpose. A *female condom* is being tested in Britain (Figure 7.5). Shaped like a large condom, it is introduced into the vagina first and then the penis is inserted into it.

Condoms are usually one uniform size, although "snug" versions are available that stretch to fit most penises.[6] Condoms are available in drugstores, campus stores, coin-operated dispensers in public restrooms, and by mail. They are sold rolled up in individual sealed packages. Explicit instructions are enclosed. Unopened condoms remain good for two years if stored away from heat. They should not come into contact with vaseline or other petroleum-based products. Spermicidal agents, K-Y jelly, and other water-soluble lubricants do not affect them. Fancier versions are made in different colors, with ribbed texture and lubrication for vaginal stimulation.

[6]Barbara Seaman has suggested that just as women buy brassieres in different cup sizes, men should be able to buy condoms in different sizes—labeled "jumbo," "colossal," and "supercolossal," so that nobody has to ask for the "small" (quoted in Djerassi, 1981, p. 17).

Condoms have a lowest reported failure rate of 4 percent used alone or 2 percent with spermicide. The addition of nonoxynal-9 enhances not only the contraceptive effect of condoms but also the protection they provide against AIDS (Chapter 5). Being manufactured to stringent specifications, they are unlikely to burst, especially if the vagina is adequately lubricated. Nonetheless, improper use results in a 12 percent typical failure rate.

The condom should be put on *before* the penis touches the female genitals. It is not safe to put it on just before ejaculation. There are two reasons: first, there is the slight risk that stray sperm may be transmitted with the Cowper's gland secretions; second, the man may lose control and ejaculate before he intends to. However, a condom put on too early in foreplay may get damaged. To avoid leakage of semen, the penis should be withdrawn from the vagina before loss of erection, while holding on to the ring at the base of the penis. To engage in coitus again, the man should discard the used condom, wash or at least wipe his penis, and then put on a new condom. Unlike the diaphragm, condoms are not reusable, even if they are washed and dried.

The condom is virtually free of side effects, though rarely a person may be allergic to the rubber or the lubricant. Women are generally not troubled by it, especially because it relieves them from the contraceptive burden, but some miss the sensation of ejaculation in the vagina. More often, men complain that it lessens their pleasure ("like taking a shower with a raincoat on"). To increase sensation, some condoms are made of animal intestines or of thinner rubber (Japanese condoms are much thinner than those made in the United States).

Men with potency problems find that the distraction of putting on the condom may make them lose their erections. By contrast, premature ejaculators sometimes find condoms helpful. In view of the protection it offers against certain sexually transmitted diseases (especially gonorrhea and AIDS), the condom is probably the single most useful contraceptive device for men and women engaging in sex with occasional partners. It can also render good service as extra protection or as a backup in more stable sexual relationships.

Public health campaigns aimed at the prevention of AIDS have greatly expanded awareness of condoms. For the first time in the United States, condoms are now advertised on

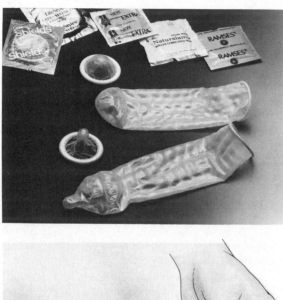

Figure 7.6   Condom rolled, unrolled, and applied. When putting on a condom without a reservoir end, leave a little space at the tip for semen.

television and in other media. The role of condoms against disease is discussed in Chapter 5.

Spermicides

Spermicides are chemical agents that kill sperm. They also provide a physical obstacle to the passage of sperm into the cervix, so they can be considered a form of barrier contraception as well. Spermicides come in the form of *creams, jellies, foam, tablets, vaginal film,* and *suppositories.* The active chemical in all of them is usually *nonoxynol-9* (Figure 7.5).

All spermicides are placed directly in the vagina, but they vary in their effectiveness and their method of application (Figure 7.7). Tablets and vaginal suppositories must be in place for 10 minutes or so before they become effective. Foam works right away, but only for half an hour. Explicit instructions are enclosed with these products, which are available on open pharmacy shelves and require no prescription.

As a group, spermicides have a lowest observed failure rate of 3 percent, but improper usage results in a much higher typical failure rate of 21 percent. Aerosol foam is the most effective form of spermicide, followed by creams and jellies; tablets and suppositories are the least effective.

There are no firmly established serious side effects of spermicidal substances. One controversial study linked them with a slightly higher rate of birth defects when women got pregnant while using them (2.2 percent for spermicide users against 1 percent for nonusers). However, this association was later refuted (Hatcher et al., 1988). There was one controversial study with evidence of a slight increase in spontaneous abortions in spermicide users. Though these findings were never found in any other scientific study, it is safer for women to stop using spermicides as soon as they suspect they may be pregnant (Jick et al., 1981; Hatcher et al., 1988).

Other drawbacks of spermicide use include occasional complaints of genital burning sensations and of allergic reactions; potential

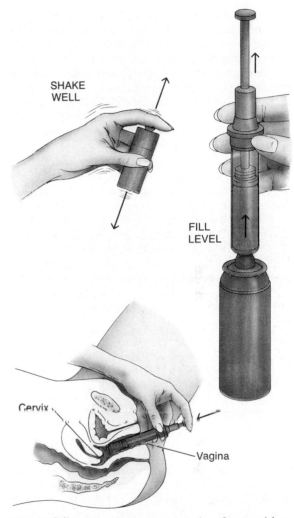

Figure 7.7   Inserting contraceptive foam with a plastic applicator. Foam must completely cover the cervical mouth.

users must first determine if they are allergic to it. Fastidious people may find them too messy. Others complain that vaginal foam makes intercourse feel "sloshy," and may make cunnilingus distasteful for the man, although some spermicides have no taste.

Nevertheless, spermicidal contraceptives have an important place, primarily in their use with barrier contraceptives. Only 4 percent of women who use contraceptives rely on spermicides exclusively; a larger number use them with a barrier method. Without cream or jelly, the diaphragm would not be safe enough to use; hence it is standard practice to combine the two. Although it has been less common to combine the use of foam with a condom, the combination would make them close to 100 percent effective and increase protection against sexually transmitted diseases as well. In cases where pregnancy must be avoided at all cost, spermicides may be combined with both diaphragm and condom used together, or some other combination. There is no reason to settle for just one method, especially when there is no risk of cumulative side effects.

Some women *douche*, or wash out the vagina, after coitus (Chapter 5). Douching is a spermicidal method of sorts, but it is so ineffective that it hardly deserves mention, no matter what fluid is used or how fast the woman rushes to use it.

Fertility Awareness Techniques

All attempts at avoiding pregnancy are contraceptive practices, yet some people distinguish between methods that are "natural" and others that are not. The Catholic church, for instance, opposes all active or intrusive methods to avoid pregnancy as immoral. But it considers it morally permissible to take advantage of a woman's transient periods of infertility during the menstrual cycle to avoid conception. Aside from morality, fertility awareness techniques provide a method of birth control that needs no external devices, has no side effects, and costs nothing.

This approach, popularly known as the *rhythm method*, depends on *fertility awareness*. This knowledge can be used not only to avoid pregnancy but to increase the chances of bringing it about. It is simply a way of knowing when a woman is fertile and when she is not.

Currently, there are three ways of determining the "unsafe" periods when a woman is fertile. There are books that explain these methods in detail (Hatcher et al., 1988), but it is better to work with a trained counselor. The

lowest reported failure rate is 2 percent, and the typical failure rate is 20 percent (Hatcher et al., 1988). Only 2 percent of sexually active women rely on fertility awareness methods, and most of them are older than 35 years (Forrest and Henshaw, 1983).

The Calendar Method   The original rhythm method is the least reliable one. Failure rates are estimated to be as high as 45 percent (Ross and Piotrow, 1974). The calendar method rests on the following three assumptions: in a regular 28-day cycle, ovulation occurs on day 14 (give or take two days); sperm remain viable for two to three days; and the ovum survives for 24 hours after ovulation.

Not all women have 28-day cycles, so the first task is to construct a personal *menstrual calendar* by noting the length of each menstrual cycle for eight months (counting the first day of bleeding in each cycle as day one). The earliest day on which the woman is likely to be fertile is calculated by subtracting 18 days from the length of her shortest cycle; the latest day she is likely to be fertile is obtained by subtracting 11 days from the length of her longest cycle. For example, if a woman has regular 30-day cycles, ovulation will take place on day 17 or within two days earlier or later, which makes days 15 to 19 unsafe. Sperm deposited on the 13th or 14th day may live to day 15, making those days unsafe. The egg may still be alive on day 19 if ovulation took place on the 17th, so cross off day 20. This calculation means that the period from day 13 to day 20 is unsafe. As an extra measure of protection, three more days may be added at each end. To be even safer, the entire preovulatory period is excluded. If the couple also refrains from sex during the several days of menstruation (for other reasons) that still leaves from 7 to 11 days that are safe to engage in coitus. Including the period of menstruation, the safe period encompasses almost half of the month.

This system is logical but it does not work, because the menstrual cycle does not function like clockwork. Many physiological and psychological factors throw it off schedule. In an extreme case, a woman with a short cycle but a long period of bleeding may ovulate while she is still menstruating, so unprotected coitus even during menstruation is not entirely safe. To predict ovulation we have to look for better evidence than the calendar.

The Basal Body Temperature Method   This approach (also called the "sympto-thermal method") depends on the fact that ovulation is accompanied by a discernible rise in the *basal body temperature* (the lowest temperature during waking hours when the body is at rest). To pinpoint the temperature rise, hence the time of ovulation, a woman must take her temperature (by a special BBT thermometer) immediately upon awakening every morning, before she gets out of bed or does anything else. She records the temperatures on a chart. An increase of at least 0.4°F (0.2°C) over the temperature of the preceding five days, which is sustained for three days, indicates that ovulation has occurred. Sometimes the temperature drops before it begins to rise, which gives some forewarning (Figure 7.8). To avoid pregnancy, the woman abstains from sexual intercourse from the end of her menstrual period until three days after ovulation (the first day the temperature rose). To have coitus anytime before there is clear evidence for ovulation is risky.

This method is subject to considerable error. Many factors other than ovulation influence body temperature. Furthermore, in 20 percent of cycles, no temperature rise accompanies ovulation, so that entire month must be considered unsafe. If women scrupulously follow the method, success rates are sometimes reported to be extremely high (99.7 percent) (Doring, 1967, in Hatcher et al., 1988). Other researchers, however, regard this method as far less reliable.

The Cervical Mucus Method   The third approach, also known as the *Billings method,* relies on changes in cervical mucus to predict ovulation. To use it, a woman must learn to identify changes in the amount and consistency of cervical mucus (Billings and Billings, 1974). For a few days before and shortly after menstruation, many women have "dry" periods, with no noticeable cervical discharge and a sensation of dryness in the vagina. These "dry" days are considered relatively safe for intercourse.

After the "dry" period following menstruation, the cervix produces a thick, sticky mucous discharge that may be white or cloudy. Gradually the mucus becomes more watery, slippery, and clear (looking like egg white). This *peak symptom* usually lasts for one or two

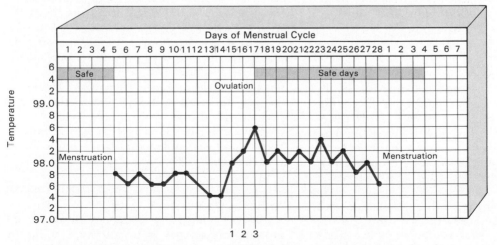

Figure 7.8   Basal body temperature variations during a sample menstrual cycle.

days. Generally, a woman ovulates about 24 hours after the last peak symptom day. The mucus then changes back to being cloudy and sticky. Intercourse is not safe from the first day in which sticky mucus is present until four days after the last peak symptom day.

To help women make these determinations, an *ovutimer* has now been developed—a plastic device that is inserted into the vagina to measure the stickiness of cervical mucus. Another device, called the *Ovulation Predictor,* uses saliva to predict ovulation five days in advance. It has not yet been approved by the FDA as a contraceptive, although it is now available over-the-counter to enhance the chance of pregnancy.

A World Health Organization review in 1978 (before some of the new devices were available) concluded that even combinations of rhythm methods were relatively ineffective. However, it ascribed failures more to risk taking (by engaging in coitus during fertile periods) than to difficulties in interpreting BBT or cervical mucus changes. Ultimately, the effectiveness of any method must rest on personal style, will power, and motivation.

If a highly reliable and easy method of predicting ovulation far enough in advance became available, it would constitute an extraordinary advance in contraception. At least those who can restrain themselves for short periods of time would have a means of avoiding pregnancy that has no side effects and raises no moral objections.

Prolonged Nursing

It has been long observed that nursing an infant seems to protect the mother from getting pregnant. In developing countries more pregnancies have been prevented by breastfeeding than by any other method; yet breastfeeding is widely regarded in the West as an unreliable form of contraception. How do we reconcile these conflicting observations?

Breastfeeding does inhibit ovulation and menstruation after childbirth, although the hormonal mechanisms underlying this effect are not fully understood. Sensory nerve endings at the nipple presumably send impulses to the brain, where they inhibit the hypothalamus from producing its releasing hormones. This depresses the secretion of LH from the pituitary, which in turn inhibits ovulation.

However, this mechanism only works if breastfeeding is done frequently and around the clock. Among certain nomadic tribes in the Kalahari desert in Africa, women on the average conceive at four-year intervals without relying on any other contraceptive. These mothers nurse their infants up to 60 times a day. The practice of giving the breast to the infant on demand remains common in many other developing countries. Even if there is no milk at a feeding, infants derive comfort from sucking, and frequent nipple stimulation keeps the mother's contraceptive mechanism active.

In recent years, breastfeeding has been rapidly losing its contraceptive function in the Third World, as it has in the West, because of the introduction of powdered milk and other food supplements that reduce the frequency of nursing. The Western practice of bottle feeding and the use of "pacifiers" in the form of rubber teats have compromised the suckling mechanism even when women do nurse their infants sporadically. However, this innate mechanism is by no means lost. A study of nursing women in Scotland has shown that if breastfeeding is carried out more than five times in 24 hours, including a night-time feed, ovulation can be inhibited for up to a year or more (Short, 1979).

Once menstruation returns, nursing exerts no further effect in preventing pregnancy. Even the absence of menstruation is not completely reliable, because 80 percent of breastfeeding women ovulate before their first period. All things considered, then, breastfeeding as currently practiced in the United States is not a reliable method of contraception (Hatcher et al., 1988).

Withdrawal

Another ancient and widely used means of avoiding pregnancy is to interrupt coitus by withdrawing the penis before ejaculation—

hence the term *coitus interruptus*. The biblical story of Onan is one of the earliest accounts of this practice (Gen. 38:9).

The major problem with coitus interruptus is that it requires a great deal of will power just at the moment when a man is most likely to throw caution to the winds. Nevertheless, this method costs nothing, requires no device, and has no physiological side effects—although some couples find it frustrating or otherwise unacceptable.

When withdrawal is the only contraceptive measure taken, the lowest observed failure rate is 7 percent, but the typical failure rate, 18 percent. Failures occur mainly because the male does not withdraw quickly enough or because some sperm seep out before ejaculation in prostatic and Cowper's gland secretions. Though withdrawal is not a method to count on, it may be, nonetheless, a useful last resort. The failure rate of withdrawal is admittedly bad, but that of not withdrawing is surely worse.

Since the unreliability of withdrawal has become well known, only 3 percent of women (which still amounts to close to a million women) continue to rely on male withdrawal for contraception. The practice is more popular among the young, where 5 percent of 15- to 19-year-olds rely on it (Forrest and Henshaw, 1983).

Sterilization

Sterilization is causing permanent (but sometimes reversible) infertility through surgery. It is the most effective method of contraception in both sexes and currently the most widely used method of birth control among married couples in the United States (Hatcher et al., 1988). Sterilization accounts for two-thirds of all methods used by Americans age 35 and older. The estimated number of couples who rely on it rose from 20 million in 1970 to 80 million in 1977; by 1975, one partner had been sterilized in a third of all married couples using contraception (Mishell, 1982). Well over 25 million people have so far undergone this procedure.

Until a few years ago women underwent sterilization in much larger numbers than men. More recently the proportion has been shifting towards a more equal distribution (Droegemueller and Bressler, 1980). However, among blacks and Mexican-Americans, many more women than men continue to be sterilized.

Three out of four women now have all the children they want by age 30, but there are another 15 to 20 years during which they remain vulnerable to unwanted pregnancy. Sterilization provides such women a reliable, safe, and simple way to be free from contraceptive concerns once and for all.

Similar considerations apply to men when they reach a point where fatherhood is no more a likely or desirable prospect. This is why older men are more likely to choose this procedure. Vasectomy is not likely to be done for men younger than 24; it accounts for 17 percent of contraceptive choices for 25- to 29-year-olds; and 23 percent for 40- to 44-year-olds (Forrest and Henshaw, 1983).

The main disadvantage of sterilization is that it may preclude having any more children. The procedure cannot always be reversed successfully. If a person remarries after a divorce or death of the spouse and wants to start a new family, it may not be possible.

Male Sterilization   The operation typically used to sterilize the male is *vasectomy*, a procedure that can be done in a doctor's office in 15 minutes (Ackman et al., 1979; Samuel and Rose, 1980). Under local anesthesia a small incision is made on each side of the scrotum to reach the vas deferens (Figure 7.9). Each vas is then tied in two places; the segment between is removed or the ends are cauterized in order to prevent them from growing together again. Sperm continues to be produced, but it now accumulates in the testes and epididymis, breaks down, and is reabsorbed.

No changes in hormonal function follow vasectomy. The testes continue to secrete testosterone into the bloodstream in normal amounts. Erection and ejaculation remain intact. Because the contribution of the testes to

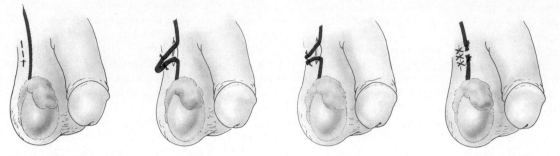

Figure 7.9   Vasectomy.

semen accounts for less than 10 percent of its volume, most men cannot detect any change in their ejaculate. The only difference is that the semen is now free of sperm.

Sperm may still be found in the ejaculate two or three months later, due to their presence in the duct system beyond the site of the vasectomy. To avoid impregnation during the immediate postvasectomy period, the male tubal system can be flushed out at the time of the vasectomy, or some other contraceptive can be used for the next three months or until two successive ejaculations are sperm-free. From then on, the man is sterile as confirmed by the absence of sperm in his ejaculate on microscopic examination.

The rare cases of failure (Table 7.2) are due to unprotected coitus in the postvasectomy

Table 7.2   First-Year Failure for Contraceptives (Percentage of women who become pregnant)

METHOD	LOWEST REPORTED RATE	RATE FOR TYPICAL USERS
Tubal ligation	0.3	0.4
Vasectomy	0.1	0.15
Birth control pills	0.1	3
Minipills (progestin only)	1	3
IUD	0.5	6
Condoms		
Plain	4	12
Spermicidal	2	—
Diaphragm with spermicide	2	18
Sponges		
No previous full-term pregnancy	14	18
Previous full-term pregnancy	28	more than 28
Cervical cap	8	18
Withdrawal	7	18
Fertility awareness	2	20
Spermicides	3	21
No method	—	89

Adapted from: Trussell, J. and Kost, K. "Contraceptive Failure in the United States: A Critical Review of the Literature." *Studies in Family Planning* V18N5 Sept/Oct 1987, pp 237–283.

period before full sterility has been achieved. Very rarely the cut ends of the vas have reunited while the wound is still fresh; current techniques make this virtually impossible.

The main disadvantage of vasectomy, as we said above, is its permanency. Despite many advances in microsurgery that permit reuniting the vas (vasovasostomy), there is only a 50 percent chance that a vasectomized man can be made fertile again (depending on the sterilization and repair methods, the rates actually vary from 5 percent to 70 percent). About one in 500 vasectomized men seek vasovasostomies, usually following a new marriage.

Before undergoing vasectomy, some men have a sample of their sperm frozen and deposited in a sperm bank, in case they want it for artificial insemination. This practice attracted considerable attention a decade ago, but enthusiasm for it has diminished. Although some pregnancies have been achieved with sperm that has been kept for several years, the success rate is much lower than with fresh sperm. Furthermore, the possibility that sperm will be damaged when stored over many years and will cause genetic defects in the offspring cannot be ruled out. Sperm banks, therefore, do not provide full fertility insurance for men who elect to undergo vasectomy (Ansbacher, 1978).

There are no serious side effects to vasectomy. The local discomfort after surgery is minimal and lasts only a few days with low risk of complications. One lingering concern is that autoimmune reactions may develop. The body may produce antibodies to the components of the sperm it reabsorbs (Shahani and Hattikudur, 1981). This reaction, in fact, is believed partly responsible for the lowered chance of fertility after vasovasostomy.

It has been found that vasectomized monkeys develop atherosclerosis, or fat deposits in the walls of blood vessels, which increases the risk of heart attacks. This effect has not been shown to hold for humans even over a decade (Clarkson and Alexander, 1980). We can never rule out the chance of finding new long-term effects, but that prospect becomes increasingly unlikely with the passage of time (Hussey, 1981).

On purely psychological grounds, vasectomy may interfere with sexual performance. A man may feel less virile, or sexually "damaged." Men who have problems with their erections to begin with or are likely to have their sense of masculinity seriously threatened may be better off avoiding this procedure. Resentment and conflict may also arise if a man feels pressured by his wife to have a vasectomy to relieve her of the burden of birth control, just as a woman would rightly feel resentful if pushed to such a choice. Despite these occasional concerns, the typical response to vasectomy is a sense of freedom and relief that leads to greater interest in sex.

Female Sterilization   Women who have their uterus removed (hysterectomy) or ovaries removed (ovariectomy) will become sterile, but these procedures are usually done for other reasons. The most common surgical procedure used for sterilizing women is tubal ligation (Tatum, 1987). Tying or severing the fallopian tubes prevents the meeting of eggs and sperm. The eggs that continue to be ovulated are simply reabsorbed by the body. The ovaries continue to supply their hormones into the blood stream normally and the menstrual cycle is not interfered with, nor is sexual interest diminished.

Female sterilization used to be a major surgical procedure, but there are now effective and inexpensive procedures involving over one hundred techniques for cutting, closing, or tying the tubes (Figure 7.10). The current trend is to use outpatient procedures performed under local anesthesia. It is possible to approach the fallopian tubes through a small incision in the vagina (culdoscopy) or in the abdominal wall (sometimes the navel) and to perform the sterilization with the help of a laparoscope (a tube with a self-contained optical system that allows the physician to see inside the abdominal cavity).

Although these techniques are all called "ligation," the tubes are not just tied but cut and cauterized. Chemicals that solidify in the tubes and lasers that destroy a portion of the tubes are among new sterilization techniques under investigation.

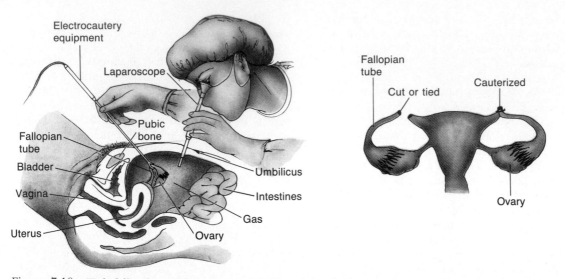

Figure 7.10   Tubal ligation with laparoscope. The abdominal cavity is slightly inflated with a harmless gas to gain easier access to the fallopian tubes.

Most methods currently used to sterilize women are virtually 100 percent effective (Table 7.2). Occasional failures in the past have been due to improper procedures. More often, the woman is already pregnant when the tubes are tied, but that fact is not known. Female sterilization, like its male counterpart, must be approached as a permanent procedure. The chances of becoming fertile again after reuniting the tubes is 10–50 percent, depending on the techniques used.

In the past, women have been generally more willing than men to undergo sterilization. The fact that women will become sterile anyway at the menopause has been one factor facilitating their decision. More importantly, women have been more willing because it is they who usually carry the burden of contraception and the risk of pregnancy. However, female sterilization is a more difficult procedure, and a sense of shared responsibility has now made more men willing to undergo vasectomy.

There is some risk of complications with female sterilization as with any other surgical procedure, but serious consequences are rare. The fatality rate for female sterilization is approximately 1 in 20,000 cases; it is virtually nil

for vasectomy (Hatcher et al., 1988). Adverse psychological reactions to sterilization are less common among women than men. Two percent of women who are sterilized regret the decision within a year (Forrest and Henshaw, 1983). More often women welcome the relief from the fear of pregnancy and freedom from the burden and side effects of contraceptives. Some women become sexually more interested and responsive as a result.

Despite the enormous expansion of contraceptive technology and use over the past two decades, we are still in the "horse and buggy days" of effective contraception, according to Alan Guttmacher, one of the pioneers in this field. The future prospects of contraception will depend on technological advances and on social attitudes influencing their use.

Future Birth Control

Contraceptive research continues to be pursued actively, and the following are some of the more likely methods to come into use in the foreseeable future.

Male Methods   Currently, there are only two reliable male contraceptive methods: the condom and vasectomy. To improve condom use we need advances in attitude, not technology. To improve vasectomy, researchers are trying to increase reversibility.

A major breakthrough would be the development of a *male pill* that would interfere with spermatogenesis, or somehow neutralize the ability of sperm to fertilize. There are interesting leads in this area, as well as formidable problems. Such a pill has to suppress the normal process of spermatogenesis that is continuously in progress. It should do so without causing chromosomal mutations, damage to other cells, or loss of sexual drive or function.

Some drugs have shown promising results in reducing male fertility. *Danazol,* a synthetic hormone with a structure similar to the androgens, prevents the release of FSH and LH from the pituitary, thereby depressing sperm production. In preliminary studies, men who were given Danazol in conjunction with testosterone for six months at a time had their sperm counts drop to 0.5-5 percent of normal levels. These men did not suffer loss of sexual function, and they regained normal fertility within five months after the end of treatment.

It had been noted that men showed decreased fertility in certain parts of China where unrefined cottonseed oil is part of the daily diet. The substance in cottonseed oil that interferes with sperm production and mobility is *gossypol,* and thousands of Chinese men have been using it now for birth control, with apparent success. The Chinese claim the method is 99.8 percent effective. The men are reported to regain fertility readily after they stop taking gossypol (Kaufman et al., 1981; Hatcher et al., 1988).

Gossypol comes in tablet form, so it would be the closest thing to a "male pill" (though it is not a hormone). However, this method is not likely to be rapidly approved or adopted in the United States, because gossypol is known to have several toxic effects, including weakness, decrease in libido, change in appetite, nausea, and sometimes serious heart problems. Extensive testing would be needed to establish its safety. A "male pill" therefore appears to be many years away (Djerassi, 1981).

Female Methods   What about research in female contraception? Most of it is aimed at the improvement of existing methods that disrupt the hormonal cycle. The same hormones, estrogens and progestins, are put to use in intravaginal and intrauterine devices, and in longer-acting forms, to provide sustained protection against fertilization (Hatcher et al., 1988).

Contraceptive vaginal rings have been shown to be 98 percent effective in preventing ovulation. These rings have an inert core of plastic around which is a layer of steroid hormones. They are slightly smaller than a regular diaphragm and more easily inserted. They are placed in the vagina on the fifth day of the menstrual period and left in place for the next 21 days. Menstruation begins a few days after the ring is removed. Each ring can be used for up to six months. The estrogens and progestins in the rings are released slowly and absorbed through the vaginal wall into the woman's bloodstream. The hormones prevent ovulation, cause changes in the endometrium, and thicken the cervical mucus. There are few side effects, and the ring does not interfere with intercourse.

Instead of taking a pill every day, women in the future may take long-acting injections of hormones. One such preparation is already being used successfully in many parts of the world but has not yet been approved for contraceptive use in the United States (Hatcher et al., 1988). The preparation, marketed as Depo-Provera (or in tablet form, Provera), contains a synthetic form of progesterone *(medroxyprogesterone),* which prevents ovulation for 90 days. It is highly effective, but side effects include weight gain, loss of sex drive, menstrual irregularities, and an unpredictable (possibly irreversible) period of infertility following cessation of the shots. It has been linked with cancer in certain animals but not in women (Gilman et al., 1985).

Similarly, implants (Norplant) that release progesterone slowly are being tested in the United States. These plastic capsules are inserted under the skin of the arm through a small incision and can provide effective contraception for up to five years, but they may be removed at any time if pregnancy is desired or side effects arise. Norplant has already been approved in several European countries (Hatcher et al., 1988).

The IUD is also being tested for improvements. *Tailless IUDs,* for instance, may greatly reduce the risk of infection. Doctors would use ultrasound to check that the IUD stays in place. The progestin-releasing IUD, which now must be replaced on a yearly basis, may be refined to slowly release a powerful progestin like norgestral over six to ten years.

An *antipregnancy vaccine* would activate the immune system in the event of conception and terminate it. Antibodies are being developed that allow the female body to be immunized to its own hCG. Women with antibodies to hCG no longer respond to the gonadotropic signals to sustain the production of ovarian hormones (Chapter 4). As a result, the immunized woman menstruates and disrupts implantation. Such a vaccine thus induces early abortion rather than preventing conception. In principle, an antipregnancy vaccine would permanently sterilize a woman, but vaccines tend to "wear off" over time and may have to be readministered. The clinical testing of antipregnancy vaccines is now under way, but it would take a decade or more for a usable vaccine to be marketed (Ory et al., 1983).

Despite all of these developments women now in their early twenties are likely to reach the menopause before any major breakthroughs occur in male or female contraceptives. The birth control methods that will be available at the end of the century are most likely to be modifications of those we have today. Government regulatory controls mean that a scientific breakthrough today would take 12 to 15 years before becoming a practical device in general use. Meanwhile, 350,000 babies are born in the world every day, and only 200,000 persons die (Djerassi, 1981, p. 225).

ABORTION

Abortion is the termination of an established pregnancy before the fetus can survive outside the uterus. Typically, a fetus is viable 28 weeks after the last menstrual period, when it typically weighs about 1000 grams (2.2. lbs.) (Chapter 6).

Abortions may be *spontaneous* or they may be *induced*—that is, brought about on purpose (Pritchard et al., 1985). Spontaneous abortions, also known as "miscarriages," typically involve defective embryos. The reason there are so few babies with serious congenital abnormalities is that over 90 percent of fetuses with abnormalities do not survive to the end of the gestation period (Lauritsen, 1982; Scott, 1986). Either because of a genetic defect or some problem in the mother, about 15 percent of all pregnancies terminate in a miscarriage. These odds are not evenly distributed—some women are more likely to miscarry than others, and some do so repeatedly.

Induced abortions may be *therapeutic* or *elective.* A therapeutic abortion must be done for medical reasons; an elective one is done when other considerations lead to the decision not to have the child. Induced abortion may be legal or illegal, depending on local laws. Complex social, moral, and legal considerations make elective abortion a highly controversial issue.

Abortion laws were liberalized in the United States in the early 1970s. The number of abortions quickly doubled, from 744,600 in 1973 to 1,553,900 in 1980. An estimated 55 million abortions per year are reported worldwide (30 abortions per 100 pregnancies) (Tietze, 1983). In the United States, a quarter of all pregnancies, and about half of all unintended ones, were terminated in 1980 by elective abortion. This rate has remained steady during the 1980s.

The majority of abortions are performed on women under 25, many of them teenagers (Hatcher et al., 1988). About eight in ten abortions are obtained by unmarried women. Blacks are over twice as likely to have abortions as whites (Ory et al., 1983). Half of unmarried women and one-third of married women say they would consider getting an abortion in the case of unwanted pregnancy, but the proportion of women willing to do so is higher (66 percent) among those with no religious affiliation than among Catholics (33 percent), Protestants (42 percent), and Jews (64 percent) (Forrest and Henshaw, 1983).

Illegal abortions are dangerous. They led to 364 deaths a year during 1958–1962 in the United States. The death rate associated with abortions had declined dramatically by 1980 and has remained very low (Table 7.1). The higher risk of illegal abortions is due to the fact that they are usually carried out by unqualified people or under unhygienic circumstances. The instruments may be unclean, leading to infection. Improper techniques may cause excessive uterine bleeding or perforation of the uterus. These risks are particularly high when women try to abort themselves with implements like knitting needles. That is why the wire coat hanger has come to symbolize the hazards of illegal abortions.

These trends have been repeated in the experience of other countries. For instance, abortion mortality rates declined 56 percent and 38 percent in Czechoslovakia and Hungary, respectively, after their abortion laws were liberalized in the mid-1950s. The opposite effect occurs when abortion laws become more restrictive: in Rumania in 1966, there was a seven-fold increase in deaths due to illegal abortion (Tietze, 1983).

Methods of Abortion

The method used for abortion is usually determined by how far the pregnancy has advanced. During the first trimester, abortion is performed by removing, through the cervix, the embryo and its associated parts (the *concepsus*). Uterine evacuation is sometimes used as late as the 20th week, but during the second trimester, abortion is frequently performed by stimulating the uterus to expel its contents— in effect, inducing labor.

Although abortion does present certain risks, the overall death rate associated with legal abortions is low, especially during the first trimester (when the great majority of abortions are done). Maternal deaths in pregnancy are 1 in 10,000. In legal abortions performed before 9 weeks, the death rate is less than or equal to 1 in 400,000; after 16 weeks, 1 in 10,000 (Table 7.1). By contrast, 1 in 3000 women die during illegal abortions (Hatcher et al., 1988).

Vacuum Aspiration   Sucking out the contents of the pregnant uterus through *vacuum aspiration* or vacuum curettage is the preferred method for first trimester abortions because it can be performed on an outpatient basis, quickly, and at a relatively low cost. Up to eight weeks of pregnancy, little or no anesthesia may be needed, and less dilation of the cervix is required.

As shown in Figure 7.11, a *suction curette* is passed through the cervical opening into the uterus to scrape and to suck out its contents. Complications of vacuum aspiration are rare but may be serious. They include perforation of the uterus, hemorrhage, uterine infection, and cervical lacerations.

Women who do not wish to face having an abortion sometimes have a *menstrual extraction* if they have missed their period for more than two weeks. The technique is basically the same as vacuum aspiration, but it is done without a pregnancy test. There is no way of knowing whether the missed period was due to pregnancy or some other cause.

Dilation and Curettage (D and C)   Before the advent of vacuum aspiration in the 1960s, the most common method of abortion was *dilation* of the cervix *and curettage* (scraping) of the uterine lining. D and C is also employed for the diagnosis and treatment of a number of uterine disorders, so if a woman has undergone this procedure it does not necessarily mean she has had an abortion. The first step, cervical dilation, can be accomplished by passing a series of progressively larger metallic dilators through the cervical opening, but in recent years a less painful but slower method has become popular, using *laminaria sticks*. These sticks, made of compressed seaweed, or synthetic cellulose, are inserted into the cervix, where they absorb cervical secretions and expand to five times their dry size in about a day. When the cervix is enlarged sufficiently, a *cur-*

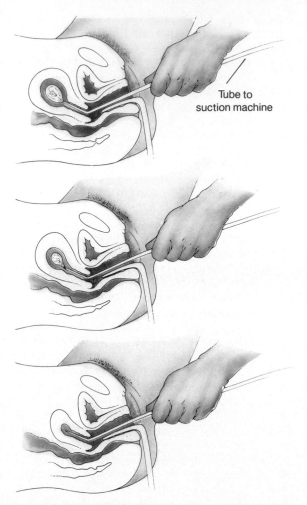

Tube to
suction machine

Figure 7.11   Abortion by vacuum aspiration. (Top to bottom) A suction curette is inserted through the cervical canal; uterine contents aspirated; uterus contracted after completion of evacuation.

rette (a bluntly serrated metal instrument) is inserted to scrape off the uterine contents. The complications of D and C are the same as those of vacuum aspiration but occur more often, which is why D and C has now been largely superseded by vacuum aspiration.

Dilation and Evacuation (D and E)   After the 12th week of pregnancy, when the uterus is softer and its walls thinner, abortion becomes a somewhat more hazardous surgical procedure, and the rate of complications increases. One abortion technique that is often used between the 13th and 20th weeks of pregnancy is *dilation and evacuation*. D and E is similar to the two previous methods, but the fetus, now being larger, is not as easily removed. After adequate cervical dilation is achieved, suction, forceps, and curettage are used to remove the uterine contents.

Saline Abortions   During the second trimester, abortion is commonly induced by the injection of a concentrated salt (or urea) solution into the uterine cavity. Known as *saline abortion,* this is a technically more difficult procedure, especially in the fourth month when pregnancy is too far along to allow a safe, simple aspiration; but the uterus is not yet large enough to allow the physician easily to locate it through the abdominal wall.

Following the injection of saline solution into the uterine cavity, contractions usually begin within 12 to 24 hours, and about 80 percent of women deliver the fetus and placenta within 48 hours. Some physicians also administer oxytocin to stimulate more vigorous uterine contractions.

Uncommon but severe complications may occur with saline abortion. Most serious is *hypernatremia* ("salt poisoning"), which can result in high-blood pressure, brain damage, and death. Other complications include uterine infection and hemorrhage. Delayed hemorrhage (days or weeks after the abortion) can occur if the fetus is aborted but part or all of the placenta remains behind.

Prostaglandin Abortions   Prostaglandins cause uterine contractions, so they can be used to induce abortion. They are usually injected directly into the uterus but can also be injected into the bloodstream or muscles. Laminaria sticks are sometimes used to dilate the cervix.

Complications with prostaglandin abortion include nausea, vomiting, and headache; at least 50 percent of women experience one or more of these side effects, which are temporary and rarely serious. Hemorrhage, infection, and uterine rupture (a danger with all types of second-trimester abortion) are infrequent with prostaglandin abortion. The likelihood of live births is higher with prostaglandin than with saline abortion, especially after the 20th week, which makes some physicians reluctant to perform abortions by this method so late.

Pregnancy Reductions   Women who take fertility pills or have multiple embryos implanted after in vitro fertilization (to ensure that at least one will survive) sometimes end up with multiple pregnancies. The number of fetuses may be as many as six or more, in which case there is virtually no chance of any of them surviving. Even if the chances of survival are better, as with quadruplets, a couple may not want to have more than one or two children. Under these circumstances, obstetricians are now aborting some of the fetuses at the woman's request. Usually only a pair of twins are left to grow to term. This still allows some reserve margin should one of the fetuses not make it; but in some cases women ask that only one fetus be spared (Kolata, 1988c).

This procedure, known as *pregnancy reduction*, involves looking at the fetuses through ultrasound (Chapter 6) when they are about 1½ inches long. The doctor selects those that are most accessible and injects potassium chloride into the chest cavities of those that are to be eliminated. After the fetus dies, it is gradually absorbed by the mother's body.

The procedure is not without its risks. It may lead to bleeding, infection, or the induction of labor, which results in the miscarriage of all the fetuses. Moreover, the practice raises various ethical and social concerns which cannot be easily resolved.

Experimental Methods   Some of the most promising advances in abortion include hormonal methods like the morning-after pill, and hormone blockers like RU 486, which we discussed earlier. Although often referred to as contraceptive devices, they are actually methods of abortion, because they disrupt the gestational process after fertilization.

If these newer methods become fully developed and widely available, a woman will be able to abort herself in the privacy of her home by simply taking a pill, or she will only need to take a pill before or after coitus to feel certain that she will not become or remain pregnant, without actually being aware of either event.

Psychological Aspects of Abortion
The psychological reactions of women and their sexual partners who are faced with the prospect of abortion are likely to be complex and varied. Much depends on their perceptions of abortion in moral and psychological terms: for some, abortion is a matter of disrupting a physiological process whose continuation is not in their best interest; for others, it is a form of infanticide. Psychological issues in abortion are usually more significant for adolescents (Melton, 1986).

Reactions to Unwanted Pregnancy   The reality of an unwanted pregnancy and the prospect of abortion elicit strong feelings. A typical initial reaction is disbelief or denial, which may lead, especially among teenagers, to delays in confirming the pregnancy and coming to terms with it. A woman may also be shocked to discover that she is further along into the pregnancy than she assumed from her missed period (by which time she was already several weeks pregnant). In other cases, there is minimal conflict. The woman knows she does not want a baby, and whatever her previous attitude towards abortion, she gets rapidly mobilized to terminate the pregnancy.

The initial disbelief gives way to a cluster of negative feelings. Distress may be compounded by guilt and recrimination. There is anger and outrage, especially if seduction, deception, or pressure was used in the sexual interaction that led to the pregnancy. At the same time there may be a deep satisfaction in being pregnant and a reluctance to terminate it, however necessary or inevitable that may be.

The men involved in a pregnancy are also likely to react strongly. They are not immune to the same yearnings for parenthood, and their own anguish, anger, guilt, and helplessness often receive insufficient recognition. Because it is usually the woman who eventually determines what is to happen to the pregnancy, men also tend to feel helpless in these situations.

The most trying time is often between the discovery of the pregnancy and the decision to abort. Once there is the prospect of action, at least the burden of uncertainty is lifted. Sometimes, though, doubts persist and women continue to feel anxious and depressed.

Deciding among the Alternatives   No two cases of unwanted pregnancy are the same. A mature woman in a secure marriage who fully intended to have a child at some future date but got pregnant unexpectedly and the teenage girl who is not even sure who got her pregnant would be obviously facing the abortion decision from quite different perspectives.

A single woman who gets unintentionally pregnant faces four choices. Traditionally, the most desirable alternative has been marriage. If it would make good sense for the woman to marry the man at this time even if she were not pregnant by him, then marriage may be the answer. Otherwise, marriage under these circumstances may turn out to be only a short-term or face-saving solution, but a long-term liability. The prospect of an unhappy marriage or a divorce with a child in the picture deserves serious thought.

The next prospect is single parenthood. Many more women decide on this alternative now than in the past, because society is more accepting and women are more independent. Raising a child on one's own is still not easy. The prospects of help from the man or other sources, the impact of single parenthood on a career, and other related considerations must be taken into account. Teenage mothers frequently do not even finish high school, economically handicapping themselves and their children.

The third alternative is to give up the child for adoption. This choice avoids both abortion, which may be unacceptable on moral or psychological grounds, and the need for dealing with the child. However, the decision to give up the child is not easy either, and a woman may not be able to make up her mind until the baby is born. If there are no adoptive parents available, the child can live in a foster home. Before the baby can be placed for adoption, the father has a right to a legal hearing.

If these three alternatives are unavailable or undesirable, then the choice comes down to abortion—which is what four out of ten pregnant teenagers choose. The earlier the decision to abort is made, the better it is for health. For this reason it is important for women to monitor their menstrual cycles. A period that is late by a day or two means nothing, but if a woman loses track of when her period should have come, then she may not know soon enough when to become concerned and get a pregnancy test. Delay in getting confirmation of pregnancy is the major difference between women who receive first-trimester abortions and second-trimester abortions.

Though time is of the essence, a woman should not be rushed or pressured into an abortion decision. All women faced with this situation need information, support, and an opportunity for counseling. The discussions should review her moral values, life situation, aspirations for the future, feelings about her partner, and the conflicts she is experiencing in making a choice. In addition to traditional sources of guidance, such as physicians, organizations like Planned Parenthood provide such services in many communities.

Reactions to Abortion   You may have found it upsetting just to read about abortion methods. A woman's reaction to undergoing the procedure itself depends on the length of the pregnancy and the sensitivity with which she is treated. The later the abortion, the more upsetting it is likely to be. Women who have had children are startled to discover that induced uterine contractions feel just like labor pains, and these memories can be disturbing. Abortions that produce an alive fetus (that cannot survive) are particularly distressing.

Women do not take abortion lightly, but some suffer more than others. The loss of the fetus may feel like the death of a child, and the woman may grieve over it. Psychological burdens of guilt and regret may lead to depression. It is important to provide post-abortion counseling to women who need it.

These emotions are much influenced by circumstances. Having an abortion is by no means a cause for ostracism in all groups (*Boston Women's Health Collective*, 1986). A woman's husband or lover, family, and friends can provide much emotional support. With help a woman can cope with the experience rather than being overwhelmed by it (Freeman, 1978).

The aftermath of an abortion is usually followed by a sense of relief. The mechanisms of denial and repression help to bury the experience in the past, yet painful doubts may also linger. Sooner or later these feelings begin to recede: by the end of several months, many of these women consider the matter resolved. In some cases women require more extensive counseling (Nadelson, 1978).

The law and large segments of professional and public opinion now endorse the right of each woman to make decisions regarding her own pregnancy. Because the fetus is part of the woman's body, it is argued that she ought to choose what can and cannot be done with it. This view may be unfair to the man, assuming he is known, is available, and cares about the outcome of the pregnancy; after all, he has feelings about the fetus, and society expects him to have a vital interest in the future of the child he has helped create. Does his fatherhood, with all its rights and responsibilities, start only after the moment of birth? On the other hand, to allow the father to decide would place an intolerable burden on the woman. Is she to bear a child that she does not want and then be primarily responsible for rearing that child as well?

There are no ready answers to these dilemmas. Abortion involves the sort of problem that is better prevented than dealt with after the fact. The need for abortion may never be eliminated entirely, but the effective use of contraception would certainly spare millions of people the experience of abortion and save them much grief.

REVIEW QUESTIONS

1. Why do people use and refuse to use contraceptives?

2. List the various types of contraception, grouping them by their methods of action. Compare their effectiveness and their other advantages and disadvantages.

3. Prescribe a contraceptive plan for an unmarried and sexually active woman from the age of 15 to 50 years.

4. Prescribe a contraceptive plan for an unmarried and sexually active man from the age of 15 to 80.

5. Discuss the methods of abortion, with their positive and negative aspects.

THOUGHT QUESTIONS

1. Respond to the argument that because pregnancy involves women's bodies, they should be responsible for contraception.

2. How would you choose between a contraceptive method that has a 0.1 percent failure rate but significant health risks, and another that has a 10 percent failure rate but no health risks?

3. As a physician, would you abort a woman because she would rather have a child of the opposite sex than that of the embryo she is carrying? As a legislator, would you pass a law prohibiting a physician from carrying out such an abortion?

4. What sort of program would you set up to help prevent teenage pregnancy?

5. What measures would you take to ensure that you do not cause or experience an unwanted pregnancy under any circumstances?

SUGGESTED READINGS

Denney, M. (1983). *A matter of choice: An essential guide to every aspect of abortion*. The biological, emotional, legal, social, religious, and political aspects of abortion are discussed clearly, candidly, and objectively.

Djerassi, C. (1981). *The politics of contraception*. New York: Norton. An authoritative, broad, and well-written account of the background, current status, and future of contraception.

Hatcher, R. A., et al. (1988). *Contraceptive technology 1988–89*. New York: Irvington. Detailed and clear descriptions of all contraceptive methods with instructions for their use.

Hatcher, R. A. (1982). *It's Your Choice*. New York: Irvington. Detailed instructions for choosing a contraceptive method and using it safely and effectively.

Melton, G. B. (Ed.) (1986). *Adolescent Abortion*. Lincoln, Neb.: University of Nebraska Press. A collection of articles on the psychological, legal, and ethical aspects of adolescent abortion.

Population Council Fact Books. New York: The Population Council. In-depth reports and review articles on ongoing research in contraceptive technology.

Population Reports. Baltimore: The Population Information Program of Johns Hopkins University.

Sexual Dysfunction and Therapy

Sexual disorders, which we discussed in Chapter 5, clearly belong to the biological side of sexuality. That is only partially true for sexual dysfunctions. There are many forms of sexual dysfunction where there is no demonstrable physical or organic cause. Nonetheless, sexual dysfunctions are included in this volume because in a significant proportion of cases, sexual dysfunctions result from a wide variety of organic causes. Their symptoms are often physical. And, some forms of treatment of sexual dysfunction entail physical methods, such as drugs and surgery. Finally, the changes in the sexual response cycle which occur with aging are basically of a biological nature.

However, because psychological and interpersonal factors are also highly significant in sexual dysfunction, this chapter deviates somewhat from previous ones by dealing, in some detail, with issues that are not entirely of a biological nature.

TYPES OF SEXUAL DYSFUNCTION

No human function works flawlessly all of the time, and sex is no exception. It is difficult to define a level of sexual activity as inadequate; there is so much variation in ordinary sexual function, and so much of sexual satisfaction is subjective. The prevalence and frequency of sexual activity also vary with age so the same standards cannot be applied across the life cycle.

Nonetheless we can identify some significant failures of sexual function and satisfaction as *sexual dysfunction* (*dys* means faulty or difficult). Such dysfunction is different from diseases of sex organs (such as STDs), reproductive failure (such as sterility), gender disorders (such as transsexualism), and atypical sexual behaviors (paraphilias). The main problems in sexual dysfunction are disturbances of sexual desire (such as lack of interest) or of sexual performance (such as the inability to become sexually aroused or reach orgasm).

Definitions
Until recently, most male sexual dysfunctions were lumped together as *impotence* and female

disorders as *frigidity*. Both terms are now considered inadequate. "Frigidity" has been discarded, but "impotence" continues to be used. Other terms for both male and female dysfunctions have been coined by various sex researchers and therapists (Masters and Johnson, 1970; Kaplan, 1974). Official terms in the Diagnostic Manual of the American Psychiatric Association also keep changing with the refinement of diagnostic categories. As a result, the profusion of terms can be confusing.

The terms that we will use in this chapter are from the 1987 edition of the *Diagnostic Manual (DSM-III-R)* (American Psychiatric Association, 1987). It classifies sexual dysfunction into four major types: *sexual desire disorders, sexual arousal disorders, orgasm disorders,* and *sexual pain disorders.* Let us clarify what a diagnosis of dysfunction means. There are two approaches. One is to consider all disturbances of sexual function, regardless of cause, as a form of sexual dysfunction. The second is to distinguish between sexual problems that are *symptoms* of some underlying disease (and consider them as part of that disease) and sexual problems that are independent. The Diagnostic Manual follows the latter model, but in practice most sex therapists and clinicians deal with all sexual problems as dysfunctions, whatever their cause.

Many physical illnesses (such as diabetes) can disrupt sexuality. An underlying condition could be affecting a key system, such as the blood vessels or the nerves supplying the genital organs. Similarly, a sexual problem can be the symptom of an underlying psychological illness (such as severe depression). Here too, the sexual disturbance is just one part of a larger disorder.

When the sexual problem is the whole story, it is considered *psychogenic*—caused by psychological or emotional factors. The cause may be intrapsychic (due to inner psychological conflicts), relational (due to conflicts with the partner), or cultural (due to social attitudes toward sex). Whatever the source, sexual dysfunctions today are generally treated as illnesses. Some people object to this model. They would rather deal with the sexual dysfunctions as forms of faulty sexual learning.

To qualify as a dysfunction, the sexual problem must be recurrent or persistent: occasional failure does not count. It is normal to fluctuate in sexual desire. Anyone's performance may occasionally falter. No one is expected to be able to engage in sex at will, with anyone, anywhere, anytime, under any circumstances.

Sexual dysfunctions are disturbances of basic sexual functions. In Kaplan's *triphasic model,* these are desire, arousal, and orgasm (Chapter 3). In physiological terms they reflect problems in the two processes underlying the sexual response cycle: vasocongestion and myotonia, the components of Masters and Johnson's *biphasic model* (Chapter 3). Vasocongestive disturbances account for problems of erection in the male and sexual arousal in the female; disturbances in myotonia are linked to difficulties in orgasm (Kaplan, 1974).

Finally, we need to distinguish between *primary* and *secondary* sexual dysfunctions. In the first case, the condition has always existed. A man has never been able to have an erection, or a woman has never experienced orgasm. In secondary dysfunctions, the person used to be healthy. A man who was once potent can no longer have an erection; a woman who was orgasmic is no longer able to reach orgasm. We discuss the importance of this distinction later.

The study of sexual dysfunctions has focused on problems during coitus, and their treatment is modeled on heterosexual couples. Remember as you read that virtually everything in this chapter applies equally to homosexuals.

Prevalence

As we noted at the beginning of the chapter, sexual dysfunction is thought to be quite prevalent. How do you find out? Do you ask people, or do you observe them? Is the subjective sensation of sexual pleasure the criterion, or objective measures of performance? How do you measure "sexual desire"? When is coitus "long enough"? What if a woman is able to reach orgasm with one man but not another? What if a man has no problem in getting an erection when masturbating but cannot do so in coitus?

No wonder accurate statistics on sexual dysfunction are hard to come by. Studies use different diagnostic criteria and call the same entity by different names, so they are hard to compare.

We cannot determine the true prevalence of sexual dysfunction from those who seek sex therapy, because the availability of such help and the willingness of people to use it vary widely. More useful are surveys of more general populations, such as couples seen at a medical clinic. One such study involved 100 couples who were predominantly white, well educated, and middle class. Eighty percent of these couples claimed to have happy marriages; none of the couples were in marital therapy or counseling at the time, although 12 percent had had such help in the past. The prevalence of sexual dysfunction in this group is shown in Table 8.1. Over half the women and over one-third of the men have had some sexual problem (Frank et al., 1978). In a similar study on a general clinic population, lack of sexual desire was reported by 27 percent of women and 13 percent of men; women were also more dissatisfied with the frequency of intercourse (23 percent) than men (18 percent). Premature ejaculation (14 percent) and erection difficulties (12 percent) were the other main male complaints, whereas the women mentioned inability to reach orgasm (25 percent) and pain during coitus (20 percent) (Ende et al., 1984). A review of 22 general sex surveys shows a greater variability in the prevalence of sexual dysfunctions (Nathan, 1986).

SEXUAL DESIRE DISORDERS

When new methods of sex therapy became established in the late 1960s, those who sought help mainly suffered from disturbances of sexual arousal and orgasm. By the end of the decade, sex therapists began to see a new kind of problem—men and women whose sexual functions were not disturbed but who complained of having lost interest in sex or satisfaction from it. These conditions, known as disorders of sexual desire (Kaplan, 1979), have come to

Table 8.1 Prevalence of Sexual Dysfunction in "Normal" Couples

	WOMEN (%)	MEN (%)
Dysfunctions		
Difficulty getting excited/getting erection	48	7
Difficulty maintaining excitement/erection	33	9
Reaching orgasm/ejaculation too soon	11	36
Difficulty reaching orgasm/ejaculation	46	4
Inability to have orgasm/ejaculation	15	0
Other Problems		
Partner chooses inconvenient time	31	16
Inability to relax	47	12
Attraction to person(s) other than mate	14	21
Disinterest	35	16
Attraction to person(s) of same sex	1	0
Different sex practices or habits	10	12
"Turned off"	28	10
Too little foreplay	38	21
Too little tenderness after intercourse	25	17
Sexual Satisfaction		
"How satisfying are your sexual relations?"		
Very satisfying	40	42
Moderately satisfying	46	43
Not very satisfying	12	13
Not satisfying at all	2	2
"How satisfactory have your sexual relations with your spouse been in comparison to other aspects of your marital life?"		
Better than the rest	19	24
About the same	63	60
Worse than the rest	18	16
"Do you have [sexual dissatisfaction] in your marriage?"		
Yes	21	33
No	79	67

Reprinted with permission from *The New England Journal of Medicine*, 299:111–115, 1978.

Hypoactive Sexual Desire Disorder

Hypoactive sexual desire (Latin *hypo*, "under") is still widely called *inhibited sexual desire* (ISD) and other terms, such as *sexual apathy*.

In its mildest form, sexual apathy takes the form of simple indifference. It is like lacking appetite for food. A person just does not feel like having sex, despite favorable circumstances and a willing partner. Sometimes the person is so preoccupied with other activities that sex is almost forgotten for a while.

Judgments of apathy tend to be subjective and relative. Suppose that one-third of the couples in a study had intercourse fewer than two or three times a month. Is that too little, too much, or just right? To answer, people are likely to compare the present against the past ("I am not as interested in sex as I used to be"); the level of interest of one partner as against another ("My spouse is not as interested in sex as I am"); experiences of others ("We think we are not having as much sex as our friends"). In practice, help is usually sought because of discrepancies in sexual desire between steady partners; the unattached are more likely to "solve" the problem by looking for other partners.

To qualify as a disorder of sexual desire, the problem has to be more serious and persistent than normal ups and downs. In *primary* hypoactive desire disorders, a person has never achieved the level of sexual interest typically expected of most healthy adult men and women, although criteria for this level tend to be arbitrary (Schover et al., 1982). Their friendships and intimate relationships are singularly lacking in sexual interest; nor do they resort to masturbation or atypical forms of sexual arousal. This condition seems to be rare. More often, in *secondary* hypoactive desire disorders, a person who used to enjoy sex loses interest. It is these individuals who are more likely to perceive a problem and seek help. Kaplan (1979) further differentiates between *global* inhibition of sexual desire, which involves a total lack of sexual interest in all situations, and *situational* lack of interest with one partner but not another; the causes here are more likely to be relational.

occupy center stage in the field of sex therapy, with various theories to explain them. They take two main forms: *hypoactive sexual desire disorder* and *sexual aversion*.

Sexual Aversion

Sexual aversion goes beyond a passive lack of interest: it involves an active avoidance of all sexual activity. Some people feel distaste. Others feel intense fear and the avoidance takes the form of a *phobia*; confronted with the prospect of sex, the person suffers acute anxiety.

The sexual aversion may be primary or secondary, global or situational. The underlying assumption once again is that there is no compelling or rational reason why a person would want to avoid sex. Sexual apathy or aversion often derives from traumatic sexual experiences (such as rape) or lack of orgasmic response; after all, the less one enjoys an activity, the less the incentive to repeat it. Apathy may also exist in cases where orgasmic ability is not disturbed or is restored through sex therapy, as illustrated in the following example, involving a 32-year-old woman married for six years:

> At the time of evaluation, both husband and wife expressed frustration with the state of their sexual relationship. This frustration led to increasingly strained and infrequent sexual encounters. During the course of her pregnancy, intercourse was discontinued altogether. Although it initially appeared that her diminished interest in sex was related to the primary anorgasmia, her subsequent progress in therapy indicated that this was not the case. Despite the fact that a program of guided self-stimulation exercises was successful in permitting Mrs. O to achieve orgasm, her interest in sex with her husband showed little change and, if anything, declined over the course of treatment. It became painfully clear to both partners that her inability to achieve orgasm at the outset of therapy had little bearing on her more fundamental lack of sexual interest (*Journal of Sex Research*, May 1987, *23*:2).

Hyperactive Sexual Desire

The concept of hypoactive sexual desire implies that there is a standard for a normal level of sexual desire. In that case, there should also be a *hyperactive* counterpart in sexual desire disorders (Latin *hyper*, "beyond"). Though there is currently no such diagnosis, the Diagnostic Manual does refer to *nonparaphiliac sexual addiction*.

Although the term is new, the concept of sexual addiction goes back to the idea that "too much" sex is maladaptive if it distresses the individual or others. Whether such behavior is a problem or a form of paraphilia or a disorder of sexual desire is a matter of how you choose to define it.

SEXUAL AROUSAL DISORDERS

Desire normally leads to sexual arousal, and vice versa. Arousal is both psychological (feeling "turned on") and physiological (the excitement phase).

The primary physical sign of sexual arousal in the male is *erection* of the penis; in the female it is *vaginal lubrication* with accompanying changes (Chapter 3). In disorders of sexual arousal a man is unable to attain or maintain erection, and a woman, lubrication. The subjective sensations of sexual arousal usually go together with the physiological changes. But the reverse is not always true; a person may feel psychologically aroused without the physical signs of arousal. Because erection is under the control of spinal reflexes that are part of the autonomic, not the voluntary nervous system, a man cannot "will" an erection, but can only "let it happen"; the same is true for vaginal lubrication. This point is important for understanding arousal disorders.

Male Erectile Disorder

The inability to attain and maintain erection is the most incapacitating of all male coital dysfunctions. In its *primary form* a man has never been able to have coitus. In its *secondary form* a previously functional person develops the problem; this form is ten times more common.

Occasional inability to have an erection is exceedingly common, especially as a man gets older; it is not sexual dysfunction. Nor is there any absolute scale against which to evaluate the length or strength of an erection. Obviously, a man must keep his erection long enough to enter the vagina if he is to have coitus. Beyond that, judgments are relative. For clinical purposes a man is considered functionally impotent if his attempts at coitus fail in one out of four instances (Masters and Johnson, 1970).

Erectile dysfunction affected about one of every hundred males under 35 years of age in the Kinsey et al. (1948) sample, but it seriously incapacitated only some of them. At 70 years of age about one in four males was impotent. Most men progressively lose some erectile function with age, but some men retain their potency well into old age. Therefore getting older is not enough to explain erectile dysfunction. Among the subjects studied by Masters and Johnson, 31 percent had problems of potency, of which 13 percent were primary and 87 percent of the secondary type (Masters and Johnson, 1970).

For many men, it is difficult to imagine a more humiliating problem. "Impotence" means "powerlessness," and its consequences go far beyond the loss of sexual pleasure. Male notions of masculinity and personal worth are so closely linked to potency that serious dysfunction is damaging to a man's self-esteem. In most cultures impotent men have been objects of derision. The classical Arabian love manual, *The Perfumed Garden* has this to say about sexually inadequate men:

> When such a man has a bout with a woman, he does not do his business with vigor and in a manner to give her enjoyment. . . . He gets upon her before she has begun to long for pleasure, and then he introduces with infinite trouble a member soft and nerveless. Scarcely has he commenced when he is already done for; he makes one or two movements, and then sinks upon the woman's breast to spend his sperm; and that is the most he can do. This done he withdraws his affair, and makes all haste to get down again from her. . . . Qualities like these are no recommendation with women (Nefzawi, 1964 ed., p. 110).

More sensible and compassionate views put sexual potency in a less negative light.

There is more to sexual satisfaction than erection and ejaculation. A less genital focus fosters a broader perception of sexuality, with an emphasis on pleasure rather than performance (Zilbergild, 1978). There are handicapped men with spinal cord injuries, for instance, who are incapable of coitus yet able to engage in other pleasurable forms of sexual relations without thinking of themselves as being less of a man (Box 8.1).

Female Sexual Arousal Disorder

Disorders of sexual arousal in women have received little attention until recently (Musaph and Abraham, 1977; Kolodny et al., 1979), and still attract much less attention than those in men (Leiblum and Pervin, 1980). This fact in part is due to the general neglect of female sexuality. Also, problems with female arousal are physically harder to detect; and unlike male erection, they do not interfere with a woman's ability to engage in coitus, although it may be lacking in enjoyment.

For these and related reasons, problems of female sexual arousal have taken second place to orgasmic problems. The term "frigidity," which implies emotional coldness, has been indiscriminately applied to both sets of problems and does justice to neither. Whether women are affectionate, aloof, or "cold" may have no bearing on their sexual arousability or responsiveness.

At the psychological level, female and male consciousness of being aroused basically appears to be the same (although women may be less genitally focused than men). At the level of physiology, vaginal lubrication is less consistently linked to sexual arousal than erection among males, although the underlying vasocongestion is the same. For instance, although erection usually remains fairly steady, lubrication may decrease during the plateau phase (Chapter 3). The level of female arousal cannot always be gauged by the level of vaginal lubrication, especially after the menopause (Chapter 4). Some women become highly aroused and reach orgasm with no perceptible vaginal lubrication for reasons that are unclear.

Box 8.1

SEX AND THE HANDICAPPED

Many people think that the physically handicapped are neither capable of nor interested in sex, or that the lack of sex should be the least of their problems. That is not the case (Boyle, 1986). Much can be done to sexually rehabilitate the handicapped and to educate the public.

Disabilities take many forms, affecting physical, sensory, and mental functions. Those with visual and hearing impairments, for instance, must overcome special challenges to communication. The mentally handicapped may struggle to interact in ways that others take for granted. That does not mean that they do not have the same needs for affection and sex as everyone else.

Especially compelling is the plight of over 100,000 persons in the United States who have sustained spinal cord injuries leading to paralysis of the legs (*paraplegia*) or of the legs and arms (*quadriplegia*). They tend to be young, more often male (85 percent) than female (15 percent) (Higgins, 1979). The most common causes of such injuries are accidents and war injuries.

Paraplegics usually lack all sensation below the level of the injury and lose normal bladder and bowel control. Sexual functions are seriously disrupted but by no means always absent. Two-thirds of men with cord lesions retain some erectile response to physical stimuli (without being able to feel it). Ejaculatory disturbance is more severe; usually fewer than 5 percent can ejaculate, and they are usually infertile (Bors and Comarr, 1960). Paraplegic women likewise lack genital responsiveness and orgasm, but they are more likely to retain fertility. They may attain orgasm instead through stimulation of the nipples, lips, and other erogenous areas. Both males and females can experience erotic dreams with orgasm ("phantom orgasm").

Some seriously handicapped persons give up sex; yet others manage to maintain an active and joyful interest. They learn to use whatever parts of their bodies allow them to give and receive sexual pleasure. Sexuality for them takes on a broader meaning than the mere coupling of genitals. As one man put it, "I can't always be genital, but that gives me more permission to be gentle" (Cole, 1975).

The handicapped have the same need for sex and affection as the non-handicapped.

The following guidelines intended for the physically handicapped hold a larger lesson for all of us.

A stiff penis does not make a solid relationship, nor does a wet vagina.

Absence of sensation does not mean absence of feeling.

Inability to move does not mean inability to learn.

The presence of deformities does not mean the absence of desire.

Inability to perform does not mean inability to enjoy.

Loss of genitals does not mean loss of sexuality. (Anderson and Cole, 1975).[1]

[1]For a list of resources and bibliographic sources on sex and disability see *Siecus Report*, Vol. 14, No 4, March 1986. Various organizations provide information and assistance to the disabled and their families: *Clearinghouse on the Handicapped*, telephone (202) 472–3796; and *Parents Helping Parents*, telephone (408) 272–4774.

ORGASM DISORDERS

Despite the efforts of sex educators and therapists to steer people away from a genital-orgasmic focus to a wider body-pleasure orientation, orgasm appears to be the central sexual experience for many people, both physically and psychologically. Moreover, male intravaginal orgasm is the primary (although no longer the exclusive) means of impregnation. Disorders of orgasm are therefore a matter of much concern.

Inhibited Female Orgasm

Persistent or recurrent delay in, or absence of, orgasm in a female following a normal sexual excitement phase during sexual activity constitutes *inhibited female orgasm* (also called *anorgasmia*). Some women are able to experience orgasm during noncoital clitoral stimulation, but are unable to experience it during coitus in the absence of manual clitoral stimulation. Because over half of all women seem to require clitoral stimulation during coitus to reach orgasm (Kaplan, 1974; Hite, 1976), most sex therapists do not consider the need for clitoral stimulation an indication of orgasmic dysfunction. Furthermore, at least some women who are unable to reach orgasm do not suffer from a sexual dysfunction but a sexually inept partner. Other women feel that orgasm is not a necessary condition for sexual satisfaction every time they engage in coitus (Kaplan, 1974). Unlike men, not only are these women able to engage in sex and conceive without orgasm, but at least some say that they enjoy coitus even without it. Whether there are physiological and psychological explanations for this, or women are still simply willing to settle for less, is not clear.

All these qualifications aside, anorgasmia is reported to be by far the most common form of female sexual dysfunction. Most studies put the prevalence of primary coital anorgasmia (no orgasm ever) at 10 percent, with another 10 percent of women having coital orgasm sporadically (Kolodny et al., 1979). There are some indications, however, that the problem is

significantly receding. In the Hunt (1974) survey, 53 percent of women who had been married for 15 years reported that they always or nearly always reached orgasm in coitus; the corresponding figure in the Kinsey et al. (1953) survey two decades earlier was only 45 percent. Likewise, the proportion of wives who never reached orgasm had gone down from 28 percent in the Kinsey sample to 15 percent in the Hunt sample. A 1983 *Family Circle* poll shows some 85 percent of wives basically satisfied with their sex lives, but that does not mean that all of these women were orgasmic (*Time*, Jan. 31, 1983, p. 80).

Premature Ejaculation

Premature ejaculation is the most prevalent form of orgasmic disorder in the male. It consists of ejaculation which occurs with minimal sexual stimulation before, upon, or shortly after penetration and before the person wishes it. The condition must be recurrent and persistent to qualify as a dysfunction. It must take into account factors that affect duration of the excitement phase, such as age, novelty of the sexual partner or situation, and frequency of sexual activity (American Psychiatric Association, 1987).

Armed with the finding that three out of four men reach orgasm within two minutes of intromission, and that most male animals do so even sooner, Kinsey made light of premature ejaculation as a form of sexual dysfunction. However, a significant number of men (and their partners) complain of the inability to delay ejaculation until sufficient mutual enjoyment has been obtained; it is small comfort to them to learn that subhuman primates ejaculate even faster. Our evolutionary forbears were vulnerable to attack while engaged in copulation, so there was, maybe, something to say back then for the "survival of the fastest" (Hong, 1984), but not today.

Although premature ejaculation is a less frequent reason than impotence for seeking help, Masters and Johnson believe it to be the most common male sexual dysfunction in the general population: an estimated 15–20 percent of men have significant difficulty controlling ejaculation (although less than 20 percent of this group seek help) (Masters and Johnson, 1970; Kolodny et al., 1979).

Attempts at defining premature ejaculation have run into many problems. The official definition does not really set a time criteria. Early researchers attempted to specify a time (ranging from 30 seconds to 2 minutes) or a number of thrusts. Later, prematurity was redefined in terms of the partner's needs. For instance, Masters and Johnson (1970) have defined it as the inability to control ejaculation long enough to satisfy a normally functional female partner in at least 50 percent of coital encounters. However, just as a woman should not be declared anorgasmic when her partner does not provide enough stimulation, a man's capacity to delay organism should not be judged by the partner's responsiveness, which can vary widely. Some sex therapists do not consider this issue to be a problem if a couple agrees that the quality of their sexual relations is satisfactory, whenever the male ejaculates (LoPiccolo, 1977).

Inhibited Male Orgasm

Inhibited male orgasm (also called *retarded ejaculation*) is the opposite of premature ejaculation. It consists of persistent or recurrent delay in, or absence of, orgasm in a male during coitus. The failure to achieve orgasm is usually restricted to an inability to reach orgasm in the vagina, with orgasm possible with other types of stimulation, such as masturbation. It is the least frequent of all male sexual dysfunctions, accounting for 1–2 percent of most clinical samples (Apfelbaum, 1980). Many males occasionally experience a temporary inability to ejaculate during coitus that is overcome by more vigorous thrusting or some other form of heightening arousal. Those with retarded ejaculation overcome it with much more difficulty if at all. Those who are totally unable to reach orgasm in the vagina are said to suffer from *ejaculatory incompetence* (Masters and Johnson, 1970). The majority of these men can have orgasm through masturbation, but in

some 15 percent orgasm does not occur outside the vagina either. In these cases arousal subsides slowly without the climactic release of orgasm. Ejaculatory disorders are sometimes further delineated by timing, level of pleasure, and so on (Vandereycken, 1986).

If a couple wants to prolong coitus, this condition might seem a mutual blessing; yet beyond a certain point, coitus ceases to be enjoyable for either partner. The man experiences a sense of failure and frustration; the woman may feel responsible for not being sufficiently exciting, or irritated by having to go on beyond the point of enjoyment. Inhibited male orgasm must not be confused with *priapism*, which consists of prolonged erections without sexual arousal. This condition is due to a variety of physical ailments; it is not a sexual dysfunction, although it may result in the loss of erectile ability if not treated promptly (Kolodny et al., 1979).

SEXUAL PAIN DISORDERS

For most people, pain ruins sexual arousal and enjoyment. Discomfort during coitus is a far more frequent complaint among women, but it also occurs in men. Painful intercourse is frequently cited as a female sexual dysfunction (Sandberg and Quevillon, 1987). It is estimated that 15 percent of women experience pain during sex at various times, and 1–2 percent have it as a chronic problem (Brody, 1988).

Dyspareunia

Dyspareunia is recurrent or persistent genital pain in either a male or a female before, during, or after sexual intercourse (American Psychiatric Association, 1987).

Female Dyspareunia   Women feel coital pain more often than men for many reasons. Any ailment in the genital organs and the pelvic region can cause pain during coitus. Women are more vulnerable to these conditions because their reproductive system is more exposed to infection (Chapter 5), traumatized during childbirth (Chapter 6), and subject to hormonal influences (Chapter 4).

The most common (and most easily treated) cause of coital pain among women is vaginal dryness, which causes irritation of the vaginal wall during coitus. Though mainly encountered among postmenopausal women (Chapter 4), vaginal dryness may also be present in women who do not naturally lubricate sufficiently during excitement; nursing women; women making love right after a menstrual period, during radiation therapy, or when using antihistamines (such as decongestants); women under undue stress; and sometimes women following a strenuous exercise program.

Infection is another common source of pain. Those with vaginitis or PID may experience a burning sensation or even bleeding during coitus. Douches, deodorant sprays, tampons, and vaginal contraceptives may also cause irritation. Injuries during childbirth that have healed poorly, tender episiotomy scars, tears in the anal region, and similar conditions are other potential sources of pain.

Male Dyspareunia   Men experience pain during coitus usually because of genital or urinary infections, especially prostatitis. Arthritis of the hip and lower back problems may also cause pain during pelvic thrusts. There is also a rare condition called *Peyronie's disease*, in which deposits of fibrous tissue in the penis lead to the curvature of the penis, painful erections, and erectile dysfunction.

Vaginismus

When a woman feels pain or expects it, she tenses up. As the muscles surrounding the vaginal introitus contract, they cause *vaginismus*, which makes penetration difficult and painful, if not impossible. Vaginismus occurs more often in response to psychological than physical causes. It affects 2–3 percent of adult women (Kolodny et al., 1979). Generally these women have no difficulty with sexual arousal and can attain orgasm through noncoital means.

Men too may experience genital muscle spasms, which cause pain during ejaculation. It may be intense enough to be disabling and may last for minutes or hours. As with vaginismus, the cause is usually psychogenic.

CAUSES OF SEXUAL DYSFUNCTION

Sexual function can be disrupted by physical, psychological, or cultural influences. In principle, all three variables interact in every case; in practice, we choose one as the cause. In most cases there is no specific link between cause and effect, especially where causes are psychological. In other words, the same psychological problem may cause sexual apathy in one woman and anorgasmia in another, erectile dysfunction in one man and premature ejaculation in another. For organic causes the connection is more often evident. For example, a drug that disrupts parasympathetic function will cause impotence; one that interferes with sympathetic function will affect orgasm.

Organic Causes

Most cases of sexual dysfunction have psychogenic causes. However, in a significant number of cases (perhaps as high as one-third), the sexual disorder has a physical basis (Munjack and Oziel, 1980). Some conditions are more apt to have physical causes. For example, erectile dysfunction is more often due to organic causes (perhaps in as high as 40 percent of cases) than premature ejaculation; deep pelvic pain during coitus is much more likely to have a physical basis than sexual apathy (Kolodny et al., 1979). Even when there are physical causes, psychological factors may still be at work. Unless the physical ailment is incapacitating, whether or not it will lead to dysfunction, or how severe the dysfunction will be, may largely depend on psychological factors. Aging is not in itself a sufficient biological reason for dysfunction.

The sexual organs are well designed to fit (Figure 8.1). Disparities in size and shape of the penis and vagina are almost never a cause

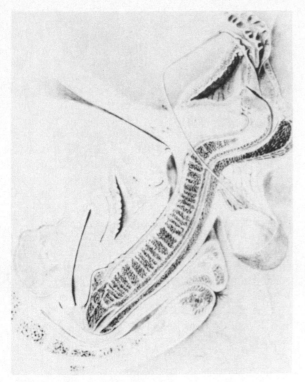

Figure 8.1   Male and female sex organs during coitus.

of coital difficulty. Like an elastic structure, the relaxed and lubricated vagina expands to accommodate the erect penis, whatever its size.

Sexual functions can suffer as a result of *acute* (short-term) or *chronic* (long-term) ailments, which may be *local* (affecting only the genital region) or *systemic* (affecting a whole system, especially the circulatory, nervous, or endocrine system). Debilitating illness such as cancer, degenerative diseases, severe infections, or systemic disorders affect sexual function indirectly. Local disturbances, such as pelvic infections, interfere more directly. The common element in a lot of these conditions is that they cause *pain*, which has a dampening effect on sexual interest, arousal, and enjoyment. Two-thirds of patients with chronic pain report a deterioration of their sexual relations (Maruta and McHardy, 1983). The most common causes of pain were discussed earlier.

Trauma   Injuries from accidents or surgery may interfere with the blood and nerve supply to the pelvis. Among males, certain forms of prostate surgery can damage the nerves in the region. A patient thinking of surgery for an enlarged prostate should ask about risk to sexual function. Newer surgical techniques reduce this risk, especially among men who are younger than 70, even in drastic surgeries for prostate cancer (Sullivan, 1987).

Female surgical procedures that damage the sexual organs (such as poorly repaired episiotomies) often cause pain during coitus. Particularly important are radical hysterectomies (where the ovaries are removed in addition to the uterus) in premenopausal women. The loss of sexual desire that affects some of these women can have a serious effect, especially if they have not been forewarned (Wigfall-Williams, 1987).

A possible consequence of giving birth is the stretching and weakening of the muscles around the vagina. This change interferes with the myotonic response and may inhibit orgasm.

Endocrine Disorders   Radical hysterectomies remove the main sources of female hormones, the ovaries. Other endocrine disorders of significance include low output from the pituitary, the gonads, or the thyroid glands (hypopituitarism, hypogonadism, and hypothyroidism); diabetes; and liver diseases like hepatitis and cirrhosis. Testosterone deficiency is now reported to be a greater cause of sexual dysfunction than previously suspected. In one group of 105 consecutive cases of impotence, 35 percent had abnormalities of the hypothalamic-pituitary-gonadal axis (Sullivan, 1987).

Neurological Disorders   Neurological disorders that may seriously influence sexual functions in both sexes include diseases of the frontal and temporal lobes of the brain, such as tumors or blood clots; nerve injuries; and disturbances in the spinal cord such as birth defects, degenerative conditions, and injury (Box 8.1). Epilepsy usually does not cause sexual problems (Kolodny et al., 1979).

The spinal cord and pelvic nerves may be injured in car accidents or war (Comfort, 1978). They can be injured also by blows to the groin, athletic mishaps (such as landing on another player's knee), or even riding on a stationary bicycle with a seat that is too high (Solomon and Cappa, 1987). In the last case, function returns promptly when the cause is eliminated; but in cases where the spinal cord is damaged, its nerves do not regenerate. Chronic lower back pain, from which many people suffer, usually does not interfere with sexual function except when there is acute pain.

Circulatory Disorders   Any interference with the blood supply to the pelvic region may result in sexual dysfunction. The buildup of cholesterol or high blood pressure will damage the arteries supplying the genital organs. The major causes of cardiovascular disease are also turning out to be associated with impotence, including smoking: 80 percent of the impotent men in one clinic are reported to be smokers (Goldstein, 1987).

The problem may also be in the veins. As we saw in Chapter 3, for the penis to be engorged with blood, the inflow must increase and the outflow must decrease. "Leaky" veins may not stop the outflow enough (Bookstein et al., 1987). (Think of trying to fill a bathtub when the drain is open.) This problem may occur more often in older men, accounting for their softer and briefer erections.

Sexual dysfunction may be the indirect result of some other circulatory ailment. For instance, people who have strokes may suffer a significant decline in sexual desire. The ratio of those who enjoyed sex in an older sample of stroke patients (average age 68 years) went down from 84 percent to 30 percent of the men and from 60 percent to 31 percent of the women. Even so, psychological factors remained important; the most common factor identified as causing decline in sexual activity was the fear that having sex might cause another stroke (Monga et al., 1986).

Similar concerns tend to disrupt the sexual lives of persons who have had heart attacks.

Men are more vulnerable in this respect than women. In one study, 76 percent of the men who had had a heart attack reported sexual dysfunction (42 percent was erectile dysfunction) as compared with a control group with similar health status but no heart attack (68 percent). Counseling can reduce anxiety about "coital death" and related fears (Dhabuwala et al., 1986). Under medical guidance many who have had heart attacks can live healthy and active sex lives, just as they are able to resume physical activity and exercise.

Diabetes is another case in point. Though it is an endocrine disorder (having to do with carbohydrate metabolism) it is through the damage caused to arteries and nerves that it causes erectile impairment. It is associated with a high prevalence of sexual dysfunction in men. Some forms of diabetes have little or no effect on women; other types do lessen female sexual desire and orgasmic capacity (Schreiner-Engel et al., 1987).

Alcohol   Alcohol is a common cause of sexual dysfunction: 40 percent of chronic alcoholics have problems with potency and 15 percent with orgasm (Kolodny et al., 1979). Chronic alcohol usage interferes with hormone production, liver function, and nutrition, and causes nerve damage, all of which are detrimental to sexual function. These organic effects are compounded by the psychological and social problems caused by alcoholism.

Alcohol is widely believed to enhance sexual activity, but in fact it puts a physiological damper on sexual arousal and performance. Alcohol may make this person less inhibited yet even well below levels of intoxication, alcohol has been shown to inhibit erections (Wilson and Lawson, 1978b), female arousal (Wilson and Lawon, 1978a), and orgasm (Malatesta, 1979; Klassen and Wilsnack, 1986).

Although alcohol intoxication retards orgasm, women who do reach orgasm under its influence report a heightened sense of pleasure; men report decreased arousal and less pleasurable orgasm. Alcohol can also be the trigger for psychogenic impotence; a man will fail to have an erection after having too much to drink; then he will feel anxiety and experience failure even when sober (O'Farrell et al., 1983).

Drugs   Drugs are another important source of sexual difficulty—especially *sedatives* (such as barbiturates) and *narcotics* (such as heroin) because of their effects on the central nervous system.

Marijuana, like alcohol, has a widespread reputation as an enhancer of sexual experience, but marijuana use has been linked with erectile problems, lowered testosterone production, and disturbance in sperm production (Kolodny et al., 1979). Men who smoke four or more marijuana cigarettes per week have significant decreases in testosterone production. The decrease in testosterone is related to the amount smoked; the heavier the usage the lower the levels of hormone production. *Nicotine* in cigarettes causes constriction of blood vessels and may thus interfere with genital blood supply.

Another class of drugs that may impair libido and sexual response is the *antiandrogens*, which include estrogen, adrenal steroids such as cortisone, and ACTH (often used to treat allergies and inflammatory reactions). Other antiandrogens like Depo-Provera and cyproterone acetate are used in connection with the suppression of certain compulsive sexual behaviors.

Other drugs block nerve impulses to the genitals. Drugs that block the effects of the parasympathetic system interfere with arousal and erection; those which block the effect of the sympathetic system interfere with orgasm (Chapter 3). It is on this basis that drugs for high blood pressure (*antihypertensives*) often impair sexual function in both sexes, but more often cause erectile dysfunction (Smith and Talbert, 1986). This effect may be avoided by shifting from one type of drug to another. Patient and doctor together should weigh the benefits of such drugs against the loss of potency.

Drugs used in the treatment of psychosis (*antipsychotic agents*) are more likely to cause ejaculatory problems, including retrograde ejaculation. Such patients have less choice in using the drug. Antidepressant drugs may also cause impairments of orgasm, which compounds the problem often caused by the depression itself (Harrison et al., 1986). Commonly used *tranquilizers* like Valium and Librium are less likely to cause problems; they may actually help sexual function by allaying anxiety.

Once the use of drugs is discontinued, their effects subside. The main thing to remember is that whenever a person is using a drug—any drug—and develops sexual problems, the drug is the first possible cause of the difficulty.

Organic or Psychogenic?

Though organic and psychogenic factors usually interact in causing sexual dysfunction, we need to determine the role of each in a given case for effective treatment. It is therefore essential that any significant organic cause be identified or ruled out first in all cases of sexual dysfunction. This procedure speeds treatment of physical conditions that have a remedy; and when that is not possible, it lets the person and the clinician know the limits to which psychological treatment is likely to help and set their therapeutic expectations accordingly.

The attempt to tell organic from psychogenic dysfunction begins with taking the history of the symptoms (Segraves et al., 1987). For example, psychogenic impotence comes on gradually over months or years. The pattern of dysfunction also tends to be different. Organic impotence tends to be more consistent and global, affecting all forms of sexual activity. Psychogenic impotence is more inconsistent; it may occur at one time but not another, during coitus but not masturbation, with one partner but not another. Even the timing of the symptom may be different. A woman who experiences pelvic pain following orgasm is more likely to have a physical ailment than another who complains of coital pain when penetration has barely occurred (unless she has an infection right at the vaginal opening).

The circumstances under which the problem starts may also help to decide the issue. If sexual dysfunction follows a significant event like divorce or death of the spouse, it is more likely to be psychogenic. Evidence of conflict between a couple or signs of emotional distress point in the same direction. However, such distress may be the result and not the cause of dysfunction. The partner of a person with an organic problem is apt to feel inadequate, guilty, or angry if the physical basis of the disorder is not suspected.

Ultimately, the decision has to be based on concrete physical evidence. Consider for instance the problem of impotence. Virtually all healthy males have erections during REM sleep. Absent or seriously deficient nocturnal erections suggest an organic cause. If *nocturnal penile tumescence* (NPT) is normal, psychogenic factors are much more likely to be the underlying problem (Marshall et al., 1987; Bohlen, 1981).

This process can be tested in some simple ways. A ring of stamps is attached at the base of the penis before going to sleep. If the ring is found broken in the morning, the man has had an erection. A more reliable device is the Snap-Gauge, which consists of Velcro straps with three connectors, each of which breaks at a defined penile pressure (Bradley, 1987; Condra et al., 1987). Most reliable is a complete penile tumescence recording carried out in a sleep laboratory, where the pattern of sleep can be followed with an electroencephalographic recording and monitoring of eye movements. Penile erections occurring during sleep are detected by special instruments such as strain gauges at the base of the penis that accurately record the magnitude, duration, and pattern of erection. Portable units now allow such monitoring while the person sleeps at home (see Box 3.2).

More specialized studies investigate the blood flow patterns in the penis. For instance, radio-opaque dyes (which show up on X-rays) are injected. Their flow provides specific data on how well the arteries and veins within the corpora of the penis are working (Bookstein et al., 1987). This procedure is called *cavernoso-graphy*. If papaverine (which dilates the arteries) is injected as well, a sonogram can show further problems in blood flow (Trapp, 1987).

Psychogenic Causes

Psychogenic factors are more difficult to identify and classify than organic causes, especially if they are deep-rooted intrapsychic conflicts. In the past, treatment of sexual dysfunction mainly focused on fundamental personality problems. In current forms of the "new sex therapy" the emphasis is on more immediate psychological factors.

Immediate Causes   The fear of failure is possibly the most common immediate cause of psychogenic impotence. Other sources of sexual problems are the demand for performance and the excessive need to please the partner. Such attitudes elicit resentment and anger, which interfere with sexual enjoyment and function. Another important cause of difficulty is "spectatoring"—anxiously and obsessively watching your own reactions during sex (Masters and Johnson, 1970). For satisfactory sex people must lose themselves in the interaction. Spectatoring distracts the person and prevents sexual responses from building up to orgasm (Kaplan, 1974).

The failure to communicate forces a couple to guess what is desired and what is ineffective or objectionable in the sexual interaction. Communication that is clear and appropriate to the occasion is necessary to provide both information and reassurance. Even when failure to communicate is not the primary cause of the dysfunction, it helps to perpetuate other problems (Fay, 1977).

The deeper causes of sexual malfunction in both sexes are internal conflicts related to past experiences. When these conflicts dominate a person's sex life to the extent that inadequate performance is the rule, then the causes can be considered to be primarily *intrapsychic*. On the other hand, when the sexual problem seems to be part of a conflict between two particular people, it is more convenient to label it as *interpersonal*. This distinction, though arbitrary, has practical merit in treatment. Intrapsychic causes must be dealt with. As veterans of successive divorces discover, even though marital partners change, the conflicts remain the same. Interpersonal conflicts often may be worked out without deeper therapy.

Psychological conflicts occur unpredictably at any time in adult life; certain developmental stages have more predictable stresses. These stresses may predispose the person to sexual dysfunction. The "midlife crisis" is a good example of such a stressful period. There is some value, therefore, in evaluating sexual problems in a broader life cycle perspective (Fagan et al., 1986).

Deeper Causes   Learning theorists have proposed a variety of models to explain the genesis of sexual malfunction which are extensions of the more basic theories of sexual learning. Central to many models is conditioning, in which the unpleasant feelings associated with an experience determine future reactions to a similar situation. This is the same basic process through which children are socialized sexually. Sometimes the antecedents of the experience are easy to trace. For instance, if a man suffers a heart attack during coitus, thereafter the very thought of sex may make him anxious and unable to perform. Similarly, after a rape, a woman may find coitus difficult even with a loving partner (Wolpe and Lazarus, 1966).

More often the causes are a complex series of long-forgotten learning experiences. The transmission of certain sexual attitudes and values to children—like sex being dirty or dangerous—is one example. A person may not remember specific or implied parental warnings and punishments, but their damaging effects persist.

Psychoanalysts explain sexual malfunctioning by infantile conflicts that remain unconscious. For instance, conflicts arising from unresolved oedipal wishes may be major causes of difficulties in both sexes. Castration anxiety is another common explanation for failures of potency, just as it is for paraphilias. When a man is impotent with his wife but not with a prostitute, he may be unconsciously equating his wife with his mother. Men who distinguish between "respectable" women (to be loved and respected) and "degraded" women (to be enjoyed sexually) are said to have a "madonna-prostitute complex."

The female counterparts to these conflicts involve the father. As certain types of men may be unconsciously identified with the father, coitus with that type of man, or any man, elicits guilt and dysfunction. By failing to enjoy the experience, the woman feels less guilty about her unconscious incestuous wishes.

Another deep psychological factor is the threat of loss of control. As orgasm implies a certain self-abandonment, some men and women are afraid that aggressive impulses will also be triggered. Still another fear is that the erect penis will tear apart the vagina or that the penis will be trapped and choked by the vagina. Such concerns may be experienced consciously but more often are unconscious: the man simply fails to have an erection, or the woman fails to reach orgasm, neither quite realizing why.

Relational Problems   Intrapsychic problems usually spill over into interpersonal conflicts, but sometimes problems arise only in a particular type of relationship. Intense disappointment, muted hostility, or open anger obviously poison sexual interactions. Subtle insults are just as detrimental. Women, for instance, are sensitive about being "used." If a man seems to be interested predominantly in a woman's body and neglects her person, she will feel that she is being reduced to the level of an inanimate object. Some women associate coitus with being exploited, subjugated, and degraded and rebel against it by failing to respond.

The attitudes most detrimental to the male's enjoyment are those that threaten his masculinity. Lack of response on the part of the woman, nagging criticism, and open or covert derision lessen male enjoyment.

Other relational causes include contractual disappointments. When people establish sexual partnerships or get married, their sexual expectations are seldom clearly communicated or negotiated. There is much room for misunderstanding and anger when the partner does not live up to what was expected. Also, as people change, new needs and new preferences alter the original relationship, and the couple may have trouble adjusting.

In power struggles, many forms of sexual sabotage can operate: one partner will pressure the other into sex at inopportune moments; sex is withheld as punishment; people make themselves undesirable or even repulsive by neglecting their bodies, cultivating the wrong physical image, or behaving in ways that are irritating and embarrassing to the other. Sabotage may even interfere with the progress of couples in sex therapy. As one partner begins to improve in sexual performance, the other (often without being aware of it) may attempt to stall or make things difficult; a significant change in the partner would alter the old pattern of the relationship, in which he or she may have a vested interest.

Cultural Causes   Societies influence sexual function by the way children are raised and by the contexts in which they let sexual relations occur. Feelings of sexual inadequacy commonly stem from ignorance, misconceptions, distorted views of the opposite gender, unwarranted fears, unresolved guilt, and unrealistic expectations of what sex has to offer (Jacobs, 1986). These attitudes all have historical and cultural roots which we cannot delve into here.

Cultural patterns also influence intimate relationships. One of the major sources of problems in heterosexual relationships has been the *double standard of morality*—different, often contradictory expectations of how men and women should behave sexually. The general effect of the double standard has been to inhibit female sexual responsiveness and relegate women to passive and resistive roles in sexual interactions. These attitudes have now been overcome by many men and women. As a result, women are more likely to have more positive attitudes toward sex and be more active and assertive sexual partners (Koblinsky and Palmeter, 1984).

Some people fear that these more liberal and egalitarian attitudes have generated new problems. Disorders of sexual desire have emerged prominently at a time when our society has become much more open about sexual topics, and some of those complaining of sexual apathy may be among the most sexually liberated. The problem may be related to the stress faced by women trying to combine career and family goals, which continues to be much less of an issue for men. In a study of 218 couples who had sought help at the Masters and Johnson Institute, married women who were pursuing careers outside the home were twice as likely to present a primary complaint of inhibited sexual desire (and had a higher ratio of vaginismus) than married women who were working at a job with no particular career prospects, or those who were not employed outside the home (Avery-Clark, 1986b). The husbands of the women who worked outside the home—either at a "career" or at a "job"—presented almost half as often primary complaints of inhibited sexual excitement and desire as the husbands of unemployed women (Avery-Clark, 1986a).

Another concern in the popular media is that the greater assertiveness of women is having a negative effect on male sexual functioning (the "new impotence"); however, there is no proof of such claims (Gilder, 1973).

Since the sexual dark ages, women have pretended (or "faked") orgasm to please their partners. Surely, such deception should no longer be needed or called for; yet the practice of pretending orgasm is reportedly increasing. In a survey of 805 professional nurses nearly two out of three said that they had pretended orgasm at one time or another (Darling and Davidson, 1986). Those who had become sexually active at a younger age and had been sexually more explorative were more likely to have pretended orgasm than others. The explanation given is that because female orgasm is now expected, women feel a greater need to live up to the standard, even if they have to fake it.

The premium we put on competence and success, combined with an overemphasis on sex, can create a formidable hurdle to its enjoyment. Orgasm becomes a challenge, rather than the natural climax of coitus. Inability to achieve it or a certain form of it (multiple, mutual, and so on) becomes not only a signal of sexual incompetence but also a reflection of personal inadequacy. As we become freer about sex, we may be thus generating new problems—excessive demands for performance and wholly unrealistic expectations of what sex can be and do.

TREATMENTS

Sexual dysfunction may be mild and transient, requiring no therapy, or it may present formidable challenges to treatment. Its remedies range from fairly simple short-term programs to highly specialized, intensive treatments, involving psychological methods, drugs, and surgery.

Until the late 1960s, sexual dysfunctions were treated by physicians, psychologists, and marriage counselors. Following the work of Masters and Johnson (1970) new methods of treating sexual dysfunctions were developed that are more effective and less time-consuming. The applications of these newer methods constitute the *new sex therapy*. Its focus is on the elimination of sexual symptoms, rather than on personality change or on restructuring relationships outside the sexual realm. Although all forms of treating sexual dysfunction could be thought of as "sex therapy," the use of the term is usually restricted to specific approaches focused on the elimination of sexual symptoms.

Sex Therapy

At present, sex therapy has no distinctive theoretical base of its own. Its practitioners may be physicians or clinical psychologists with many perspectives. The basic methods of sex therapy are derived from behavior modification techniques developed in psychology (Caird and Wincze, 1977; Jehu, 1979). These techniques have been refined and adapted for dealing with sexual dysfunction (LoPiccolo and LoPiccolo, 1978; Leiblum and Pervin, 1980).

Physicians combine medical approaches with various aspects of sex therapy. For instance, Helen Kaplan (who is also a psychoanalyst) integrates psychodynamic principles with sex therapy approaches in her work (Kaplan, 1974). Urologists who implant penile prostheses primarily approach these problems from a surgical angle.

Sex therapy has clear merits, but definite shortcomings. It has as yet no solid conceptual base and a young clinical tradition. It is likely to be most effective within a broader program treating psychological problems and interpersonal conflicts (Arentewicz and Schmidt, 1983).

The PLISSIT Principle   Common sense dictates that simpler and briefer methods of therapy be tried first. This principle is embodied in the PLISSIT model for treating sexual dysfunction. The acronym stands for Permission, Limited Information, Specific Suggestions, and Intensive Therapy (Annon, 1976). Therapists take these approaches either sequentially or more or less concurrently. *Permission* takes the form of reassurance, reinforcement of positive values, and encouragement to behave sexually according to the needs and values of the individual. *Limited information* provides facts and self-knowledge about the issues that are specific to a person's sexual problem. *Specific suggestions* are the sort of techniques originally developed by Masters and Johnson (1970) and modified by others. *Intensive therapy* goes beyond the techniques of behavior modification and delves into the person's intrapsychic conflicts with some form of intensive therapy. Kaplan (1974) deals with these issues through *psychosexual therapy*. In this approach, explorations of psychological background as individuals is combined with a focus on the sexual problem to gain insights into the unconscious conflicts underlying the sexual problem.

The Masters and Johnson Model   The Masters and Johnson (1970) sex therapy format, though modified by others, still defines the basic approach in sex therapy. It has a number of general procedures for the treatment of all

Box 8.2

PSYCHOSEXUAL TREATMENT OF A SEXUALLY DYSFUNCTIONAL COUPLE*

The couple consisted of a handsome and successful 42-year-old real estate broker and his 40-year-old wife, who was a teacher. When they applied for treatment, they had been married for 18 years and had three daughters.

Their chief complaints were the husband's impotence and the wife's inability to attain coital orgasm. Neither spouse had had psychiatric treatment previously. On the surface, both functioned well and neither had ever experienced significant psychiatric symptoms. It became apparent early on, however, that both husband and wife had personality problems. . . .

Basically, their marital relationship was good. Despite their difficulties, they respected, loved, and were committed to one another. At the time of the initial consultation, however, there was a great deal of anger and hostility between them. She would often explode in violent tantrums and he would respond to these outbursts by withdrawing. . . .

The patient's history revealed that he had always had some sexual difficulty. He had been shy with girls in his teens, so that he did not have his first coital experience until he was in his early twenties. In college he had several dates with a girl after she had indicated interest in him, and had become extremely aroused when they petted. However, he experienced erectile difficulties when he attempted coitus and was unable to consummate the act.

Shortly afterward he met his wife, who also took the initiative in their relationship. And again he was very excited by kissing and petting, and had frequent and urgent erections in her presence, although they did not take off their clothes. They had known each other for a year when they decided to get married. No attempt was made to have premarital sexual relations, because both felt that this would not be "proper."

On their honeymoon, when faced with the inevitability of intercourse, the patient was unable to achieve an erection in his wife's presence. However, he stimulated her clitorally and she was very responsive. . . .

Occasionally, with a great deal of stimulation,

he was able to have intercourse. She never reached orgasm on these occasions, although she continued to be responsive to clitoral stimulation. Frequency of sexual contact was limited to approximately once every two or three months. On such occasions the husband always took the initiative. . . . Since he was "in charge," he never allowed himself a passive role in sex, and felt guilty about not pleasing his wife. . . .

Treatment began with the prohibition of coitus. As usual the patient and his wife were instructed to take turns gently caressing each other and to tell one another what each found especially pleasing. This experience produced an erection in him, while he reveled in his role as the passive recipient of pleasure, a role he had previously denied himself because it wasn't "manly." He especially enjoyed it when his wife gently played with his penis.

She, on the other hand, was furious and weepy. She enjoyed the experience when it was her turn to receive his caresses, but felt fatigue first, and then rage, when it was her turn to pleasure him and she saw his erectile response. . . . She perceived her "obligation" to him, particularly the fact that she was required to "service" his penis, as a humiliation which evoked feelings of persecution. On the other hand, because she was an extremely intelligent and basically stable person, she was struck by the intensity and irrational quality of her reactions. . . .

The next obstacle to treatment was created by the patient in the form of obsessive self-observation and doubt regarding his sexual competence. He was able to respond to and enjoy their mutual caresses, but when coitus seemed imminent he would begin to "turn himself off" with fears, e.g., "It won't work," "She'll be mad at me. . . ," etc. This difficulty was surmounted by the usual instructions that he consciously avoid such thoughts and focus his attention on sexual sensations. In addition, he was advised to immerse himself in fantasy if he could not control the tendency by conscious effort. . . .

*From Helen Singer Kaplan (1974), *The new sex therapy.*

This couple's sexual relationship improved considerably as a result of treatment, but the patient did not achieve complete erectile security. When treatment was terminated, the couple was having intercourse approximately twice a week, and occasionally the husband still lost his erection. However, the couple regarded these experiences as minor "setbacks," and accepted them with equanimity; consequently, they did not have a detrimental effect on their subsequent lovemaking.

---

dysfunctions, supplemented by specific strategies for particular conditions. After organic factors are ruled out, the key task is to have the dysfunctional couple accept sex as a natural function that requires no heroic effort, but only a relaxed, accepting attitude. The focus is on the couple, not the individual. Neither partner is at fault or sick; there are merely inhibitions to overcome. Each must learn to give, as well as to receive, sexual pleasure, to get actively involved rather than be a spectator or a passive participant.

Treatment progresses along two complementary tracks. The first involves discussions between the couple and two cotherapists (one male, one female). Detailed sexual histories are taken from each patient and roundtable sessions explore past experiences, conflicts, feelings, and attitudes on both sexual and nonsexual matters that have a bearing on the dysfunction, successes and failures in the ongoing treatment are analyzed; and so on. Concurrently, the couple goes through a sequence of sexual assignments in private that eventually culminate in mutually satisfactory sexual intercourse, if the treatment is successful. Because of the focus on the couple, Masters and Johnson provided female *partner surrogates* to men without partners early in their work. Legal and ethical considerations have made this practice a problem although some sex therapists continue to rely on surrogate partners in treating select cases. The practice of using a pair of therapists (which is expensive) is not considered essential by others (Kaplan, 1974). The effectiveness of treatment does not seem to depend on the number of therapists (LoPiccolo et al., 1985).

Of importance in the early phase of therapy are *sensate focus* exercises, which prepare the ground for tackling specific problems. These exercises overcome the common immediate causes of sexual dysfunction: anxiety, spectatoring, demand for performance, noncommunication. They start with activities focused on touch awareness and pleasuring rather than sexual arousal. The couple is comfortably positioned so that the helping partner has easy access to the body of the person with the primary problem: when the symptom is female anorgasmia, the man is seated behind the woman (Figure 8.2); in impotence, the position is reversed. What follows next depends on the nature of the problem.

Treatment of Erectile Dysfunction In cases of erectile dysfunction, the woman takes the initiative to touch, caress, or gently fondle the man's body, initially keeping away from the genitals. The man guides his partner by his verbal or nonverbal responses as well as by leading her hand (*handriding*). In subsequent stages, the couple alternates in pleasuring one another, with more direct communication to guide each other's actions.

After the couple becomes adept at general physical enjoyment, they move on to more explicitly erotic techniques, with the woman now stimulating the man's genitals, along with the rest of his body (Figure 8.3). The interaction continues in a relaxed, nondemanding manner; the man is not expected to have an erection at any particular moment, and neither person is allowed to rush to coitus. When erections do occur in the natural course of events (which is often quite soon in treatment), they are allowed to come and go to instill in the man the conviction that he does not "will" an erection but lets it happen; and should it subside, there need be no sense of failure. Intercourse

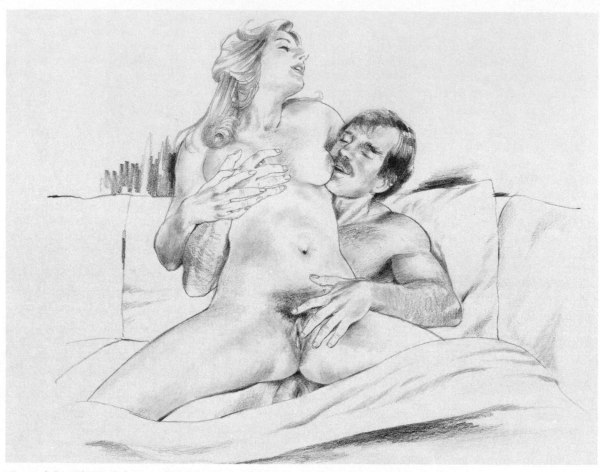

Figure 8.2   The training position for the treatment of female sexual dysfunction.

is forbidden at this stage; it is to be attempted only when erections become frequent and stable during sensate focus exercises.

The transition from sensate focus exercise to coitus is gradual. When the man has gained enough confidence, the woman positions herself astride him and inserts his penis into her vagina, relieving him of taking responsibility and risking fumbling and failure. Intercourse then proceeds with the primary aim of providing the man with adequate satisfaction: the man is encouraged to be "selfish" during these periods; the sharing of pleasure will follow later. In sum, the basic aim in the treatment of erectile dysfunction is to reduce anxiety so that physiological reflexes can take their natural course.

Treatment of Anorgasmia   The procedure for dealing with female anorgasmia is basically the same. This time the man takes the initiative with early sensate focus exercises; but when it comes to coitus, again the woman assumes the woman-above position. This position lets her be in charge, counteracting her fears of being hurt, used, or put down. It also enables her to gauge the depth of intromission and the force and frequency of coital thrusts, all to suit her needs and help achieve orgasm.

Other strategies focus on more particular

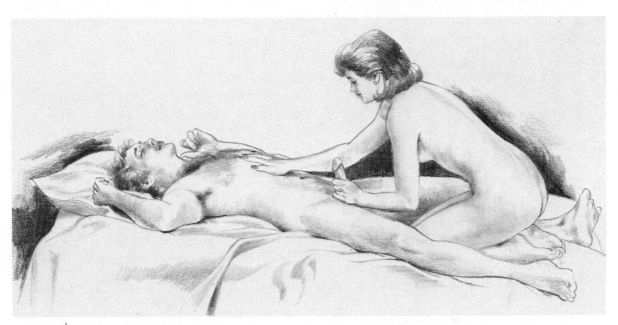

Figure 8.3   The training position for the treatment of male sexual dysfunction.

issues. For instance, women are more likely than men to have concerns over their physical attractiveness, to be inhibited in allowing themselves sexual pleasure, and to be reticent about exposing their erotic needs. Such issues are dealt with both in discussion and practice. A woman who has never experienced orgasm must be made to have that experience first, however it is attained. She may be encouraged to stimulate her own body and genitals or be brought to orgasm by the partner manually or orally (some also advocate the use of vibrators). Once the inhibiting barrier is broken, the woman will know what it feels like to have an orgasm and can be gradually helped to transfer the experience to coitus. Other "bridging" techniques consist of clitoral stimulation by the man or woman during coitus. The basic aim in treating failures of female sexual arousal and orgasm is to help the woman let go of the inhibitions that block orgasm.

Treatment of Male Inhibited Orgasm   The first goal in the treatment of male ejaculatory incompetence is similarly to cross the orgasmic

barrier. The man is helped to reach orgasm first through solitary masturbation; then he does it in the partner's presence, or she masturbates him (having him ejaculate at her genitals). Coitus is attempted last, with her bringing him to a high pitch of excitement and inserting the penis as he is about to ejaculate. This is the only instance of the man-above position. It lets him enter while she continues to stimulate the base of his penis manually.

Treatment of Premature Ejaculation   The basic aim in treating premature ejaculation is to train the man to anticipate orgasm and modulate his level of arousal accordingly. The woman assumes the stimulating position (Figure 8.3) and gently brings him to erection. When the man says that ejaculation is imminent, she uses the *squeeze technique* to avert it—squeezing the glans between thumb and forefinger or putting pressure at the base of the penis (Figure 8.4). She must be gentle yet firm in applying pressure at the front and back of the glans for three to four seconds, being careful not to hurt. Why this maneuver works is

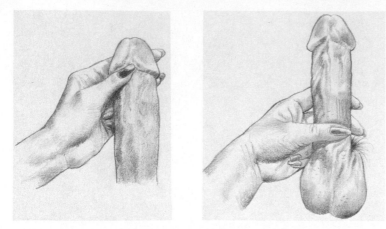

Figure 8.4   Two versions of the "squeeze technique."

not known. It results in partial loss of erection, which is then brought back with further stimulation, and the cycle is repeated. The basic procedure in this approach was reported originally by Semans (1956).

Coitus takes place in the woman-above position. During it, the squeeze cycle is repeated a number of times; as he senses the approach of ejaculation she takes out the penis, squeezes the glans, and reinserts it. This maneuver is followed by slow thrusting, and the cycle is repeated until the man wishes to ejaculate.

To avoid coital disengagement, the woman either puts the pressure on the base of the penis or tries a *stop-and-go* technique: as excitement mounts, thrusting stops; after a few moments of rest, when ejaculatory urgency is past, thrusting resumes.

Treatment of Vaginismus   The basic aim in treating vaginismus is the elimination of the vaginal spasm. The treatment utilizes *desensitization*. The patient is taught the nature of the problem and is made aware of the muscular tension at the vaginal opening. Then in a relaxed and safe setting (with the male partner present), a small plastic dilator is gently inserted in her vagina, as she watches in a mirror. The woman then uses dilators, moving from the smallest to the largest, which is about the size of an erect penis, leaving each in place

for short periods of time. When the couple moves on to coitus, she take charge, inserting the penis herself, controlling the thrusts, and so on.

These treatments strategies sound straightforward, but they only work in a broader therapeutic context. Their success calls for skillful therapists, motivated patients, and resourceful and compassionate sexual partners; these tasks are not for the squeamish. Sex therapy is still a new field, and its methods are being refined. There are still important questions to be answered about its effectiveness (Box 8.3).

Psychotherapy and Behavior Therapy

Psychotherapeutic methods of treating sexual dysfunction long antedate sex therapy. Their practitioners are a diverse group of specialists with different theoretical orientations: psychoanalysts, psychiatrists, clinical psychologists, psychiatric social workers, and marriage and family therapists. The common element in their approach is the reliance on verbal interchange to resolve personality problems and interpersonal conflicts.

Psychotherapy   Psychotherapists frequently deal with sexual issues, but most of them do not think of themselves as sex therapists; the methods they use to treat sexual problems are

Box 8.3

TREATMENT OUTCOMES OF SEX THERAPY

How well does sex therapy work? The results originally reported by Masters and Johnson were spectacular. Their experience with a total of 1872 cases showed an overall success rate of 82 percent (85 percent for males; 78 percent for females), ranging from 99 percent for vaginismus and 96 percent for premature ejaculation to a relatively more modest 67 percent for primary impotence, 78 percent for secondary impotence, 76 percent for ejaculatory incompetence, 79 percent for primary anorgasmia, and 71 percent for situational anorgasmia (Masters et al., 1982).

Moreover, these results were achieved over periods of treatment lasting a matter of weeks, using methods that seemed so beguilingly simple that almost anyone could master them. Almost everyone tried to do so in a surge of therapeutic optimism that led, during the 1970s, to the emergence of a whole new field with numerous practitioners.

A decade after Masters and Johnson inaugurated sex therapy, questions began to be raised about the effectiveness of their methods and the replicability of their reported results (Zilbergeld and Evans, 1980). In other hands, the same techniques have produced less spectacular results. One factor in this shift is that the more superficial types of sexual dysfunction now get taken care of through educational and self-help methods, leaving the harder cases, with more deep-seated personality problems, in sex therapy.

Despite these concerns, the results of sex therapy continue to be impressive. In a systematic study of sex therapy outcomes, using modifications of the Masters and Johnson techniques, Arentewicz and Schmidt (1983) have reported the following cures, according to the therapists:

Orgasmic dysfunction	19%
Vaginismus	67%
Ejaculatory dysfunction	49%
Premature ejaculation	40%
Overall rate	49%

Patients and their partners agreed that their sexual problems were much better: the overall rate was 65 percent for patients, 58 percent for partners; but only 8 percent of patients and 11 percent of their partners said the symptoms were fully removed.

Are the improvements in sex therapy sustained over time? Follow-up studies show that after three years some forms of sexual dysfunction tend to return, although the person may feel less distressed about it. Sexual desire dysfunction appears to be particularly resistant to sustained change (DeAmicis et al., 1985).

It may be some time before the treatment of sexual dysfunction occupies its rightful place among respected medical and behavioral specialties; but that day will surely come, since sexual dysfunction deserves to be attended to as much as any other form of human suffering.

---

the same methods they use to treat any other kind of problem. *Psychoanalysis*, the most intensive form of psychotherapy, involves analyzing erotic fantasies, dreams, events, and past and present sexual relationships. Improvements in the patient's condition depend on insight into the origins of the difficulties. A scaled-down version of psychoanalysis is *insight-oriented therapy*, which pursues more limited goals (DeWald, 1971). It is the basis of Kaplan's psychosexual therapy.

The Systems Approach   Rather than focusing on the past, the systems approach focuses on the current relationship of an individual with the sexual partner (LoPiccolo and Stock, 1986). The aim is to learn what function the sexual symptom serves in their relationship. Once the couple can deal with the underlying issue, the sexual dysfunction is next to resolve.

Behavior Therapy   Behavior therapy is another form of psychotherapy, but for historical reasons it is more often referred to as *behavior modification*. Its focus is the problem behavior itself, which it aims to change by techniques derived from learning theory.

Behavior therapy subsumes a variety of techniques. *Systematic desensitization* is based on the principle of *reciprocal inhibition*—the fact that contrary emotional states are mutually exclusive. For instance, you cannot be simultaneously anxious and calm; if anxiety interferes with a certain function, its effects can be counteracted by relaxation.

Anxiety-producing situations cause distress at various gradations. For example, a woman may find it very difficult to have intercourse with her husband in the nude, although coitus when partially clothed may be more tolerable. The behavior therapist and the patient draw up a detailed list of situations in order from the most unbearable to the comfortably tolerable (for instance, being kissed on the cheek).

The patient is then trained in the techniques of deep-muscle relaxation (which is incompatible with anxiety) and desensitization proceeds. When the woman is fully relaxed, she is asked to fantasize the least fearful of the anxiety-provoking situations on the list; if she gets anxious, she switches back to the relaxation routine; otherwise she tackles the next troublesome scene. This process is repeated until the patient is able to confront in fantasy the anxiety-provoking situation in full (coitus in the nude). The next step is to transfer this mastery of the anxiety-producing stimulus to real-life situations. The same process may be used in connection with other techniques without the use of relaxation.

Group Therapy

The therapies we have described so far treat individual patients or couples. In *group therapy* a small number of individuals (usually about six to eight) work with one or two therapists. The members of a group may be chosen on the basis of having similar or different problems; they may be unrelated or couples. In addition to the economy in the therapist's time,

this approach allows members of the group to share and learn from their problems and lend each other emotional support. On the other hand, this format does not allow for as much individual attention, and especially with a topic like sexual dysfunction, a person may be less open or candid.

Helen Kaplan (1974) has worked effectively with groups of dysfunctional couples, and Lonnie Barbach (1975) has reported good results working with groups of five to seven women with orgasmic problems. These women met with two female therapists for discussions and also carried out daily assignments aimed at "getting in touch" with their own body and sexual feelings. Orgasm is initially attained through masturbation and eventually through coitus (the sexual partners were not directly involved in the treatment program). After five weeks, 93 percent of these women could reach orgasm through masturbation, and three months later, half were orgasmic during coitus.

Self-Help

The outpouring of books on various aspects of sexuality, including dysfunction, has reached a mass market. During the 1970s, best-seller lists virtually always included one or more books on sex. Many of these were "how-to" sex manuals with explicit texts and illustrations. Such books, films, and videos have been helpful to some people with varying degrees of sexual dysfunction. However, some of them contain incorrect information and foster unrealistic goals and expectations.

More specifically aimed at the dysfunctional are instructional books that approach these issues in more thoughtful and systematic ways, including graduated exercises for overcoming inhibitions and gaining sexual satisfaction (Barbach, 1975).

Sexual experience itself probably remains the greatest teacher and therapist of all. With a serious problem, however, practice does not make perfect—it makes matters worse by failure compounding failure. Couples merely trying harder may dig themselves deeper into

Box 8.4

GETTING HELP FOR SEXUAL DYSFUNCTION

We usually seek help when pain or distress becomes more than we can bear, or when we fear that it will get worse. Sexual dysfunction does not hurt in the same way; so it is harder to decide when to seek help, especially if the problem is not sexually incapacitating, such as total impotence or anorgasmia.

If you experience a fairly rapid and consistent change in sexual function, it is important to get a medical checkup. Because symptoms of sexual dysfunction may be the early signs of an underlying illness, there may be more at stake than sex.

On the other hand, the fact that a sexual problem has hung around for many years does not mean that you should simply accept it and learn to live with it. If you wonder whether the dysfunction warrants therapy, you can always seek a consultation. Seeing a physician or sex therapist does not necessarily mean committing yourself beforehand to treatment.

Sex therapy, like other forms of specialized care, is, in principle, best sought not directly but through your doctor or a counselor at a student health center.

Unlike other health care professions, the practice of sex therapy is still unregulated. Anybody entitled to see patients or clients can set up shop as a sex therapist; as a result there are people in practice who simply lack the skills and training to do sex therapy well.

It is generally better to seek help from individuals and clinics affiliated with hospitals, medical schools, and universities. Even if such institutions themselves do not provide such care, they may refer you to reputable therapists. Similar advice may be obtained from medical and psychological societies and the directories of professional organizations in this field, such as the American Association of Sex Educators, Counselors and Therapists (AASECT); 11 Dupont Circle, Suite 220, Washington, D.C. 20036.

Membership requirements in such groups are not stringent, so you must continue to exercise good judgment before making a final choice. Deal with therapists who have a graduate degree from a reputable institution, who show clear evidence of having received serious training in the treatment of sexual dysfunction, and who are willing to discuss openly their qualifications, methods, time schedules, and fees. If there is any hint that sexual intimacies with the therapist are expected, go elsewhere.

The choice of a competent therapist is important, but the success of sex therapy largely depends on the individuals or couples who seek it. Going through the motions of seeking help to placate a spouse, or hoping that things are somehow going to change without serious effort, are futile exercises. Enthusiasm may be too much to expect from those contemplating therapy, but a certain measure of motivation and determination is essential for success.

---

the problem. On the other hand, the experience of having a caring, compassionate, and competent partner goes a long way in sorting out many sexual problems and imparting confidence. When these approaches prove insufficient, it is time to seek help from a sex therapist. Box 8.4 addresses that choice.

Treatment with Drugs

Drugs are among the oldest forms of treatment. Their application to sexual dysfunction is one facet of the search for aphrodisiacs. However, until modern times, the drugs available to physicians to treat sexual disorders were neither effective nor safe. A well-known example is *Spanish fly,* a powder made from dried cantharis beetles (hence also called *cantharides*). It causes urethral irritation and penile vasocongestion, which may result in erection but also has dangerous side effects.

Yohimbine (from the bark of the African yohimbe tree) is a nervous system stimulant (Gilman et al., 1985). Though long claimed to be a sexual stimulant, its "prosexual" effects have only been recently established in experimental animals (Kwong et al., 1986). There is also some evidence that it can be effective in treating impotence.

The great chemical hope for the treatment of sexual dysfunction, especially impotence, has resided in sex hormones—a hope that has been largely unfulfilled. Given their indispensable role in the organization and activation of sexual functions, it was reasonable to expect that hormones would enhance sexual drive and potency. To this end, testicular extracts began to be used way back in the 19th century. Furthermore, since sexual decline and aging are closely associated in people's minds, such treatments also held high hopes for rejuvenation, or regaining youthfulness. Though there has never been any convincing proof that they work, extracts of testicular, placental, and embryonal tissue from animals continue to be used in various parts of the world in futile attempts to fight the effects of aging.

The popularity of testosterone treatment has waxed and waned over the years. There is little question that it can be effective in restoring libido and potency in males if there is androgen deficiency. There is also some evidence that testosterone may enhance sexual function in cases of psychogenic impotence by giving the body a temporary physiological boost (Kaplan, 1974). Testosterone is therefore a legitimate drug for the treatment of select cases, but there is no convincing evidence that it counteracts the effects of aging. Furthermore, the use of testosterone by older persons poses the risk of prostatic hypertrophy and cancer (Chapter 5).

The effect of estrogens on female libido remains unclear (Chapter 4). Estrogen in menopausal women can improve sexual function, counteracting the changes in the vaginal lining and the drying out of the lubricatory response (Chapter 4). The effect of androgens on women's sex drive is ambiguous, and their virilizing side effects so unacceptable that testosterone is not used to treat female sexual dysfunction.

Various drugs have occasionally appeared on the scene with aphrodisiacal side effects, but none of them is reliable and safe for treating sexual dysfunction. Examples are L-dopa, which is widely used in treatment of Parkinson's disease (a neurological disorder); PCPA (p-chlorophenylalanine), an experimental psychiatric drug; and vitamin E (used in some cases of infertility) (Gessa and Tagliamonte, 1974; Hyppa et al., 1975).

During a conference of urologists in Las Vegas in 1983, a British researcher, presenting a paper on the effect of a drug injected into the penis, drove the point home by unzipping his pants to show the audience his own induced erection. The drug in question was a smooth muscle relaxant that causes the penile tissues and arteries to relax, increasing the blood flow into the penis.

A number of substances have since been used with similar effects in causing vasodilation. Most commonly used is a mixture of *papaverine* and *phentolamine*, which must be injected directly into the penis (the pain is said to be minimal). It gives many impotent men a firm erection that lasts more than an hour. The drug will not work on those with severe blood vessel problems. Nonetheless, 14,000 men in the United States are now reportedly using it, many by injecting themselves at home (Stipp, 1987).

In about half of the cases of organic impotence and even a greater proportion of cases with psychogenic impotence, a single injection will be followed by spontaneous erections, in some cases for several weeks (Kiely et al., 1987). In 5 percent of cases the drugs cause sustained erections (priapism) that will permanently damage penile tissues if not relieved. Reducing the dosage often avoids a recurrence.

Physical Treatment Methods

Kegel Exercises   It is not uncommon for women, following childbirth, to develop *stress incontinence*—involuntary loss of spurts of urine when they cough or strain (raising abdominal pressure on the bladder). The cause is loss of tone of the pubococcygeal (PC) muscle (Chapter 2). To correct this problem, a set of exercises were devised whereby squeezing and relaxing the vaginal orifice gradually strengthened the muscles in the region. In the 1950s, the gynecologist Arnold Kegel discovered that these exercises helped improve his patients' sexual responsiveness and orgasm ability. These *Kegel exercises* have become part of various sex therapy programs (Kegel, 1952).

One of the merits of Kegel exercises is their simplicity. The woman first learns to control the PC muscle by squeezing on a finger inserted in the vagina, or by alternately interrupting and releasing the stream of urine. She is then instructed to tighten and relax these muscles in a prescribed manner (such as tighten for 3 seconds, relax for 3 seconds, in sets of 10, three times a day) (Kline-Graber and Graber, 1978). The use of a *perineometer* (see Box 3.2) or similar devices allows for more accurate assessments of vaginal muscle tone (McKey and Dougherty, 1986). It also provides an aid to Kegel exercises by allowing the woman to monitor the effect of contracting her PC muscle.

Despite much clinical evidence of the value of strengthening the perineal muscles, data on the correlation of vaginal muscle tone with orgasmic capability have been lacking. One study with normal subjects did not find pubococcygeal strength to be positively correlated to frequency or self-reported intensity of orgasm (Chambless et al., 1982). Another study, though, has shown Kegel exercises to enhance the sexual arousal of a group of normal women (Messe and Geer, 1985).

Masturbation Training Masturbation, with or without vibrators, is recommended by some therapists, especially in cases that do not yield to more standard approaches. Its main purpose is to allow a woman with primary anorgasmia to experience what it feels like to have an orgasm, which then serves as the goal. As more powerful sources of stimulation, vibrators may bring about orgasm when other means fail.

Various regimens have been devised for such masturbation training, with the purpose of eventually enabling anorgasmic women to attain coital orgasm (LoPiccolo and Lobitz, 1977). These attempts are often successful (LoPiccolo and Stock, 1986). It can be argued that as long as a woman reaches orgasm, by any means, she need not be considered dysfunctional. In this view, the attainment of masturbatory orgasm alone would be a successful outcome.

Surgical Methods

The use of splints to support a limp penis has been known in many cultures. These splints have been in the form of flat rods or hollow dildos housing the flaccid penis.

Advances in surgery have made it possible to implant *penile prostheses*. These devices do not cure impotence; they are mechanical aids that make intercourse possible. Mainly intended for cases of impotence that are due to irreversible organic causes, such as a severed spinal cord, such surgery is currently used for psychologically caused impotence as well. They are also a last resort for cases that do not respond to sex therapy and otherwise would have no hope of coital function. Some 30,000 prostheses are now implanted per year in the United States, mostly with good results. In select cases, the procedure can even be done on an outpatient basis (Small, 1987).

One type of penile prosthesis is a fixed or flexible plastic rod. Implanted in the penis, it provides enough rigidity to allow coitus. A more sophisticated prosthesis is the inflatable variety (Figure 8.5). This device has a reservoir filled with a fluid, which is implanted in the abdomen. It is connected by tubes to a small pump lodged in the scrotal sac, which in turn has tubes leading to the inflatable cylinders implanted in the penis. To attain erection, the man pumps the fluids into the cylinders; to return the penis to a flaccid state, a valve in the pump is released, sending the fluid in the penile cylinders back to the reservoir. The disadvantages of the inflatable device include the greater technical difficulty of installing it. Complications that occasionally arise include infection or leakage of pump fluid (Fishman,

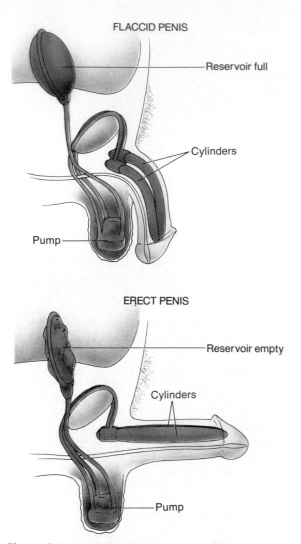

Figure 8.5   An inflatable penile prosthesis.

1987); yet in 98 percent of cases, the implanted cylinders remain intact after a year, and in 92 percent of cases after three years (Gregory and Purcell, 1987). If a prosthesis is removed because of infection or malfunction, reimplantation of another device at a later date is not always possible (Gasser et al., 1987), but success rates are now quite high.

A mechanically inflated penis does not feel exactly the same as a naturally erect one, but these surgical methods restore a level of func-

tion and mutual satisfaction that would be otherwise unavailable to these men and their partners. Ninety percent of men with penile implants say they would do it over again (Steege et al., 1986).

Prevention of Sexual Dysfunction

An ounce of prevention is worth a pound of cure—the truism applies to sexual dysfunction as to any other problem. Unfortunately, we do not yet have a clear enough understanding of the causes of sexual dysfunction to speak with confidence about their prevention. Nonetheless, we do have some guidelines (Qualls et al., 1978).

Sex is a physiological function whose full enjoyment requires a healthy and vigorous body. Preventing and treating disease, exercising, eating right, and following proper sexual hygiene all help avoid sexual dysfunction.

Sex is also a psychological function that is highly vulnerable to anxiety, depression, and interpersonal conflict. There is no way to banish these painful emotions forever from your life, but you can face your feelings and your partner's feelings with understanding and sensitivity. A healthy relationship, like a healthy body, increases the chances of a healthy sex life. Therefore sex must be seen in this larger context. Also, the more you can keep sex free from other conflicts, the better off you will be sexually. Some couples treat sex as a "demilitarized" zone: they can retreat to it as a haven from conflict. At least one important facet of their lives remains satisfying and heartens them to keep working on the rest. But sex is not a balm to be used indiscriminately to smooth over all difficulties.

Sex is also highly influenced by culture. Shame and guilt serve certain legitimate social purposes, but they can also reflect unwarranted antisexual prejudice that will cripple sexual development and enjoyment. You do not have to stick to all the assumptions you grew up with. It is healthy to think them through. At the moment, our culture does not have a clear common understanding of how to raise children who will be sexually healthy and fulfilled adults; you can help develop such understanding.

One of the most important requirements for preventing sexual dysfunction is sexual activity itself. To keep at it, to work and play at its enhancement, to deal with its shortcomings are what it takes to remain sexually alive. Equally important is to place sexual fulfillment in a realistic perspective for each stage of life. The best way to get sexual satisfaction is not to chase it constantly like a rainbow. Sex is one of the greatest joys in life, yet sex alone cannot make life fulfilling. On the other hand, even people with inhibited and faltering sexual lives need to remember that it is not sex that gives up on people, but people who give up on sex.

REVIEW QUESTIONS

1. What are the main types of sexual dysfunction?
2. What are the main physical and psychological causes of sexual dysfunction?
3. How would you tell if a sexual dysfunction is mainly organic or psychogenic?
4. Should sexual interactions between therapist and patient or client be considered a form of sex therapy or a breach of professional ethics?
5. What educational programs would you institute to help prevent sexual dysfunctions?

THOUGHT QUESTIONS

1. Why was the recognition and treatment of sexual dysfunctions neglected for so long?
2. What are the advantages and disadvantages of using the illness model in approaching sexual dysfunctions?
3. What do you think about using surrogate partners in the treatment of sexual dysfunction?
4. How does the new sex therapy differ from traditional psychological methods of treating sexual dysfunctions?
5. How are sexual dysfunctions treated medically and surgically?

SUGGESTED READINGS

Barbach, L. G. (1982). *For each other: Sharing sexual intimacy*. New York: Doubleday. A well-written book aimed at helping women to fulfill their sexual potential.

Belliveau, F., and Richter, L. (1970). *Understanding human sexual inadequacy*. New York: Bantam. An abbreviated and simpler account of Masters and Johnson's *Human sexual inadequacy* that established the modern field of sex therapy.

Kaplan, H. S. (1974). *The new sex therapy*. New York: Quadrangle. A comprehensive and clearly presented work on sex therapy, with excellent illustrations.

Leiblum, S., and Pervin, L. (Eds.) (1980). *Principles and practice of sex therapy*. London: Tavistock. Valuable contributions on sexual dysfunctions and their treatment.

Zilbergeld, B. (1978). *Male sexuality*. Boston: Little, Brown. An interesting and insightful account of male sexuality, including attitudes that limit male sexual fulfillment.

Glossary

abortion   The ending of a pregnancy before the fetus is mature enough to survive outside the uterus; may be induced (done intentionally) or spontaneous (see **spontaneous abortion**).

abstinence   Voluntarily not engaging in sexual acts, such as intercourse.

Acquired Immunodeficiency Syndrome (AIDS)   Primarily a sexually transmitted disease that destroys the immune system's ability to fight infections; however, can also be passed by needle sharing among intravenous drug users or contaminated blood products.

adolescence   The term used to describe the psychosocial development of a child into an adult.

adrenal glands   A pair of endocrine glands, located above the kidneys, that produce and secrete (among other things) androgens.

afterbirth   The placenta and amniotic sac that are expelled from the mother's body right after giving birth.

amenorrhea (ah meh NOR ee auh)   The absence of menstruation; see **menstruation.**

amniocentesis (AM nee oh sen TEE sis)   A procedure in which amniotic fluid is removed from a mother's uterus and tested to see if the fetus has genetic disorders or other abnormalities.

amniotic fluid (AM nee ot ik)   The fluid within the amniotic sac that surrounds the fetus and cushions it from shock until birth.

anal intercourse   Sexual behavior in which one person's penis is inserted into another's anus.

androgen (ANN dro gin)   A generic term used to describe a class of hormones (such as testosterone) that are primarily responsible for

the development and functioning of the male reproductive system.

androgen insensitivity   A condition in which a genetic male develops female genitals and breasts because of his body's failure to respond to testosterone; sometimes called testicular feminization syndrome.

androgyny (ann DRODGE ih nee)   The blending of stereotypically masculine and feminine characteristics in the same person.

anorgasmia (ann ore GAZ mee ah)   A female sexual dysfunction characterized by a persistent inability to reach orgasm; also called inhibited female orgasm.

aphrodisiac (AF ruh DIZ ee ak)   A substance thought to increase a person's sexual desire, sexual performance, and/or sexual pleasure.

areola (AIR ee oh lah)   The dark, circular area of skin that surrounds the nipple.

artificial insemination   The placement of semen into a woman's vagina or uterus by means other than sexual intercourse so as to induce pregnancy.

autonomic nervous system (AW toe NAWM ik)   The part of the nervous system that regulates the glands and organs of the body, including sexual reflexes.

Bartholin's glands (bahr TOE linz)   Two small glands located on either side of the vaginal opening which are thought to secrete fluid to lubricate the vagina during sexual arousal.

basal body temperature method (BAY sul)   A method of birth control in which the changes in a woman's temperature are recorded in order to determine when she ovulates.

Billings method   A method of birth control in which a woman monitors the changes in the amount and consistency of her cervical mucus, thus identifying when she ovulates.

birth control   See **contraception.**

bisexual   A person who feels sexually attracted to or has had sexual contact with both males and females.

blastocyst (BLAS toe sist)   A small mass of cells, formed from the morula, that attaches itself to the uterine wall during the first week of pregnancy.

bonding   The process by which a parent and child become emotionally attached to each other.

candidiasis (KAN dih DYE ah sis)   A vaginal infection that is caused by the excessive growth of a yeastlike fungus that occurs normally in the body.

castration (kas TRAY shun)   The surgical removal of the testes.

cervical cap (SIR vih kull)   A contraceptive device, made of plastic or rubber, that fits over the mouth of the cervix and acts as a barrier to sperm.

cervix (SIR vix)   The lower portion of the uterus that opens into the vagina.

cesarian section (see SAIR ee uhn)   A surgical procedure whereby the infant is delivered through an incision in the uterus and abdominal wall.

chancre (SHAN ker)   A hard, raised sore (usually painless) that characterizes the first stage of syphilis.

chancroid (SHAN kroid)   A bacterial sexually transmitted disease that causes soft, painful ulcers to appear on the genitals.

chlamydia (clah MID ee uh)   A bacterial sexually transmitted disease that in men causes painful urination, itching, and discharge, but in women may only cause a slight vaginal discharge.

chorionic villi sampling (CVS)   A procedure in which a sample of tissue is taken from the membrane that surrounds the fetus to test for possible birth defects.

chromosome (KROW moe sohm)   The genetic material found in the nucleus of every human cell; each cell has 46 chromosomes, except those in sperm and eggs which carry 23.

cilia (SIL eh uh)   The numerous hairlike structures that line the fallopian tubes and propel the egg to the uterus.

circumcision (SIR come SIH zhun)   The surgical removal of the foreskin (**prepuce**) of the penis.

climacteric (KLI mack tur ik)   A midlife period for both men and women during which many physiological changes take place due to the transition from fertility to infertility; also see **menopause.**

clitoris (KLI tore iss)   In females, a small highly sensitive genital organ; its function is to provide erotic pleasure when stimulated.

coitus (COY tus)   The technical term for the insertion of the penis into the vagina; commonly referred to as sexual intercourse.

colostrum (cuh LAW strum)   A watery substance, rich in protein and antibodies, secreted from the mother's breasts immediately following delivery; precursor to breast milk.

conception   The beginning of a new life, marked by fertilization.

condom (KON dum)   A male contraceptive device—usually a thin, latex sheath—that fits over the penis; also used to prevent sexually transmitted diseases.

congenital adrenal hyperplasia   A condition in which the adrenal glands of the female fetus produce an excess amount of androgen, resulting in masculinized genitals upon birth; also referred to as adrenogenital syndrome.

contraception   Techniques, devices, or drugs used to prevent pregnancy; also called birth control.

copulation (cop you LAY shun)   Sexual intercourse as a means of reproducing; applies to animals (and humans).

corona (cor OH nah)   The sensitive rim of the glans.

corpora cavernosa (CORE pour ah kah ver NOH sa)   Two parallel masses of erectile tissue, located in the shaft of the penis and the clitoris, that become engorged with blood during sexual arousal.

corpus luteum (CORE pus LOO tee um)   The part of the ovarian follicle that is left in the ovary after the egg has been released; its primary function is to secrete progesterone.

corpus spongiosum (CORE pus spun gee OH

sum)   Within the penis, a column of spongy, erectile tissue that contains the urethra and becomes engorged with blood during sexual arousal; also called spongy body.

couvade (koo VAHD)   The male experience of symptoms that mimic pregnancy and/or childbirth.

Cowper's glands (KOW perz)   Two small structures, attached to the urethra in the male, that secrete a few drops of clear, sticky fluid during sexual arousal.

cremasteric muscle (KRE mah ster ik)   A muscle, located under the scrotum, that involuntarily elevates the testes in response to cold, sexual excitement, and other stimuli.

cremasteric reflex   The reflexive contractions of the cremasteric muscle when the inner thigh is stroked.

critical period   A point during development when key events must take place or development is forever altered.

cryptorchidism (krip tor KID ism)   A condition in which the testes fail to descend from the abdominal cavity into the scrotal sac.

culture   A society's characteristic behavior patterns, such as language, customs, beliefs, and artistic accomplishments.

cunnilingus (KUN ih LING gus)   The oral stimulation of the female genitals.

cystitis (sis TIE tiss)   An inflammation of the urinary bladder causing painful and frequent urinations.

desensitization   A form of behavior modification therapy in which the patient while relaxed and in a safe environment, is gradually exposed to the anxiety-provoking situation until his or her anxiety extinguishes.

diaphragm (DYE uh fram)   A dome-shaped rubber contraceptive device that is inserted into the vagina to block the cervical opening.

dihydrotestosterone (die HIE dro tes TOSS tur ohn)   A hormone, similar to testosterone, that is responsible for the embryonic development of the male's external genitals.

dilation and curettage (D & C) (CURE eh taj)   A process in which the cervix is dilated and then an instrument scrapes the uterine lining, used to treat uterine disorders or abort fetuses.

dilation and evacuation (D & E)   A method of abortion in which the cervix is dilated and the fetus is removed from the uterus with forceps, curettage, and suction; it is generally used in the second trimester.

douch (DOOSH)   To rinse out the inside of the vagina with water or chemical solutions.

Down's syndrome   A chromosomal disorder causing mental retardation and defects of the internal organs.

dysmenorrhea (dis men OR ee uh)   Pain experienced before or during menstruation—typically backaches, abdominal cramps, and aches in the thighs.

dyspareunia (DIS par OO nee ah)   Persistent pain during or after sexual intercourse.

ectopic pregnancy (EK top ik)   A pregnancy in which the fertilized egg implants itself in a site other than the uterus, most often a fallopian tube.

ejaculation (ee JACK you LAY shun)   The expulsion of semen from the penis, usually during orgasm.

ejaculatory ducts (ee JACK you la TORE ee)   Two tubelike structures that serve as a pathway for sperm and seminal fluid to move from the prostate gland to the urethra.

ejaculatory incompetence   A male sexual dysfunction characterized by an inability to ejaculate during intercourse.

embryo (EM bree oh)   The term used to refer to the unborn child during the first eight weeks of pregnancy.

endocrine glands (EN doe krin)   Ductless glands that produce and secrete hormones into the bloodstream.

endometrium (en doe MEE tree um)   The inner lining of the uterus, which is partially shed during menstruation.

epididymis (ep ih DID ih mus)   A tightly coiled tube, located at the upper back of each testis, in which sperm mature.

episiotomy (eh PIZ ee ot uh me)   An incision made, during delivery, between the vagina and anus to give the baby's head more room to emerge and to avoid injury to vaginal tissues.

erectile dysfunction   See **impotence.**

erogenous zones (eh RAW jeh nus)   Areas of the body particularly responsive to sexual stimulation.

Eros   According to psychoanalytic theory, erotic love or libidinal instinct.

estrogen replacement therapy (ES tro gin)   A treatment before and during menopause in which a woman takes supplemental estrogen.

estrogens   A group of "female" hormones—produced primarily by the ovaries—responsible for sexual maturation, regulation of the menstrual cycle, and maintenance of the uterine lining.

estrus (ES truss)   The period in which many nonhuman female mammals are ovulating or sexually active; also referred to as "being in heat."

eunuch (YOU nik)   A castrated male.

excitement phase   The first phase of the human sexual response cycle, in which the person begins to feel physically and/or psychologically aroused: heart beat quickens, muscles tense, and genitals become engorged with blood.

fallopian tubes (fah LOW pee un)   The two narrow tubes that serve as a pathway for the eggs (ova) to move from the ovaries to the uterus; the place where fertilization usually takes place.

fellatio (feh LAY shee oh)   The oral stimulation of the male genitals.

fertility symbols   Primitive art, used for magical-religious purposes, that was thought to promote human and animal fertility; art objects included spears, clubs, erect penises, and exaggerated breasts, buttocks, and stomachs.

fertilization   The moment when the sperm and ovum unite.

fetal alcohol syndrome   A birth disorder due to the mother's heavy drinking while pregnant; abnormalities may be as serious as mental retardation.

fetus (FEE tus)   The term used to refer to the unborn child from the ninth week of pregnancy to birth.

follicle (FALL ih kul)   The group of cells in the ovary that contain an egg.

follicle-stimulating hormone (FSH)   A pituitary hormone that stimulates the maturation of ovarian follicles in the female and sperm production in the male.

foreplay   Kissing, touching, genital stimulation, and other forms of physical contact between two people that leads to sexual intercourse; see **petting.**

foreskin   See **prepuce.**

frigidity (fri JID ih tee)   An outdated term once used to describe most sexual dysfunctions common to females.

gametes   See **germ cells.**

gender identity   The view of one's self as being either male or female.

gender roles   The behaviors and traits that a given culture deems appropriate for males and females.

gene   The DNA molecules, found on chromosomes, responsible for hereditary transmission from parent to child.

genitals (JEN ih tulz)   The sex organs of males and females, typically the penis, testes, and scrotum in males and the vulva in females.

genital tubercle (TOO ber kul)   A small bud of fetal tissue that develops into either a penis or a clitoris.

genital warts   Warts on the genitals, caused by a sexually transmitted virus.

germ cells   Sperm or egg cells.

glans   The tip of the penis or clitoris, rich with nerve endings.

gonadotropin releasing hormone (GnRH) (goe NAD o TROE pin)   A hormone, produced by the hypothalamus, that controls the output of LH and FSH by the pituitary.

gonadotropins   Pituitary hormones that stimulate the gonads (testes and ovaries) to secrete their own hormones.

gonads (GOE nads)   The testes or ovaries—reproductive glands.

gonorrhea (GONE or REE ah)   A venereal disease caused by bacteria, causing urethral and vaginal discharge.

Graafian follicle   The mature follicle; it houses the egg that is released during ovulation.

Grafenberg spot (GRAY fen berg)   An erotically sensitive area presumably located in the anterior wall of the vagina; sometimes called the G-spot.

granulosa cells (gran oo LOW suh)   The cells that line the ovarian follicle, and later become the corpus luteum.

hepatitis (HEP ah TIE tis)   An inflammation

of the liver caused by a virus, which may be transmitted through sexual contact.

hermaphrodite (her MAF roe dite)   A person who has both ovarian and testicular tissue, and thus has genital features of both sexes.

herpes (HER peez)   Painful blisters of the genitals or mouth that are caused by a virus, transmitted through sexual contact.

heterosexuality   A sexual orientation in which a person is sexually, emotionally, and socially attracted toward members of the opposite sex.

homosexuality   A sexual orientation in which a person is sexually, emotionally, and socially attracted toward members of the same sex.

hormones (HOR mohnz)   Chemical substances, secreted by the endocrine glands into the bloodstream, that have an effect on physiological functioning and psychological behavior.

hot flashes   A common symptom of menopause in which the woman experiences sudden sensations of warmth, sweating, or chilliness.

human chorionic gonadotropin (KORE ee ON ik goe NAD o TROE pin)   A hormone produced by the placenta and similar to pituitary gonadotropin. Normally found only in the urine of pregnant women.

human immunodeficiency virus (HIV) (ih MEW no dee FIH chun see)   The retrovirus that causes AIDS.

hymen (HI men)   The delicate tissue partially covering the vaginal opening, usually until a female first has intercourse.

hypoactive sexual desire disorder   A sexual dysfunction characterized by a persistent absence of both sexual fantasies and desire for sex.

hypothalamus (hie poe THAL uh mus)   The part of the brain that regulates (among other things) sexual functions by controlling the production and release of gonadotropic hormones.

hysterectomy (HISS tur EK tuh me)   A surgical procedure in which a woman's uterus is removed.

impotence (im PUH tence)   The inability to have or maintain a firm erection despite stimulation; also called erectile dysfunction.

incest   Sexual activity between close relatives, for example, between fathers and daughters.

infertility (in fur TILL ah tee)   The inability of a man, woman, or couple to conceive a child, usually after one year of trying.

inhibited female orgasm   See **anorgasmia.**

inhibited male orgasm   See **retarded ejaculation.**

interstitial cells (in tur STISH ee ul)   The cells, located between seminiferous tubules in the testes, that produce most of the androgens in the male; also called Leydig's cells.

intrauterine device (IUD) (IN tra YOU ter in)   A small plastic device inserted into the uterus for birth control.

in vitro fertilization (IVF) (in VEE trow)   A procedure in which a woman's mature eggs are extracted from her body, fertilized with sperm in a laboratory, and then implanted in her uterus.

Kegel exercises   A series of exercises aimed at strengthening the muscles surrounding the vagina and urethra.

Klinefelter's syndrome   A sex chromosome disorder in which a male is born with an extra X chromosome, resulting in an XXY pattern; these men have small testes and penises and are infertile.

labia majora (LAY bee uh MUH jor ah)   The two, outer, elongated folds of skin surrounding the labia minora, the clitoris, and the urethral and vaginal openings.

labia minora (LAY bee uh mi NOR uh)   The two, small, inner folds of skin surrounding the urethral and vaginal openings.

labioscrotal swelling   The fetal tissue that develops into the scrotum in the male or the labia majora in the female.

labor   The process involved in giving birth: rhythmic uterine contractions, cervical dilation, the baby's birth, and expulsion of the placenta.

lactation (lak TAY shun)   The production of breast milk in the female that begins about two to three days after giving birth.

Lamaze method (la MAHZ)   A method of prepared childbirth emphasizing muscle relaxation and controlled breathing.

lesbian (LEZ bee un)   A female homosexual.

leukorrhea (LOO kor REE ah) Whitish or puslike vaginal discharge caused by various infections or irritating chemicals.

Leydig's cells (LEE digz) See **interstitial cells.**

libido (lih BEE doe) The psychoanalytic term for sexual drive or energy.

limbic system (LIM bik) A set of brain structures that regulates sexual and emotional behavior.

lochia (LOH key ah) A reddish-brown uterine discharge lasting several weeks following childbirth.

luteinizing hormone (LH) (LOO tuh nye zing) A pituitary hormone that stimulates the corpus luteum in the ovaries to produce progesterone, and the Leydig cells in the testes to produce androgens.

mammary glands (MAM ah ree) Milk-producing organs in the female breast.

mastectomy (mass TEK toe me) The surgical removal of a breast.

meiosis (my OH sis) Germ cell division in which the number of chromosomes is reduced by half.

menarche (MEN ark) A pubescent girl's first menstrual period.

menopause (MEN oh pause) The gradual ending of a woman's menstrual cycles, usually beginning in her late forties.

menorrhagia (men or HAG ee ah) An increase in the amount or duration of menstrual bleeding.

menstruation (MEN stroo a shun) A bloody vaginal discharge caused by the sloughing off of the uterine lining, occurring at about monthly intervals. Also called menses or period.

midwife A woman who assists in another's childbirth.

mitosis (my TOE sis) A type of cell division in which the cell splits into two identical cells.

monogamy (mah NAW guh mee) A type of marriage in which a person has only one spouse at a time.

mons pubis (monz PEW bis) The soft, fatty tissue over the female pubic bone that becomes covered with pubic hair during puberty; also called mons veneris or mound of Venus.

morning sickness The nausea upon awakening, followed by an aversion to food, commonly experienced by women during their first eight weeks of pregnancy.

morula (MOHR oo lah) A round mass of cells developing from the zygote within a few days of fertilization.

Mullerian ducts (mew LAIR ee un) The parts of the female fetus that develop into the fallopian tubes, uterus, and upper portion of the vagina.

myotonia (MY oh TONE ee ah) Increased muscle tension that occurs during the sexual response cycle.

nature versus nurture The theoretical controversy over whether nature (biology) or nurture (environment) is primarily responsible for behavior—including sexuality.

nipple The prominent tip of the breast, made of smooth muscle and nerve fibers, that in females is the outlet of the milk ducts.

nocturnal emission Ejaculation while asleep; also called a wet dream.

nursing The infant's sucking of milk from a woman's breast; also called breastfeeding.

nymphomania (NIM foe MAY nee ah) Hyperactive sexual desire in women.

Oedius complex (ED ih pus) According to psychoanalytic theory, a child's sexual attraction toward the parent of the opposite sex and hostility toward the parent of the same sex.

orgasm (or GAZ um) An intensely pleasurable subjective sensation marking the sudden discharge of accumulated sexual tension.

orgasmic platform As part of the female sexual response, the expansion of the outer vaginal walls and narrowing of the vaginal opening due to vasocongestion.

osteoporosis (ah stee oh poh RO sis) A condition, common in postmenopausal women, in which the bones become thin and brittle due to calcium loss.

ovaries (OH vah reez) The pair of female reproductive glands, located on each side of the uterus, which produce eggs and sex hormones; also called the female gonads.

ovulation (ohv you LAY shun) The process by which a mature egg is released from the ovary.

ovum (OH vum)   A mature egg cell.

oxytocin (ahk see TOK sin)   A pituitary hormone that stimulates the release of breast milk as well as uterine contractions.

pair bonds   The strong, lifelong attachments that naturally take place between two animals; it includes "love" when speaking of humans.

pap smear   A routine test for cervical cancer; a small sampling of cervical cells is lightly scraped from the cervix for examination.

paraphilias (pare uh FILL ee ahz)   Variations in sexual behavior that society deems unusual or atypical.

parasympathetic nervous system   One of two subdivisions of the autonomic nervous system associated primarily with the vasocongestive response in sexual arousal (among other effects on the body). See **sympathetic nervous system.**

pediculosis pubis (peh DIK you LOW sis PYOU bis)   Lice infesting pubic hair, usually acquired through sexual contact but sometimes through infected clothing or bedding; commonly called crabs.

pedophilia (PEH doe FILL ee ah)   Engaging in sexual activity with children as the preferred form of erotic gratification.

pelvic inflammatory disease   An inflammation of the fallopian tubes, the lining of the uterus, and/or the lower abdominal cavity as a result of various bacterial infections—gonorrhea or chlamydia, for example.

penis (PEE nis)   The male sexual organ of copulation that also expels urine from the body.

petting   Kissing, touching, genital stimulation, and other forms of physical contact between two people that does not include sexual intercourse; see **foreplay.**

Peyronie's disease (pay ROH neez)   A rare condition, caused by abnormal fibrous tissue in the penis, resulting in painful erections.

phallic stage (FAHL ik)   According to psychoanalytic theory, the psychosexual stage of development in which the child's sexual energies are focused on its genitals.

phallic symbols   Objects that symbolize the penis, such as sticks, knives, and cylindrical monuments.

phallus   The penis.

pheromones (FARE oh mohnz)   Chemical substances that send out messages, including sexual ones, through the sense of smell.

pituitary gland (pih TEW ih tare e)   A small endocrine gland at the base of the brain, that secretes several hormones important to sexual development and functioning—like FSH and LH.

placenta (pluh SEN tah)   The organ attached to the uterine wall through which the fetus receives nutrients and oxygen from the mother and through which waste products are expelled.

plateau phase   Following excitement, the phase of human sexual response in which sustained sexual tension precedes orgasm.

pleasure principle   According to psychoanalytic theory, the concept that libido is motivated to seek immediate gratification, regardless of realistic constraints; also see **reality principle.**

postpartum depression (post PAR tum)   The mild to moderate depression that women may experience after giving birth; also called "the baby blues."

postpartum period   The six-to-eight-week period following childbirth in which a woman is adjusting to many physiological and psychological changes.

premature ejaculation   A common male sexual dysfunction in which ejaculation uncontrollably occurs before or shortly after intercourse begins.

premenstrual syndrome (PMS)   A combination of physical and psychological symptoms (fatigue, depression, irritability) that some women experience prior to menstruation.

prepuce   The fold of skin over the clitoris (also called the clitoral hood) or the loose, movable skin that covers the penis (also called the foreskin).

priapism (PRY ah PIZ um)   Prolonged erection independent of sexual arousal.

primary sexual characteristics   The internal sex organs and the external genitalia.

primary sexual dysfunction   A sexual dysfunction which has always existed; for example, a man has never been able to have an erection or a woman has never experienced an orgasm; see **secondary sexual dysfunction.**

progesterone (Pro JES tur own)   Primarily a "female" hormone—produced by the corpus luteum—that maintains the uterine lining during pregnancy.

prolactin (pro LACK tin)   A pituitary hormone that stimulates milk production in lactating women.

prostaglandins (pro stah GLAN dinz)   Hormones, produced by the prostate glands and other tissues, that cause uterine contractions, especially during labor; sometimes used to induce abortion.

prostate gland (PROS tate)   A gland, located at the base of the bladder, that produces most of the seminal fluid in a male.

prostatitis (PRO state I tis)   An inflammation of the prostate gland that causes frequent and painful urination in men.

prostitution   Engaging in sexual activity for money.

pseudocyesis (sue doe CYE ah sis)   A false pregnancy, whereby the female experiences symptoms—morning sickness, tender breasts, abdominal sensations—but she is not pregnant.

pseudohermaphrodite (sue doe her MAF roe dite)   A person with the sex chromosomes and gonad of one sex but the external genitals of the other sex.

psychoanalytic theory   A psychological theory, originated by Sigmund Freud, in which unconscious processes and infantile sexuality receive a particularly strong emphasis.

puberty (PEW burr tee)   The stage of life between childhood and adulthood during which the reproductive system matures and secondary sexual characteristics develop.

pubococcygeus muscle (pew bow cock SEE gee us)   The muscle that surrounds the vaginal opening.

reality principle   According to psychoanalytic theory, the concept that the rational, conscious self operates in harmony with the real world; see also **pleasure principle.**

refractory period   In males only, the period of time immediately following an orgasm in which it is physiologically impossible to experience another orgasm.

resolution phase   The final phase of the human sexual response in which the body returns to its initially unaroused state—muscles relax, breathing slows down, and sex organs return to their usual color and size.

retarded ejaculation   A male sexual dysfunction in which ejaculation occurs (if at all) only after lengthy intercourse and effort; sometimes called inhibited male orgasm.

retrograde ejaculation   A condition in which semen is expelled into the urinary bladder during orgasm instead of out of the penis.

Rh incompatibility   A condition in which the mother's antibodies destroy the red blood cells of the fetus, leading to anemia, jaundice, and even death.

saline abortion   A method of abortion, used in the second trimester, in which a strong salt solution is injected into the uterus to induce labor.

scabies (SKAY bies)   A highly contagious infection, caused by tiny parasitic mites, that may be sexually transmitted.

scrotum (SKRO tum)   The loose sac of skin that lies behind the penis and contains the testes.

secondary sexual characteristics   The physical characteristics, other than genitals, that indicate sexual maturity in a female or male, such as breasts, pubic hair, facial hair, and a deepened voice.

secondary sexual dysfunction   A sexual dysfunction caused by a physiological and/or psychological problem that has not always existed; see **primary sexual dysfunction.**

semen (SEE men)   The milky white fluid, containing sperm, that is ejaculated from the penis during orgasm.

seminal vesicles (SEM in nul VES ih kulz)   Two saclike organs that contribute secretions to seminal fluid.

seminiferous tubules (sem ih NIF er us)   Thin coiled tubes, located in the testes, that produce and store billions of sperm.

sensate focus   A sex therapy exercise in which people touch each other with the goal of giving pleasure and not with the expectation of subsequent intercourse; the purpose is to reduce anxiety and thereby reduce sexual dysfunctions.

sex chromosomes A single pair of chromosomes that determine the sex and related characteristics of each individual.

sex flush A rashlike redness of the skin that appears on the chest and/or breasts during sexual arousal; more common in women.

sexism The institutionalized and/or culturally based discrimination against women; for example, sexist language or the exaggeration by the media of stereotypic masculine or feminine behaviors.

sex skin The pink or red coloration of the labia minora (minor lips) when a woman is sexually aroused to the point of orgasm.

sex therapy A variety of general procedures (such as couple-counseling) and specific strategies (such as sensate focus) used to treat sexual dysfunctions.

sexology A general term encompassing sex research, education, and therapy.

sexual dimorphism (die MORE fih zum) The differences between male and female anatomical and physiological differences and sexual behavior.

sexual dysfunction A disturbance with sexual desire (such as lack of interest) or sexual performance (such as inability to reach orgasm) for physical, psychological, interpersonal, and/or cultural reasons.

sexual identity The view of one's own sexual characteristics (degree of sexiness), sexual orientation (heterosexual, homosexual, bisexual), sex values, and gender identity.

sexual intercourse See **coitus.**

sexually transmitted disease A contagious infection (caused by a bacteria, virus, or protozoa) that is transmitted primarily through sexual contact; also called venereal disease.

smegma (SMEG muh) A yellowish substance that is secreted under the foreskin of the penis or the hood of the clitoris.

society A network of human relationships organized in such a way to bind a group of people together; also see **culture.**

Spanish fly An alleged aphrodisiac derived from the dried powder of beetles. See **aphrodisiac.**

spectatoring Anxiously watching or judging one's own sexual performance.

sperm The male reproductive cell which contains half of the chromosomes necessary for fertilizing an egg.

spermatogenesis (SPUR mat oh GEN neh sus) Sperm production.

spermicides (SPUR mih sides) Contraceptive chemicals that kill sperm.

spontaneous abortion An abortion due to medical problems with the fetus or mother; sometimes called a miscarriage; also see **abortion.**

squeeze technique A technique used to reduce the tendency for rapid ejaculation; it consists of squeezing the glans or the base of the penis.

sterilization A surgical procedure performed to make a person incapable of reproducing.

steroids (STARE oidz) A group of chemical substances, including the sex hormones—estrogen, progesterone, and testosterone.

sublimation (sub lih MAY shun) According to psychoanalytic theory, the process by which libido is channeled into socially acceptable activities, like art or athletics.

surrogate mother A woman who carries a baby for a couple because the wife is sterile; she contractually agrees to relinquish the child to them following birth.

surrogate partner A member of a sex therapy team who engages in sexual activity with a client as part of the therapy process.

swinging The two-couple practice of exchanging marital partners for sex, sometimes referred to as wife-swapping.

sympathetic nervous system One of two subdivisions of the autonomic nervous system associated primarily with the orgasmic response (among other effects on the body). See **parasympathetic nervous system.**

syphilis (SIH fil lis) A venereal disease caused by a microorganism; if left untreated, it progresses through three stages—primary, secondary, and tertiary—with serious consequences.

teratogen (teh RAT oh jin) A substance, such as a chemical or drug, that causes birth defects.

testes (TESS teze) The pair of male reproductive glands, located in the scrotum, which produce sperm and sex hormones; also called the male gonads.

testosterone (tess TOSS tur ohn)   Primarily a "male" hormone, secreted by the testes in the male and the adrenal cortex in both sexes.

toxemia   An abnormal condition that occurs in pregnant women; symptoms include high blood pressure, protein in the urine, and fluid retention.

toxic shock syndrome   An illness associated with bacteria present in highly absorbent menstrual tampons.

transsexualism   A condition in which a person feels persistently uncomfortable about his (or her) assigned sex, and desires to change his (or her) anatomy so as to live as a member of the opposite sex.

trichomoniasis (TRIH ko mon I ah sis)   A common vaginal infection, characterized by a smelly, yellowish discharge and vaginal itching; caused by a parasite.

tubal ligation (lie GAY shun)   A surgical procedure in which a woman is sterilized by cutting or tying her fallopian tubes.

Turner's syndrome   A sex chromosome disorder in which a female is missing a chromosome, resulting in an XO pattern; these women are infertile, may have webbing between the fingers and toes or neck and shoulders, and usually have organ defects.

ultrasonography   A procedure in which sound waves are converted into a photographic image of the fetus to detect possible fetal abnormalities or to forecast complications in delivery.

umbilical cord   The cord that connects the fetus to the placenta.

urethra (ur REE thrah)   A tube through which urine passes from the bladder to outside the body; in men, it is also the passageway for semen.

urethral meatus (you REE thral ME a tus)   The opening of the urethra to the outside.

urogenital folds   The fetal tissue that develops into the penis and urethra in the male and the labia minora in the female.

uterus (YOU ter us)   A hollow, muscular organ in which a woman nourishes a fetus until birth; the womb.

vacuum aspiration   The preferred method of abortion during the first trimester in which the cervix is dilated and the contents of the uterus are extracted through a suction tube.

vagina (vah JYE nah)   The female organ of copulation through which menstrual blood is passed and babies are born.

vaginal sponge   A circular, highly absorbent sponge, treated with spermicide, that is placed in front of the cervical opening to block sperm.

vaginismus (VAH jih NIS mus)   A female sexual disorder marked by involuntary spasms of the muscles surrounding the vagina; it makes intercourse painful if not impossible.

vaginitis (VAH jih NIT is)   An inflammation of the vagina caused by any of a number of vaginal infections; symptoms include itching, pain, discharge, and discomfort during intercourse.

vas deferens (vas DEH fur renz)   Ducts that carry sperm from the testes to the base of the urethra.

vasectomy (vah SEK tuh mee)   A surgical procedure in which a man is sterilized by cutting or tying the vas deferens.

vasocongestion (VAH so con JES chun)   In response to sexual arousal, the accumulation of blood in the vessels and tissues of various body parts, especially the genitals.

vasovasectomy (VAH so vah SEK tuh mee)   A surgical procedure in which the vas deferens are reattached after a vasectomy; see **vasectomy.**

venereal disease (vah NEAR ee ul)   See **sexually transmitted disease.**

voyerism (VOY yer ism)   Obtaining sexual gratification from spying on people while they are undressing, nude, or engaging in sexual activity.

vulva (VUL va)   A collective term for the external genitals of the female—the mons pubis, the labia majora and minora, the clitoris, and the urethral and vaginal openings.

Wolffian ducts (WOOL fee un)   The tissue in the male fetus that develops into the epididymus, vas deferens, and seminal vesicles.

womb   See **uterus.**

Yohimbine   An alleged aphrodisiac derived from the bark of the African Yohimbe tree. See **aphrodisiac.**

zygote (ZY goat)   A single cell created by the union of an egg and sperm.

References

Abel, G. (1981). The evaluation and treatment of sexual offenders and their victims. Paper presented at St. Vincent Hospital and Medical Center, Portland, OR, October 15. Cited in Crooks, R. & Baur, K. *Our sexuality*, 3d ed. Menlo Park, CA: Benjamin/Cummings.

Abplanalp, J. M., Haskett, R. F., & Rose, R. M. (1980). The Premenstrual Syndrome. *Advances in Psychoneuroendocrinology*, 3, 327–347.

Abplanalp, J. M., Rose, R. M., Donnelly, A. F., & Livingston-Vaughan, L. (1979). Psychoendocrinology of the menstrual cycle: II. The relationship between enjoyment of activities, moods, and reproductive hormones. *Psychosomatic Medicine*, 41, 605.

Ackman, C. F. D., MacIsaac, S. G., & Schual, R. (1979). Vasectomy: Benefits and risks. *International Journal of Gynecology and Obstetrics*, 16, 493–496.

Addiego, F., Belzer, E. G., Comolli, J., Moger, W., Perry, J. D., & Whipple, B. (1981). Female ejaculation: A case study. *The Journal of Sex Research*, 17(1), 13–21.

Altman, L. D. (1982, July 22). Measuring the benefits of the pill. *San Francisco Chronicle*, p. 23.

Altman, L. K. (1987, June 30). AIDS mystery: Why do some infected men stay healthy? *New York Times*.

Altman, L. K. (1986, November 20). Global program aims to combat AIDS "disaster." *New York Times*, p. 4.

Alzate, H., & Hoch, Z. (1986). The "G spot" and "female ejaculation": A current appraisal. *Journal of Sex and Marital Therapy*, Fall, 12(3), 211–220.

American Psychiatric Association. (1987). *Diagnostic and statistical manual (DSM III-R)*. Washington, DC: American Psychiatric Association.

Andersch, B., Wendestam, C., Hahn, L., Ohman, R., & Goteborgs, U. Premenstrual complaints: I. Prevalence of premenstrual symptoms in a Swedish urban population (1986). *Journal of Psychosomatic Obstetrics and Gynecology*, March, 5(1), 39–49.

Anderson, C. L. (1975). What are psychology departments doing about sex education? *Teaching of Psychology*, 2(1), 24–27.

Anderson, T. P., & Cole, T. M. (1975). Sexual counseling of the physically disabled. *Postgraduate Medicine*, 58, 117–123.

Annon, J. S. (1976). *The behavioral treatment of sexual problems: Brief therapy*. New York: Harper & Row.

Ansbacher, R. (1978). Artificial insemination with frozen spermatozoa. *Fertility and Sterility*, 29, 375–379.

Apfelbaum, B. (ed.). (1980). *Expanding the boundaries of sex therapy*. Berkeley, CA: Berkeley Sex Therapy Group.

Arentewicz, G., & Schmidt, G. (eds.). (1983). *The treatment of sexual disorders*. New York: Basic.

Arey, L. B. (1974). *Developmental anatomy*, 7th ed. Philadelphia: Saunders.

Arms, K., & Camp, P. S. (1987). *Biology*, 3d ed. Philadelphia: W. B. Saunders.

Arms, S. (1975). *Immaculate deception*. Boston: Houghton Mifflin.

Arsdalen, V. K. N., Mallow, T. R., & Wein, A. J. (1983). Erectile physiology, dysfunction and evaluation. *Monographs in Urology*, 136–156.

Avery-Clark, C. (1986a). Sexual dysfunction and disorder patterns of husbands of working and nonworking women. *Journal of Sex-Marital Therapy*, Winter, 12(4), 282–296.

Avery-Clark, C. (1986b). Sexual dysfunction and disorder patterns of working and nonworking wives. *Journal of Sex-Marital Therapy*, Summer, 12(2), 93–107.

Baird, D., & Wilcox, A. (1985). Cigarette smoking associated with delayed conception. *Journal of the American Medical Association*, 253, 2979–2983.

Bancroft, J. (1986a). Reproductive hormones and sexual function. Abstracts: Conference on the scientific basis of sexual dysfunction. Bethesda, MD: National Institutes of Health.

Bancroft, J. (1986b). The role of hormones in female sexuality. *Proceedings of the 8th International Congress of Psy-

chosomatic Obstetrics and Gynecology*. Amsterdam: Excerpita Medica.

Bancroft, J. (1983). *Human sexuality and its problems*. Edinburgh: Churchill-Livingstone.

Bancroft, J. (1980). Endocrinology of sexual function. *Clinics in Obstetrics and Gynecology*, 7(2), 253–281.

Bancroft, J. (1974). *Deviant sexual behavior*. Oxford: Clarendon Press.

Bancroft, J., & Wu, F. C. W. (1983). Changes in erectile responsiveness during androgen therapy. *Archives of Sexual Behavior*, 12, 59–66.

Bandura, A. (1986). *Social foundations of thought and action—A social cognitive theory*. Englewood Cliffs, NJ: Prentice-Hall.

Barbach, L. G. (1975). *For yourself: The fulfillment of female sexuality*. New York: Doubleday.

Barclay, D. (1987). Benign disorders of the vulva and vagina. In M. Pernoll & R. Benson (eds.), *Current obstetric and gynecologic diagnosis and treatment 1987*. Los Altos, CA: Appleton & Lange.

Beach, F. A. (1976). Cross-species comparisons and the human heritage. *Archives of Sexual Behavior* 5(5), 469–485.

Beach, F. A. (1971). Hormonal factors controlling the differentiation, development, and display of copulatory behavior in the ramstergig and related species. In E. Tobach, L. R. Aronson, & E. Shaw (eds.), *The biopsychology of development*, pp. 249–295. New York: Academic.

Beach, F. A. (1947). A review of physiological and psychological studies of sexual behavior in mammals. *Physiological Review*, 27(2), 15.

Behrman, R. E., & Vaughan, V. C. (1983). *Nelson textbook of pediatrics*, 12th ed. Philadelphia: W. B. Saunders.

Belzer, E. J., Jr. (1981). Orgasmic expulsions of women: A review and heuristic inquiry. *The Journal of Sex Research*. 17, 1–12.

Benson, G. S., McConnell, J. A., & Schmidt, W. A. (1981). Penile polsters: Functional structures or atherosclerotic changes? *Journal of Urology,* *125*(6), 800–803.

Bentler, P. M., & Peeler, W. H. (1979). "Models of female orgasm." *Archives of Sexual Behavior, 8,* 405–423.

Benton, D., & Wastell, V. (1986). Effects of androstenol on human sexual arousal. *Biology of Psychology,* April, *22*(2), 141–147.

Berkeley, S. F., Hightower, A. W., Broome, C. V., & Reingold, A. (1987). The relationship of tampon characteristics to menstrual toxic shock syndrome. *Journal of American Medical Association,* August 21, *258*(77), 917–920.

Bermant, G., & Davidson, J. M. (1974). *Biological bases of sexual behavior.* New York: Harper & Row.

Bernstein, R. (1981). The Y chromosome and primary sexual differentiation. *Journal of the American Medical Association 245*(19): 1953–1956.

Billings, E. L., & Billings, J. J. (1974). *Atlas of the ovulation method.* Collegeville, MN: The Liturgical Press.

Billstein, S. Human lice. (1984). In K. Holmes, P. Mardh, P. Sparling, & P. Wiesner (eds.), *Sexually transmitted diseases.* New York: McGraw-Hill.

Bleier, R. (1984). *Science and gender.* New York: Pergamon Press.

Boffey, P. M. (1988, March 6). Masters and Johnson says AIDS spread is rampant. *New York Times,* p. 14.

Bohlen, J. G. (1981). Sleep erection monitoring in the evaluation of male erectile failure. *Urological Clinics of North America, 8*(1), 119–134.

Bohlen, J. G., Held, J. P., & Sanderson, M. O. (1980). The male orgasm: Pelvic contractions, measured by anal probe. *Archives of Sexual Behavior, 9,* 403–521.

Bohlen, J. G., Held J., Sanderson, M., & Ahlgren, A. (1982). The female orgasm: Pelvic contractions. *Archives of Sexual Behavior, 2*(5).

Bonsall, R. W., & Michael, R. P. (1978). Volatile odoriferous acids in vaginal fluid. In E. S. E. Hafez & T. N. Evans (eds.), *The human vagina,* pp. 167–177, New York: Elsevier.

Bookstein, J., Valji, K., Parsons, L., & Kessler, W. (1987). Penile pharmacocavernosography and cavernosometry in the evaluation of impotence. *Journal of Urology,* April, *137*(4), 772–776.

Bors, E., & Comarr, A. E. (1960). Neurological disturbances of sexual function with special reference to 529 patients with spinal cord injury. *Urological Survey, 10,* 191–222.

Bowen, S. (1988). *Sexually transmitted diseases and society.* Stanford, CA: Stanford University Press.

Boyle, P. S. (1986). Sexuality and disability. *SIECUS Report, 14,* 1–3.

Bradford, J. M. (1986). The use of a bioimpedance analyzer in the measurement of sexual arousal in male sexual deviants. *Canadian Journal of Psychiatry,* February, *31*(1), 44–47.

Bradley, W. E. (1987). New techniques in evaluation of impotence. *Urology,* April, *29*(4), 383–388.

Brandt, A. M. (1987). *No magic bullet.* New York: Oxford University Press.

Brecher, E. M. (1969). *The sex researchers.* Boston: Little, Brown.

Brobeck, J. R. (ed.). (1979). *Best and Taylor's physiological basis of medical practice,* 10th ed. Baltimore: Williams and Wilkins.

Brody, J. E. (1988, March 10). Personal health. *New York Times,* p. 16.

Brozan, N. (1987). "How effective is breast self-examination?" *New York Times,* May 4.

Brozan, N. (1985, March 9). Fetal health: New early diagnosis studied. *New York Times.*

Budoff, P. (1980). *No more menstrual cramps and other good news.* New York: Putnam.

Caird, W., & Wincze, J. P. (1977). *Sex therapy: A behavioral approach.* Hagerstown, MD.: Harper & Row.

Calhoun, L. et al. (1981). The influence of pregnancy on sexuality: A review of current evidence. *Journal of Sex Research, 17*(2), 139–151.

Camp, S. L., & Speidel, J. (1987). *The international human suffering index* (pamphlet). Washington, DC: Population Crisis Committee.

Campbell, N. A. (1987). *Biology.* Menlo Park, CA: Benjamin/Cummings.

Capraro, V., Ridgers, D., & Rodgers, B. (1983). Abnormal vaginal discharge. *Medical Aspects of Human Sexuality. 17*(8).

Carmichael, M. S., Humbert, R., Dixen, J., Palmisano, G., Greenleaf, W., & Davidson, J. M. (1987). Oxytocin increase in human sexual response. *Journal of Clinical Endocrinology and Metabolism.*

Cates, W. Jr., & Holmes, K. (1986). Sexually transmitted diseases. In W. Cates, Jr. & K. Holmes (eds.), *Public health and preventive medicine,* 12th ed., pp. 257–295. Atlanta: U.S. Public Health Service.

Chambless, D., et al. (1982). The pubococcygeus and female orgasm: A correlational study with normal subjects. *Archives of Sexual Behavior, 11*(6).

Chang, J. (1977). *The Tao of love and sex.* New York: Dutton.

Chayen, B., Tejani, N., Verman, U. L., & Gordon, G. (1986). Fetal heart rate changes and uterine activity during coitus. *Acta-Obstet-Gynecol-Scand., 65*(8), 853–855.

Christenson, C. V. (1971). *Kinsey: A biography.* Bloomington: Indiana University Press.

Clarkson, T. B., & Alexander, N. J. (1980). Long-term vasectomy: Effects on the occurrence and extent of atherosclerosis in rhesus monkeys. *Journal of Clinical Investigation, 65*(1), 15–25.

Cohen, H. D., Rosen, R. C. & Goldstein, L. (1976). Electroencephalographic laterality changes during human sexual orgasm. *Archives of Sexual Behavior, 5*(3), 189–199.

Cole, T. M. (1975). Sexuality and physical disabilities. *Archives of Sexual Behavior, 4,* 389–401.

Comfort, A. (1978). *Sexual consequences of disability.* Philadelphia: George F. Shickley.

Condra, M., Fenemore, J., Reid, K., Phillips, P., Morales, A., Owen, J., & Surridge, D. (1987). Screening assessment of penile tumescence and rigidity. Clinical test of Snap-Gauge. *Urology,* March, *29*(3), 254–257.

Connell, E. B. (1979). Barrier methods of contraception: A reappraisal. *International Journal of Gynecology and Obstetrics, 16,* 479–481.

Cooper, J. F. (1928). *Technique of contraception.* New York: Day-Nichols.

Cooper, P. T., Cumber, B., & Hartner, R. (1978). Decision-making patterns and post-decision adjustment of child-free husbands and wives. *Alternative Lifestyles, 1*(1), 71–94.

Corey, L. (1984). Genital herpes. In K. Holmes, P. Mardh, P. F. Sparling, & P. Wiesner (eds.), *Sexually transmitted diseases.* New York: McGraw-Hill.

Crapo, L. (1985). *Hormones.* New York: W. H. Freeman.

Cutler, W. B., Preti, G., Huggins, G. R., Erikson, E., & Garcia, C. R. (1985). Sexual behavior frequency and biphasic ovulatory type menstrual cycles. *Physiological Review, 34,* 805–810.

Dalton, K. (1979). *Once a month.* New York: Hunter.

Dalton, K. (1972). *The premenstrual syndrome.* London: William Heineman Medical Books.

Dalton, K. (1969). *The menstrual cycle.* New York: Pantheon.

Daly, M., & Wilson, M. (1978). *Sex evolution and behavior.* Belmont, CA: Wadsworth.

Darling, C. A., & Davidson, J. K. Sr., (1986). Enhancing relationships: Understanding the feminine mystique of pretending orgasm. *Journal of Sex and Marital Therapy,* Fall, *12*(3), 182–196.

Davenport, W. H. (1977). Sex in cross-cultural perspective. In F. A. Beach (ed.), *Human sexuality in four perspectives*, pp. 115–163. Baltimore, MD: Johns Hopkins University Press.

Davidson, J. M. (1980). The psychology of sexual experience. In J. M. Davidson & R. J. Davidson (eds.), *The psychobiology of consciousness*. New York: Plenum.

Davidson, J. M., Kwan, M., & Greenleaf, W. J. (1982). Hormonal replacement and sexuality in men. *Clinics in Endocrinology and Metabolism*, 11(3), 599–623.

Davidson, J. M., et al., (1979). Effects of androgen on sexual behavior in hypogonadal men. *Journal of Clinical Endocrinology and Metabolism*, 48(6).

Dawson, D. A., Meny, D. J., & Ridley, J. C. (1980). Fertility control in the United States before the contraceptive revolution. *Family Planning Perspectives*, 12(2), 76–86.

DeAmicis, L., Goldberg, D., LoPiccolo, J., Friedman, J., & Davies, L. (1985). Clinical follow-up of couples treated for sexual dysfunction. *Archives of Sexual Behavior*, 14, 467–489.

Debrovner, C. H. (1983). Premenstrual syndrome. *Medical Aspects of Human Sexuality*, 215–216.

DeGroat, W. C. (1986). Organization of the reflex pathways mediating penile erection. In *Abstracts: Conference on the scientific basis of sexual dysfunction*. Bethesda, MD: National Institutes of Health.

DeGroat, W. C., & Booth, A. M. (1980). Physiology of male sexual function. *Annals of Internal Medicine*, 92(2), 329–331.

DeJong, F. H., & Sharpe, R. M. (1976). Evidence for inhibin-like activity in bovine follicular fluid. *Nature*, 263, 71–72.

Dennerstein, L., et al. (1980). Hormones and sexuality: Effect of estrogen and progestogen. *Obstetrics and Gynecology*, 56(3), 316–322.

DesJarlais, D. C. (1987, June 4). Quoted in *New York Times*, p. 1

DeWald, P. A. (1971). *Psychotherapy: A dynamic approach*, 2d ed. New York: Basic.

Dhabuwala, C. B., Kumar, A., & Pierce, J. M. (1986). Myocardial infarction and its influences in male sexual function. *Archives of Sexual Behavior*, 15(6), 499–505.

Diamond, M. (1965). A critical evaluation of the ontogeny of human sexual behavior. *Quarterly Review of Biology*, 40, 147–175.

Dickinson, R. L. (1949). *Atlas of human sex anatomy*, 2d ed. Baltimore, MD: Williams & Wilkins.

Dick-Read, G. (1932). *Childbirth without fear*. New York: Harper & Row.

Djerassi, C. (1981). *The politics of contraception*. New York: Norton.

Doty, R. L. et al. (1975). Changes in the intensity and pleasantness of human vaginal odors during the menstrual cycle. *Science*, 190, 1316–1318.

Dowdle, W. R. (1987, October 16). Quoted in *New York Times*, p. 11.

Draper, N. E. E. (1976). Birth control. In *Encyclopaedia Britannica*, 2, 1065–1073. Chicago: Benton.

Droegemueller, W., & Bressler, R. (1980). Effectiveness and risks of contraception. *Annual Review of Medicine*, 31, 329–343.

Dunn, H. G. et al. (1977). Maternal cigarette smoking during pregnancy and the child's subsequent development: II. Neurological and intellectual maturation to the age of 6½ years. *Canadian Journal of Public Health*, 68, 43–50.

Durfee, R. B. (1987). Obstetric complications of pregnancy. In M. Pernoll, & R. Benson (eds.), *Current obstetric and gynecologic diagnosis and treatment 1987*, 6th ed. Los Altos, CA: Appleton & Lange.

Easterling, W. E., & Herbert, W. N. P. (1982). The puerperium. In D. N. Danforth (ed.), *Obstetrics and gynecology*, 4th ed. pp. 787–799. Philadelphia: Harper & Row.

Eckholm, E. (1986, October 28). Heterosexuals and AIDS: The concern is growing. *New York Times*, p. 25.

Ehrhardt, A. A., & Meyer-Bahlburg, H. (1981). Effects of prenatal sex hormones on gender-related behavior. *Science*, 211(4488), 1312–1318.

Ehrlich, P., & Ehrlich, A. (1968). *Population resources, environment*, 2d ed. San Francisco: Freeman.

Ellis, H. (1942). *Studies in the psychology of sex*. New York: Random House.

Ende, J., Rockwell, S., & Glasgow, M. (1984). The sexual history in general medicine practice. *Archives of Internal Medicine*, 144, 558–561.

Eschenbach, D. (1986). Pelvic infections. In D. Danforth & J. Scott (eds.), *Obstetrics and gynecology*, 5th ed. Philadelphia: J. B. Lippincott.

Essex, M. (1985, August 12). Cited in *Time*, p. 44.

Evans, J. R. et al. (1976). Teenagers: Fertility control behavior and attitudes before and after abortion, childbearing or negative pregnancy test. *Family Planning Perspectives*, 8, 192–200.

Fagan, P., Meyer, J., & Schmidt, C., Jr. (1986). Sexual dysfunction within an adult developmental perspective.

Journal of Sexual and Marital Therapy, Winter, 12(4), 243–257.

Fausto-Sterling, A. (1985). *Myths of gender*. New York: Basic Books.

Fawcett, J. T. (1970). *Psychology and population: Behavioral research issues in fertility and family planning*. New York: Population Council.

Fay, A. (1977). "Sexual problems related to poor communication." *Medical Aspects of Human Sexuality*, 3, 48–62.

Federman, D. D. (1968). *Abnormal sexual development*. Philadelphia: W. B. Saunders.

Fisher, C. et al. (1983). Patterns of female sexual arousal during sleep and waking: Vaginal thermo-conductance studies. *Archives of Sexual Behavior*, 12(2).

Fisher, S. (1973). *The female orgasm*. New York: Basic Books.

Fishman, I. J. (1987). Complicated implantation of inflatable penile prostheses. *Urology Clinic of North America*, February, 14(1), 217–239.

Ford, C. S., & Beach, F. A. (1951). *Patterns of sexual behavior*. New York: Harper & Row.

Forrest, J., & Henshaw, S. (1983). Contraception in America. *Family Planning Perspectives*, 15, 154–156.

Fox, C. A., & Fox, B. (1969). Blood pressure and respiratory patterns during human coitus. *Journal of Reproduction and Fertility*, 19(3), 405–415.

Frank, E., Anderson, C., & Rubinstein, D. (1978). Frequency of sexual dysfunction in "normal" couples. *New England Journal of Medicine*, 299, 111–115.

Freeman, D. (1983). *Margaret Mead and Samoa*. Cambridge, MA: Harvard University Press.

Freeman, E. D. (1978). Abortion: Subjective attitudes and feelings. *Family Planning Perspectives*, 10, 150–155.

Freeman, E. W. (1985). PMS treatment approaches and progesterone therapy. *Psychosomatics*, October, 811–815.

Freud, S. (1957–1964). *The standard edition of the complete psychological works of Sigmund Freud*. J. Strachey, ed. London: Hogarth Press and Institute of Psychoanalysis.

Friedland, G. H., & Klein, R. S. (1987). Transmission of human immunodeficiency virus. *New England Journal of Medicine*, 317(18), 1125–1135.

Frisch, R. E. (1974). Critical weight at menarche, initiation of the adolescent growth spurt, and control of puberty. In M. M. Grumbach et al. (eds.), *Control of the onset of puberty*. New York: Wiley.

Gagnon, J. H. (ed.). (1977). *Human sex-*

uality in today's world. Boston: Little, Brown.

Gagnon, J., & Simon, W. (1973). *Sexual conduct: The social sources of human sexuality.* Chicago: Aldine.

Gal, A., Meyer, P., & Taylor, C. (1987). Papillomavirus antigens in anorectal condyloma and carcinoma in homosexual men. *Journal of the American Medical Association,* January 16, 257(3), 337–340.

Gasser, T., Larsen, E., & Bruskewitz, R. (1987). Penile prosthesis reimplantation. *Journal of Urology,* January, 137(1), 47–47.

Gessa, G. L., & Tagliamonte, A. (1974). Role of brain monoamines in male sexual behavior. *Life Science, 14*(3), 425–436.

Ghadirian, A. M., & Kamaraju, L. S. (1987). *Premenstrual mood changes in affective disorders.*

Gilder, G. F. (1973). *Sexual suicide.* New York: Quadrangle.

Gilman, A. G., Goodman, L. S., Rall, T. W., & Murad, F. (eds.). (1985). *Goodman and Gilman's pharmacological basis of therapeutics,* 7th ed. New York: Macmillan.

Giuliano, A. (1987). The breast. In M. Pernoll & R. Benson (eds.), *Current obstetric and gynecologic diagnosis and treatment 1987.* Los Altos, CA: Appleton & Lange.

Gladue, B., Green, R., & Hellman, R. (1984). Neuroendocrine response to estrogen and sexual orientation. *Science, 225,* 1496–1499.

Goedert, J., Biggar, R., Melbye, M., Mann, D., Wilson, S., Gail, M., Grossman, R., Digioia, R., Sanchez, W., Weiss, S. et al. (1987). Effect of T4 count and cofactors on the incidence of AIDS in homosexual men infected with human immunodeficiency virus. *Journal of the American Medical Association,* January 16, 257(3), 331–334.

Gold, A. R., & Adams, D. B. (1978). Measuring the cycles of female sexuality. *Contemporary Obstetrics and Gynecology, 12,* 147–156.

Goldberg, D., Whipple, B., Fishkin, R., Waxman, H., Fink, P., & Weisberg, M. (1983). The Grafenberg spot and female ejaculation: A review of initial hypotheses. *Journal of Sex and Marital Therapy, 9,* 27–37.

Goldfoot, D. A. et al. (1976). Lack of effect of vaginal lavages and aliphatic acids on ejaculatory responses in rhesus monkeys: Behavioral and chemical analyses. *Hormones and Behavior, 7,* 1–27.

Goldstein, B. (1976). *Human sexuality.* New York: McGraw-Hill.

Goldstein, I. (1987, April 14). Quoted in the *Wall Street Journal,* p. 1.

Goldstein, I. (1986). Impact of drugs on penile smooth muscle. *Abstracts: Conference on the scientific basis of sexual dysfunction.* Bethesda, MD: National Institutes of Health.

Goldstein, S., Halbreich, U., Endicott, J., & Hill, E. (1983). Premenstrual hostility, impulsivity and impaired social functioning. *Journal of Psychosomatic Obstetrics and Gynecology,* March, 5(1), 33–38.

Gonda, A. G., & Ruark, J. E. (1984). *Dying dignified.* Menlo Park, CA: Addison-Wesley.

Gong, V. (1987). Signs and symptoms of AIDS. In V. Gong & N. Rudwick (eds.), *AIDS.* New Brunswick, NJ: Rutgers University Press.

Gordon, J. W., & Ruddle, F. H. (1981). Mammalian gonadal determination and gametogenesis. *Science, 211,* 1265.

Gorski, R., Gordon, J., Shryne, J., & Southam, A. (1978). Evidence for a morphological sex difference within the medial preoptic area of the rat brain. *Brain Research, 148,* 333–346.

Gottlieb, M. S., Schroff, R., Schanker, H. M. et al. (1981). Pneumocystis carinii pneumonia and mucosal candidiasis in previously healthy homosexual men: Evidence of a new acquired cellular immunodeficiency. *New England Journal of Medicine, 305,* 1425–1431.

Goy, R. W., & McEwen, B. S. (1980). *Sexual differentiation in the brain.* Cambridge, MA: MIT Press.

Grafenberg, E. (1950). The role of urethra in female orgasm. *International Journal of Sexology, 3*(3).

Gray, D. S., & Gorzalka, B. B. (1980). Adrenal steroid interactions in female sexual behavior: A review. *Psychoneuroendocrinology, 5*(2), 157–175.

Greenspan, F. S., & Forsham, P. H. (1986). *Basic and clinical endocrinology,* 2d ed. Los Altos, CA: Lange Medical Publications.

Gregersen, E. (1983). *Sexual practices.* New York: Watts.

Gregory, J., & Purcell, M. (1987). Scott's inflatable penile prosthesis: Evaluation of mechanical survival in the series 700 model. *Journal of Urology,* April, 137(4), 676–677.

Groopman, J. E. (1988). The acquired immunodeficiency syndrome. In J. B. Wyngaarden & L. H. Smith (eds.), *Cecil textbook of medicine,* pp. 1799–1805. Philadelphia: W. B. Saunders.

Grumbach, M. M. (1980). The neuroendocrinology of puberty. *Hospital Practice, 15*(3), 51–60.

Grumbach, M. M., & Conte, F. A. (1985). Disorders of sexual differentiation. In J. D. Wilson & D. W. Foster (eds.), *Williams textbook of endocrinology,* 7th ed., pp. 313–401. Philadelphia: W. B. Saunders.

Grumbach, M. M. et al. (1974). Hypothalamic pituitary regulation of puberty: Evidence and concepts derived from clinical research. In M. M. Grumbach, G. D. Grave, & F. E. Mayer (eds.), *Control of the onset of puberty,* Chapter 6. New York: Wiley.

Guerrero, R. (1975). "Type and time of insemination within the menstrual cycle and the human sex ratio." *Studies in Family Planning, 6*(10), 367–371.

Guyton, A. C. (1986). *Textbook of medical physiology,* 7th ed. Philadelphia: W. B. Saunders.

Hafez, E. S. E. (1980). *Human reproduction,* 2d ed. New York: Harper & Row.

Hagen, I. M., & Bech, R. K. (1980). The diaphragm: Its effective use among college women. *Journal of the American College Health Association, 28*(5), 263–266.

Hällstrom, T. (1973). *Mental disorder and sexuality in the climacteric.* Stockholm: Scandinavian University Books.

Hamerton, J. L. (1988). Chromosomes and their disorders. In J. B Wyngaarden & L. H. Smith (eds.), *Cecil textbook of medicine,* pp. 161–171. Philadelphia: W. B. Saunders.

Hansfield, H. H. (1984). Gonorrhea and uncomplicated gonococcal infection. In K. Holmes, P. Mardh, P. F. Sparling & P. Wiesner (eds.), *Sexually transmitted diseases.* New York: McGraw-Hill.

Harbison, R. D., & Mantilla-Plata, B. (1972). Prenatal toxicity, maternal distribution and placental transfer of tetrahydrocannabinol. *Journal of Pharmacology and Experimental Therapeutics, 180,* 446–453.

Harlow, H. F. (1958). The nature of love. *American Psychologist, 13,* 673.

Harrison, W. M., Rabkin, J. G., Ehrhardt, A. A., Stewart, J. W., McGrath, P. J., Ross, D., & Quitkin, F. M. (1986). Effects of antidepressant medication on sexual function: A controlled study. *Journal of Clinical Psychopharmacology,* June, 6(3), 144–149.

Hart, B. L., & Leedy, M. G. (1985). Neurological bases of male sexual behavior. In N. Adler, D. Pfaff, & R. Gay (eds.), *Handbook of behavioral neurobiology,* Vol. 7: *Reproduction.* New York: Plenum Press.

Harvey, S. M. (1987). Female sexual behavior: Fluctuations during the menstrual cycle. *Journal of Psychosomatic Research, 31*(1), 101–110.

Harvey, S. M. (1980). Trends in contraceptive use at one university:

1974–1978. *Family Planning Perspectives, 12*(6), 301–304.

Haseltine, F. P., & Ohno, S. (1981). Mechanisms of gonadal differentiation. *Science, 211,* 1272.

Hatcher, R. A. (1982). *It's your choice.* New York: Irvington.

Hatcher, R A., Guest, F. A., Stewart, F. H., Stewart, G. K., Trussel, J., Bowen, S. C., & Cates, W. (1988). *Contraceptive technology 1988–1989: AIDS, a special section.* New York: Irvington.

Hausfater, G., & Skoblick, B. (1985, January 4). Premenstrual problems may best baboons. Reported by E. Eckholm in *New York Times,* p. 19.

Hayes, R. W. (1975). Female genital mutilation, fertility control, and the patrilineage in modern Sudan: A functional analysis. *American Ethnologist, 2,* 617–633.

Haynes, D. M. (1982). Course and conduct of normal pregnancy. In D. N. Danforth (ed.), *Obstetrics and gynecology,* 4th ed. Philadelphia: Harper & Row.

Hearst, N., & Hulley, S. B. (1988). Preventing the heterosexual spread of AIDS. *Journal of the American Medical Association,* April 22, *259*(16), 2428–2432.

Heath, R. G. (1972). Pleasure and brain activity in males. *Journal of Nervous and Mental Disease, 154,* 3–18.

Heim, N. (1981). Sexual behavior of castrated sex offenders. *Archives of Sexual Behavior, 10,* 11–19.

Heim, N., & Hursch, C. J. (1979). Castration for sex offenders: Treatment or punishment? A review and critique of recent European literature. *Archives of Sexual Behavior, 8,* 281–305.

Heiman, M. (1968). "Discussion of Sherfey's paper on female sexuality." *Journal of the American Psychoanalytic Association, 16,* 406–416.

Henley, N. M., & Freeman, J. (1976). The sexual politics of interpersonal behavior. In S. Cox (ed.), *Female psychology: The emerging self.* Chicago: Science Research Associates.

Henslin, J. M., & Sagarin, E. (eds.). (1978). *The sociology of sex.* New York: Schocken.

Herbst, A. L. (1981). Clear cell adenocarcinoma and the current status of DES-exposed females. *Cancer, 48* Suppl (2), 484–488.

Hiernaux, J. (1968). Ethnic differences in growth and development. *Eugenics Quarterly, 15,* 12–21.

Higgins, G. E. (1979). Sexual response in spinal cord injuries: A review. *Archives of Sexual Behavior, 8,* 173–196.

Hill, E. (1987). Premalignant and malignant disorders of the uterine cervix. In M. Pernoll & R. Benson (eds.), *Cur-*

rent obstetric and gynecologic diagnosis and treatment 1987, 6th ed. Los Altos, CA: Appleton & Lange.

Himes, N. (1970). *Medical history of contraception.* New York: Gambut.

Hite, S. (1976). *The Hite report.* New York: Macmillan.

Hoch, Z. (1986). Vaginal erotic sensitivity by sexological examination. *ActaObstet-Gynecol-Scand, 65*(7), 767–773.

Hoffman, C. H. (1981). Sexually transmitted diseases. *Journal of the American Medical Association, 246*(15), 1709.

Hollingshead, W. H., & C. Rosse (1985). *Textbook of anatomy,* 4th ed. New York: Harper & Row.

Holmes, K., March, P. A., Sparling, P. F., & Weisner, P. J. (1984). *Sexually transmitted diseases.* New York: McGraw-Hill.

Holt, L. H., & Weber, M. (1982). *Woman care.* New York: Random House.

Hong, L. K. (1984). Survival of the fastest. *Journal of Sex Research, 20,* 109–122.

Hopson, J. S. (1979). *Scent signals: The silent language of sex.* New York: Morrow.

Houseknecht, S. (1978). Voluntary childlessness. *Alternative Lifestyles, 1*(3), 379–402.

Howard, J., Blumstein, P., & Schwartz, P. (1986). Sex, power, and influence tactics in intimate relationships. *Journal of Personality and Social Psychology,* July, *51*(1), 102–109.

Hrdy, S. B. (1981). *The woman that never evolved.* Boston: Harvard University Press.

Huelsman, B. R. (1976). An anthropological view of clitoral and other female genital mutilations. In T. P. Lowry and T. S. Lowry (eds.), *The clitoris.* St. Louis, MO: Warren H. Green.

Hull, C. L. (1943). *Principles of behavior.* New York: Appleton-Century-Crofts.

Hunt, M. (1974). *Sexual behavior in the 1970's.* Chicago: Playboy Press.

Hurtig, A., & Rosenthal, I. (1987). Psychological findings in early treated cases of female pseudohermaphroditism caused by virilizing congenital adrenal hyperplasia. *Archives of Sexual Behavior, 16*(3), 209–223.

Hussey, H. H. (1981). Vasectomy—A note of concern: Reprise editorial. *Journal of the American Medical Association, 245*(22), 2333.

Hyppa, M. T., Falck, S. C., & Rinne, V. K. (1975). Is L-dopa an aphrodisiac in patients with Parkinson's disease? In M. Sandler and G. L. Gessa (eds.), *Sexual behavior: Pharmacology and biochemistry,* New York: Raven.

Imperato-McGinley, J. (1985). Disorders of sexual differentiation. In J. B.

Wyngaarden & L. H. Smith, Jr. (eds.), *Cecil textbook of medicine,* 17th ed. Philadelphia: W. B. Saunders.

Imperato-McGinley, J., Guerrero, L., Gautier, T., & Peterson, R. (1974). Steroid 5-alphareductase deficiency in man: An inherited form of male pseudohermaphroditism. *Science, 186,* 1213–1215.

Jacobs, L. (1986). Chief complaint: Sexual inadequacy. *Medical Aspects of Human Sexuality,* May, *20*(5), 44–50.

Jacobs, P. A., Brunton, M., Melville, M. M., Britain, R. P., & McClemont, W. F. (1965). Aggressive behavior, mental subnormality, and the XYY male. *Nature, 208,* 1351–1352.

James, W. H. (1971). The distribution of coitus within the human intermenstruum. *Journal of Biosocial Science, 3,* 159–171.

Jehu, D. (1979). *Sexual dysfunction.* New York: Wiley.

Jick, H. et al. (1981). Vaginal spermicides and congenital disorders. *Journal of the American Medical Association, 245*(13), 1329–1332.

Johns, D. R. (1986). Benign sexual headache within a family. *Archives of Neurology,* November, *43,* 1158–1159.

Johnson, S., Jr., & Joseph, S. C. (1987, December 20). Pro and con: Free needles for addicts to help curb AIDS? Interview in the *New York Times.*

Johnson, W. (1985, January 16). Landers' survey does not surprise experts. *Times Tribune,* p. B-1.

Jolly, A. (1972). *The evolution of primate behavior.* New York: Macmillan.

Jost, A. (1953). Problems of fetal endocrinology: The gonadal and hypophyseal hormones. *Recent Progress in Hormone Research, 8,* 379–418.

Judd, H. L. (1987). Menopause and postmenopause. In M. L. Pernoll & R. C. Benson (eds.), *Current obstetric and gynecologic diagnosis and treatment.* Norwalk, CT: Appleton & Lange.

Judson, F. (1983). What practical advice can physicians give patients on avoiding genital herpes. *Medical Aspects of Human Sexuality, 17*(8).

Kaiser, I. (1986). Fertilization and the physiology and development of fetus and placenta. In D. Danforth & J. Scott (eds.), *Obstetrics and gynecology,* 5th ed. Philadelphia: J. B. Lippincott.

Kando, T. M. (1978). *Sexual behavior and family life in transition.* New York: Elsevier/North Holland.

Kaplan, H. S (1987, November 5). Cited in Brody, J., Changing attitudes on masturbation. *New York Times.*

Kaplan, H. S. (1979). *Disorders of sexual desire.* New York: Brunner/Mazel.

Kaplan, H. S. (1974). *The new sex therapy.* New York: Brunner/Mazel.

Kaplan, J., Spira, T., Fishbein, D., Pinsky, P., & Schonberger, L. (1987). Lymphadenopathy syndrome in homosexual men. Evidence for continuing risk of developing the acquired immunodeficiency syndrome. *Journal of the American Medical Association,* January 16, 257(3), 335–337.

Katchadourian, H. (1981). Sex education in college: The Stanford experience. In L. Brown (ed.), *Sex education in the eighties.* New York: Plenum.

Kaufman, A. et al. (1981). Recent developments in family planning in China. *Journal of Family Practice, 12(3),* 581–582.

Kegel, A. (1952). Sexual functioning of the pubococcygeus muscle. *Western Journal of Surgery, Obstetrics and Gynecology, 60,* 521–524.

Kessel, R. G., & Kardon, R. H (1979). *Tissues and organs.* San Francisco: W. H. Freeman.

Keusch, G. T. (1984). Enteric bacterial pathogens: Shigella, campylobacter, salmonella. In K. Holmes, P. Mardh, P. F. Sparling, & P. Wiesner (eds.), *Sexually transmitted diseases.* New York: McGraw-Hill.

Kiely, E., Williams, G., & Goldie, L. (1987). Assessment of the immediate and long-term effects of pharmacologically induced penile erections in the treatment of psychogenic and organic impotence. *British Journal of Urology,* February, 59(2), 164–169.

Kinch, R. A. H. (1979). Help for patients with premenstrual tension. *Consultant.* April, 187–191.

Kingsley, L., Detels, R., Kaslow, R., Polk, B., Rinaldo, C., Jr., Chmiel, J., Detre, K., Kelsey, S., Odaka, N., Ostrow, D. et al. (1987). Risk factors for seroconversion to human immunodeficiency virus among male homosexuals. Results from the Multicenter AIDS cohort study. *Lancet,* February 14, 1(8529), 345–349.

Kinsey, A. C., Pomeroy, W. B., & Martin, C. E. (1948). *Sexual behavior in the human male.* Philadelphia: Saunders.

Kinsey, A. C., Pomeroy, W. B., Martin, C. E., & Gebhard, P. H. (1953). *Sexual behavior in the human female.* Philadelphia: Saunders.

Kirkendall, L. A. (1981). Sex education in the United States: A historical perspective. In L. Brown (ed.), *Sex education in the eighties.* New York: Plenum.

Klassen, A. D., & Wilsnack, S. C. (1986). Sexual experience and drinking among women in a U.S. national survey. *Archives of Sexual Behavior,* October, 15(5), 363–392.

Klaus, M. H., & Kennell, J. H. (1976). *Maternal-infant bonding.* St. Louis, MO: Mosby.

Klein, H. G., & Altar, H. J. (1987). Blood transfusions and AIDS. In *AIDS,* Vol. 11, pp. 7–10. Chicago, IL: American Medical Association.

Kline-Graber, G., & Graber, B. (1978). Diagnosis and treatment procedures of pubococcygeal deficiencies in women. In J. LoPiccolo & L. LoPiccolo (eds.), *Handbook of sex therapy.* New York: Plenum.

Knuppel, R., & Godlin (1987). Maternal-placental-fetal unit; fetal and early neonatal physiology. In M. Pernoll & R. Benson (eds.), *Current obstetric and gynecologic diagnosis and treatment 1987.* Los Altos, CA: Appleton & Lange.

Koblinsky, S., & Palmeter, J. (1984). Sex-role orientation, mother's expression of affection toward spouse, and college women's attitudes toward sexual behaviors. *Journal of Sex Research, 20,* 32–43.

Kolata, G. B. (1988a, June 7). The evolving biology of AIDS: Scavenger cell looms large. *New York Times,* p. B5.

Kolata, G. B. (1988b, March 29). Fetuses treated through umbilical cords. *New York Times,* p. 20.

Kolata, G. B. (1988c, January 25). Multiple fetuses raise new issues tied to abortion. *New York Times.*

Kolata, G. B. (1987, October 28). Earlier U.S. AIDS incursions hinted. *New York Times.*

Kolodny, R. C., Masters, W. H., & Johnson, V. E. (1979). *Textbook of sexual medicine.* Boston: Little, Brown.

Kolodny, R. C. et al. (1979). *Textbook of human sexuality for nurses.* Boston: Little, Brown.

Koop, E. (1986). Acquired immune deficiency syndrome. *Journal of Medical Association,* November 28.

Krane, R. J., & Siroky, M. B. (1981). Neurophysiology of erection. *Urologic Clinics of North America, 8*(1).

Kwan, M., Greenleaf, W. J., Mann, J., Crapo, L., & Davidson, J. M. (1983). The nature of androgen action on male sexuality: A combined laboratory and self-report study in hypogonadal men. *Journal of Clinical Endocrinology and Metabolism, 57,* 557–562.

Kwong, L., Smith, E., Davidson, J., & Peroutka, S. (1986). Differential interactions of prosexual drugs with 5-hydroxytryptamine-sub(1A) and alpha-sub-2-adrenergic receptors. *Behavioral Neuroscience,* October, 100(5), 644–668.

Lacey, C. (1987). Premalignant and malignant disorders of the uterine corpus. In M. Pernoll & R. Benson (eds.), *Current obstetric and gynecologic diagnosis and treatment 1987.* Los Altos, CA: Appleton & Lange.

Ladas, A. K., Whipple, B., & Perry, J. D. (1982). *The G spot.* New York: Holt, Rinehart and Winston.

Lamaze, F. (1970). *Painless childbirth.* Chicago: Regnery.

Lancaster, J. B. (1979). Sex and gender in evolutionary perspective. In H. A. Katchadourian (ed.), *Human sexuality: A comparative and developmental perspective,* pp. 51–80. Berkeley: University of California Press.

Lancaster, J. B., & Lee, R. B. (1965). "The annual reproductive cycle in monkeys and apes." In DeVore, I. (ed.), *Primate behavior: Field studies of monkeys and apes.* New York: Holt, Rinehart and Winston.

Lauritsen, J. (1982). Research review: The cytogenetics of spontaneous abortion. *Research in Reproduction, 14,* 3ff.

Laws, J. L., & Schwartz, P. (1977). *Sexual scripts.* Hinsdale, IL: Dryden.

Laws, S. (1983). The sexual politics of pre-menstrual tension. *Women's Studies International Forum, 6,* 20.

Leary, W. E. (1988, July 17). Sharp rise in rare sex-related diseases. *New York Times,* p. B9.

Leboyer, F. (1975). *Birth without violence.* New York: Knopf.

Ledger, W. A. (1987). AIDS and the obstetrician/gynecologist: Commentary. In *AIDS,* Vol 2, pp. 5–6. Chicago: American Medical Association.

Lehrman, D. S. (1970). Semantic and conceptual issues in the nature-nurture problem. In L. R. Aronson & E. Tobach (eds.), *Development and evolution of behavior.* New York: Freeman.

Leiblum, S. R., & Pervin, L. A. (1980). *Principles and practice of sex therapy.* London: Tavistock.

Leibovici, L., Alpert, G., Laor, A., Kalter-Leibovici, O., & Danon, Y. (1987). Urinary tract infections and sexual activity in young women. *Archives of Internal Medicine,* February 147(2), 345–347.

Lemon, S. M. (1984). Viral hepatitis. In K. Holmes, P.Mardh, P. F. Sparling, & P. J. Wiesner (eds.), *Sexually transmitted diseases.* New York: McGraw-Hill.

Lertola, J. (1986, November 3). Illustration in *Time,* p. 69.

Lessing, D. (1962). *The golden notebook.* London: Michael Joseph.

Levin, R. J. (1981). The female orgasm: A current appraisal. *Journal of Psychosomatic Research, 25*(2), 119–133.

Levin, R. J. (1980). The physiology of sexual function in women. *Clinics in Obstetrics and Gynecology, 7*(2), 213–252.

Libby, R. W. (1976). Social scripts for sexual relationships. In S. Gordon & R. W. Libby (eds.), *Sexuality today and tomorrow*. N. Scituate, MA: Duxbury Press.

Lipkin, M., Jr., & Lamb, G. S. (1982). The Couvade syndrome: An epidemiologic study. *Annals of Internal Medicine, 96*, 509–511.

Lipson, J., & Engleman, E. (1985). Special Report on AIDS. *Stanford Medicine*, Spring, pp. 24–25.

Lisk, R. D. (1967). In L. Martinini & W. F. Ganong (eds.), *Neuroendocrinology*, Vol. 2. New York: Academic, p. 197.

London, S. N., & Hammond, C. B. (1986). The climacteric. In D. N. Danforth & J. R. Scott (eds.), *Obstetrics and gynecology*, 5th ed. Philadelphia: J. B. Lippincott.

LoPiccolo, J. (1977). Direct treatment of sexual dysfunction in the couple. In J. Money & H. Musaph (eds.), *Handbook of sexology*. New York: Elsevier.

LoPiccolo, J., Heiman, J., Hogan, D., & Roberts, C. (1985). Effectiveness of single therapists versus cotherapy teams in sex therapy. *Journal of Consulting and Clinical Psychology, 53*, 287–294.

LoPiccolo, J., & Lobitz, W. C. (1977). The role of masturbation in the treatment of orgasmic dysfunction. In J. LoPiccolo & L. LoPiccolo (eds.), *Handbook of sex therapy*. New York: Plenum.

LoPiccolo, J., & LoPiccolo, L. (1978). *Handbook of sex therapy*. New York: Plenum.

LoPiccolo, J., & Stock, W. (1986). Treatment of sexual dysfunction. *Journal of Consulting and Clinical Psychology*, April *54*(2), 158–167.

Lowry, T. P. (ed.). (1978). *The classic clitoris*. Chicago: Nelson-Hall.

Lowry, T. P., & Lowry, T. S. (1976). *The clitoris*. St. Louis, MO: Warren H. Green.

Mabie, B., and Sibai (1987). Hypertensive states of pregnancy. In M. Pernoll & R. Benson (eds.), *Current obstetric and gynecologic diagnosis and treatment 1987*. Los Altos, CA: Appleton & Lange.

MacLean, P. D. (1976). Brain mechanisms of elemental sexual functions. In B. J. Sadock, H. I. Kaplan, & A. M. Freedman (eds.), *The sexual experience*. Baltimore, MD: Williams & Wilkins.

MacLusky, N. J., & Naftolin, F. (1981). Seuxal differentiation of the central nervous system. *Science, 211*(4488), 1294–1302.

Mahler, H. (1986, December 1). Quoted in *Time*, p. 45.

Malatesta, V. J. (1979). Alcohol effects on the orgasmic ejaculatory response in human males. *The Journal of Sex Research, 15*, 101–107.

Malla, K. (1964). *The ananga ranga*. R. F. Burton & F. F. Arbuthnot, translators. New York: Putnam.

Manniche, L. (1987). *Sexual life in ancient Egypt*. London: KPI.

Marshall, J. (1987). Infertility. In M. Pernoll & R. Benson (eds.), *Current obstetric and gynecologic diagnosis and treatment 1987*. Los Altos, CA: Appleton & Lange.

Marshall, P., Surridge, D., & Delva, N. (1987). The role of nocturnal penile tumescence in differentiating between organic and psychogenic impotence: The first stage of validation. *Archives of Sexual Behavior, 10*(1).

Marshall, W. A., & Tanner, J. M. (1974). Puberty. In J. A. Douvis & J. Dobbing (eds.), *Scientific foundations of pediatrics*. London: William Heinemaun Medical Books.

Maruta, T., & McHardy, M. (1983). Sexual problems in patients with chronic pain. *Medical Aspects of Human Sexuality, 17*(2).

Masters, W., & Johnson, V. (1982). Sex and the aging process. *Medical Aspects of Human Sexuality, 16*(6).

Masters, W. H., & Johnson, V. E. (1970). *Human sexual inadequacy*. Boston: Little, Brown.

Masters, W. H., & Johnson, V. E. (1966). *Human sexual response*. Boston: Little, Brown.

Masters, W. H., Johnson, V. E., & Kolodny, R. C. (1988). *Crisis: Heterosexual behavior in the age of AIDS*. New York: Grove Press.

Matteo, S., & Rissman, E. F. (1984). Increased sexual behavior during the midcycle portion of the human menstrual cycle. *Hormones and Behavior, 18*, 249–255.

McCance, A. A., Luff, M. C., & Widdowson, E. C. (1952). Distribution of coitus during the menstrual cycle. *Journal of Hygiene, 37*, 571–611.

McCary, J. L. (1975). Teaching the topic of human sexuality. *Teaching of Psychology, 2*(1), 16–21.

McClintock, M. K. (1983). Pheromonal regulation of the ovarian cycle. In J. G. Vandenbergh (ed.), *Pheromones and reproduction in mammals*, pp. 113–149. New York: Academic Press.

McClintock, M. K. (1971). Menstrual synchrony and suppression. *Nature, 299*, 244–245.

McDougall, W. (1908). *An introduction to social psychology*. London: Methuen.

McGee, A. (1984). Gonococcal pelvic inflammatory disease. In K. Holmes, P. Mardh, P. F. Sparling, & P. Wiesner (eds.), *Sexually transmitted diseases*. New York: McGraw-Hill.

McKey, P. L., & Dougherty, M. C. (1986). The circumvaginal musculature: Correlation between pressure and physical assessment. *Nursing Resource*, September–October, *35*(5), 307–309.

McQuarrie, H. G., & Flanagan, A. D. (1978). Accuracy of early pregnancy testing at home. Paper presented at the annual meeting of the Association of Planned Parenthood Physicians, San Diego, October 24–27.

Melody, G. F. (1977). A case of penis captivus. *Medical Aspects of Human Sexuality, 11*, December.

Melton, G. B. (ed.) (1986). *Adolescent abortion*. Lincoln, NE: University of Nebraska Press.

Menning, B. (1977). *Infertility: A guide for childless couples*. Englewood Cliffs, NJ: Prentice-Hall.

Mertz, G. J. (1984). Double blind placebo-controlled trial of oral Acyclovir in first episode genital herpes simplex virus infection. *Journal of American Medical Association, 254*, 1147–1151.

Messe, M. R., & Geer, J. H. (1985). Voluntary vaginal musculature contractions as an enhancer of sexual arousal, *Archives of Sexual Behavior, 4*(1), 13–38.

Meyer, A. W. (1939). *The rise of embryology*. Stanford, CA: Stanford University Press.

Michael, R. P., Bonsall, R. W., & Zumpe, D. (1976). The evidence for chemical communication in primates. *Vitamins and Hormones, 34*, 137–186.

Michael, R. P., Bonsall, R. W., & Warner, P. (1974). Human vaginal secretions: Volatile fatty acid content. *Science, 186*, 1217–1219.

Michael, R. P., & Keverne, E. B. (1968). Pheromones in the communication of sexual status in primates. *Nature, 218*, 746–749.

Michael, R. P., & Zumpe, D. (1978). Potency in male rhesus monkeys: Effects of continuously receptive females. *Science, 200*, 451–453.

Miller, D., Rich, L., & Steinberg, C. (1987). STD talk: Students coping with sexually transmitted diseases. Brown University Health Services, Providence, RI.

Mishell, D. R. (1982). Control of human reproduction. In D. N. Danforth (ed.), *Obstetrics and gynecology*, 4th ed., pp. 252–280. Philadelphia: Harper & Row.

Mishell, D. R. (1979). Intrauterine devices: Medicated and nonmedicated. *International Journal of Gynecology and Obstetrics, 16,* 482–487.

Mitamura, T. (1970). *Chinese eunuchs.* Tokyo: C. Tuttle.

Moghissi, K. S. (1982). Nutrition in obstetrics and gynecology. In D. N. Danforth (ed.), *Obstetrics and gynecology,* 4th ed, pp. 203–215. Philadelphia: Harper & Row.

Moghissi, K. S., & Evans, T. N. (1982). "Infertility." In Danforth, D. N. (ed.), *Obstetrics and gynecology,* 4th ed. Philadelphia: Harper & Row.

Money, J. (1987). Human sexology and psychoneuroendocrinology. In D. Crews (ed.), *Psychobiology of reproductive behavior: An evolutionary perspective,* pp. 323–344. Englewood Cliffs, NJ: Prentice-Hall.

Money, J. (1973). Gender role, gender identity, core gender identity: Usage and definitions and terms. *Journal of American Academy of Psychoanalysis, 1,* 397–403.

Money, J., & Erhardt, A. A. (1972). *Man and woman, boy and girl.* Baltimore, MD: Johns Hopkins University Press.

Monga, T., Lawson, J., & Inglis, J. (1986). Sexual dysfunction in stroke patients. *Archives of Physical and Medical Rehabilitation,* January, 67(1), 19–22.

Moore, K. L. (1982). *The developing human,* 3d ed. Philadelphia: Saunders.

Moos, R. (1969). Fluctuations in symptoms and moods during the menstrual cycle *Journal of Psychosomatic Research, 13,* 37–44.

Morbidity and mortality weekly report. (1988). *Journal of the American Medical Association, 259,* 2657–2661.

Morganthau, T., & Hager, M. (1987, November 10). AIDS: Grim prospects. *Newsweek,* pp. 20–21.

Morris, D. (1967). *The human zoo.* New York: McGraw-Hill.

Morris, D. (1977). *Manwatching: A field guide to human behavior.* New York: Abrams.

Mosher, B. A., & Whelan, E. M. (1981). Postmenopausal estrogen therapy: A review. *Obstetric and Gynecology Survey, 9,* 467–475.

Munjack, D. J., & Oziel, L. J. (1980). *Sexual medicine and counseling in office practice.* Boston: Little, Brown.

Murad, F., & Haynes, R. C., Jr. (1985a). Androgens. In A. F. Gilman, L. S. Goodman, T. W. Rall, & F. Murad (eds.), *Goodman and Gilman's, the pharmacological basis of therapeutics,* 7th ed. New York: Macmillan.

Murad, F., & Haynes, R. C., Jr. (1985b). Estrogens and progestins. In A. F. Gilman, L. S. Goodman, T. W. Rall, & F. Murad (eds.), *Goodman and Gilman's, the pharmacological basis of therapeutics,* 7th ed. New York: Macmillan.

Murphy, F. K., & Patamasucon, P. (1984). Congenital syphilis. In K. Holmes, P. Mardh, P. F. Sparling, & P. Wiesner (eds.), *Sexually transmitted diseases.* New York: McGraw-Hill.

Murphy, J. (1986, December 29). The month-after pill. *Time.*

Musaph, H., & Abraham, G. (1977). Frigidity or hypogyneisms. In J. Money & H. Musaph, (eds.), *Handbook of sexology.* New York: Elsevier.

Nadelson, C. C. (1978). The emotional impact of abortion. In M. T. Notman & C. C. Nadelson (eds.), *The woman patient,* Vol. 1, pp. 173–179. New York: Plenum.

Nadler, R. D. (1977). Sexual behavior of captive orangutans. *Archives of Sexual Behavior, 6,* 457–476.

Nathan, S. (1986). The epidemiology of the DSM-III psychosexual dysfunctions. *Journal of Sex and Marital Therapy,* Winter, 12(4), 267–281.

Nefzawi, S. (1964 ed.). *The perfumed garden.* R. F. Burton, translator. New York: Putnam.

Neiberg, P., Marks, J. S., McLaren, N., & Remington, P. (1985). The fetal tobacco syndrome. *Journal of the American Medical Association, 253,* 2998–2999.

Nelson, N. M. et al. (1980). A randomized clinical trial of the Leboyer approach to childbirth. *New England Journal of Medicine, 302,* 655–660.

Newman, H. F., & Northup, J. D. (1981). Mechanism of human penile erection: An overview. *Urology, 17*(5), 399–408.

Noonan, J. T., Jr. (1967). *Contraception: A history of its treatment by the Catholic theologians and canonists.* New York: New American Library.

Nory, M. J. (1987). The normal puerperium. In M. Pernoll & R. Benson (eds.), *Current obstetric and gynecologic diagnosis and treatment 1987.* Los Altos, CA: Appleton & Lange.

Novak, E. R., et al. (1970). *Novak's textbook of gynecology,* 8th ed. Baltimore: Williams & Wilkins.

Oakley, G. (1978). Natural selection, selection bias and the prevalence of Down's syndrome. *New England Journal of Medicine, 299*(19), 1068–1069.

O'Farrell, T., Weyand, C., & Logan, D. (1983). *Alcohol and sexuality: An annotated bibliography on alcohol use, alcoholism, and human sexual behavior.* Phoenix, AZ: Oryx.

Olds, J. (1956). Pleasure centers in the brain. *Scientific American, 193,* 105–116.

Oriel, J. D. (1984). Genital warts. In K. Holmes, P. Mardh, P. F. Sparling, & P. Wiesner (eds.), *Sexually transmitted diseases.* New York: McGraw-Hill.

Orkin, M., & Maibach, H. (1984). Scabies. In K. Holmes, P. Mardh, P. F. Sparling, & P. Wiesner (eds.), *Sexually transmitted diseases.* New York: McGraw-Hill.

Ortner, A., Glatzl, J., & Karpellus, E. (1987). Clinical and endocrinologic study of precocious puberty in girls. *Archives of Gynecology, 240*(2), 81–93.

Ory, H. W., Forrest, J. D., & Lincoln, R. (1983). *Making choices.* New York: Alan Guttmacher Institute.

Osser, S., Liedholm, P., & Oberg, S. J. (1980). Risk of pelvic inflammatory disease among users of intrauterine devices, irrespective of previous pregnancy. *American Journal of Obstetrics and Gynecology, 138*(7 pt 2), 864–867.

Page, D. C. et al. (1987). The sex-determining region of the human Y chromosome encodes a finger protein. December 24, 51(6), 1091–1104.

Paige, K. E. (1978). The ritual of circumcision. *Human Nature, 1,* 40–48.

Paige, K. E. (1973, April). Women learn to sing the menstrual blues. *Psychology Today,* 41–46.

Pear, R. (1987, May 20). Three health workers found infected by blood of patients with AIDS. *New York Times.*

Pepe, F., Iachello, R., Panella, M., Pepe, G., Panella, P., Pennisi, F., Pepe, P., Salemi, F., Privitera, D., Sanfilippo, A. et al. (1987). Parity and sexual behavior in pregnancy. *Clinical Exp-Obstet-Gynecol, 14*(1), 60–65.

Perkins, R. P. (1979). Sexual behavior and response in relation to complications in pregnancy. *American Journal of Obstetrics and Gynecology, 134,* 498–505.

Pernoll, M., & Benson, R. (eds.) (1987). *Current obstetrical and gynecologic diagnosis and treatment,* 6th ed. Los Altos, CA: Appleton & Lange.

Perry, J. D., & Whipple, B. (1981). Pelvic muscle strength of female ejaculators: Evidence in support of a new theory of orgasm. *The Journal of Sex Research, 17*(1), 22–39.

Perry, J. D., & Whipple, B. (1980). Female ejaculation by Grafenburg spot stimulation. Paper presented at the annual meeting of the Society for the Scientific Study of Sex. Dallas, November, 15.

Persky, H. (1983). Psychosexual effects of hormones. *Medical Aspects of Human Sexuality, 17*(9).

Peterman, T. A., & Curran, J. W. (1986). Sexual transmission of human immunodeficiency virus. *Journal of the American Medical Association, 256,* p. 2222–2226.

Pfeiffer, E., Verwoerdt, A., & Davis, G. C. (1972). Sexual behavior in middle life. *American Journal of Psychiatry, 128,* 1262–1267.

Pincus, G. (1965). *The control of fertility.* New York: Academic Press.

Pohlman, E. G. (1969). *Psychology of birth planning.* Cambridge, MA: Shenkman.

Pomeroy, W. B. (1976). *Your child and sex: A guide for parents.* New York: Delacorte.

Pomeroy, W. B. (1972). *Dr. Kinsey and the Institute for Sex Research.* New York: Harper & Row.

Price, W., & Forejt, J. (1986). Neuropsychiatric aspects of AIDS: A case report. *General Hospital Psychiatry,* January, *8*(1), 7–10.

Pritchard, J. A., MacDonald, P. C., & Gant, N. (1985). *Williams obstetrics,* 17th ed. Norwalk, CT: Appleton-Century-Crofts.

Qualls, C. B., Wincze, J. P., & Barlow, D. H. (eds.). (1978). *The prevention of sexual disorders.* New York: Plenum.

Quinn, T. C., Mann, J. M., Curran, J. W., & Piot, P. (1986). AIDS in Africa: An epidiologic paradigm. *Science,* November 21, 955–986.

Rabkin, C., Thomas, P., Jaffe, H., & Schultz, S. (1987). Prevalence of antibody to HTLV-III/LAV in a population attending a sexually transmitted diseases clinic. *Sexually Transmitted Diseases,* January–March, *14*(1), 48–51.

Raisman, G., & Field, P. (1971). Sexual dimorphism in the preoptic area of the rat. *Science, 173,* 731–733.

Ramasharma, K., & Sairam, M. R. (1982). Isolation and characterization of inhibin from human seminal plasma. *Annals of New York Academy of Sciences, 383,* 307–328.

Raven, P. H., & Johnson, G. B. (1986). *Biology.* St. Louis, MO: C.V. Mosby.

Reichlin, S. (1963). Neuroendocrinology. *New England Journal of Medicine, 269,* 1182, 1246, 1296.

Reid, R. L. (1986). Premenstrual syndrome: A time for introspection. *American Journal of Obstetrics and Gynecology,* November, *155*(5), 921–926.

Reinhold, R. (1987, October 31). AIDS book brings power to a gay San Franciscan. *New York Times.*

Remy, J. (1979). Mutilations sexuelles: En France aussi. *L'Express,* No. 1447, 58–60.

Richart, R. (1983). Condyloma viruses that progress to invasive cancer can be identified. *Contraceptive Technology Update, 4,* 143–144.

Robbins, M. B., & Jensen, G. D. (1976). Multiple orgasms in the male. In R. Gemme & C. C. Wheeler (eds.),

Progress of sexology, pp. 323–338. New York: Plenum.

Robinson, P. (1976). *The modernization of sex.* New York: Harper & Row.

Rodgers, B. (1972). *Gay talk.* New York: Putnam.

Rousso, H. (1986). Confronting the myth of asexuality: The network project for disabled women and girls. *SIECUS Report, 14,* 4–6.

Rowan, R. L., & Gillette, P. J. (1978). *The gay health guide.* Boston: Little, Brown.

Rowell, T. E. (1972). Female reproduction cycles and social behavior in primates. *Advances in the Study of Behavior, 4,* 69–105.

Rowell, T. E. (1972). *The social behavior of monkeys.* Baltimore: Penguin.

Rubin, P. (ed.) (1987). *Clinical oncology: A multidisciplinary approach,* 6th ed. American Cancer Society.

Rubin, Z., Peplau, L. A., & Hill, C. T. (1981). Loving and leaving: Sex differences in romantic attachment. *Sex Roles, 7,* 821–836.

Rubinow, D. R. (1984). Premenstrual syndrome: Overview from a methodological perspective. *American Journal of Psychiatry,* February, 163–170.

Rubinow, D. R., Roy-Byrne, P., Hoban, M. D., Grover, G. N., Stambler, N., & Post, R. M. (1986). Premenstrual mood changes: Characteristic patterns in women with and without premenstrual syndrome. *Journal of Affective Disorders,* March–April, *10*(2), 85–90.

Ruble, D. N. (1977). Premenstrual symptoms: A reinterpretation. *Science, 197,* 291–292.

Russell, B. (1945). *A history of Western philosophy.* New York: Simon & Schuster.

Russell, K. P. (1987). The course and conduct of normal labor and delivery. In M. Pernoll & R. Benson (eds.), *Current obstetric and gynecologic diagnosis and treatment 1987.* Los Altos, CA: Appleton & Lange.

Russell, M. J., Switz, G. M., & Thompson, K. (1977). Olfactory influences on the human menstrual cycle. Delivered at the American Association for the Advancement of Science, San Francisco.

Rutledge, F. (1986). Gynecologic malignancy: General considerations. In D. Danforth & J. Scott et al. (eds.), *Obstetrics and gynecology,* 5th ed. Philadelphia: J. B. Lippincott.

Saah, A. J. (1987). Serologic tests for human immunodeficiency virus (HIV). *AIDS, 2,* 11–14.

Sadler, T. W. (1985). *Langman's medical embryology,* 5th ed. Baltimore, MD: Williams & Wilkins.

Sadock, B. J., & Sadock, V. A. (1976). Techniques of coitus. In B. J. Sadock et al. (eds.), *The sexual experience.* Baltimore, MD: Williams & Wilkins.

Salzman, L. (1968). Sexuality in psychoanalytic theory. In J. Marmor (ed.), *Modern psychoanalysis,* pp. 123–145. New York: Basic Books.

Samuel, T., & Rose, N. R. (1980). The lessons of vasectomy: A review. *Journal of Clinical Laboratory Immunology, 3*(2), 77–83.

Sandberg, G., & Quevillon, R. (1987). Dyspareunia: An integrated approach to assessment and diagnosis. *Journal of Family Practice,* January *24*(1), 66–70.

Sarrel, P. J., & Coplin, H. R. (1971). A course in human sexuality for the college student. *American Journal of Public Health, 61,* 1030–1037.

Savage, D. C. L., & Evans, J. (1984). Puberty and adolescence. In J. O. Forfar & G. C Arneil (eds), *Textbook of pediatrics,* 3d ed., Vol. 1, pp. 366–388. Edinburgh: Churchill Livingston.

Scarpinato, L., & Calabrese, L. H. (1987). Prospects for AIDS therapy and vaccine. In V. Gong & N. Rudnick (eds.), *AIDS.* New Brunswick, NJ: Rutgers University Press.

Scharfman, M. A. (1977). Birth and the neonate. In R. C. Simons & H. Pardes (eds.), *Understanding human behavior in health and illness.* Baltimore, MD: Williams & Wilkins.

Schmeck, H. M., Jr. (1987, October 20). Venereal virus strongly implicated in several cancers. *New York Times,* p. 17.

Schover, L., Friedman, J., Weiler, S., Heiman, J., & LoPiccolo, J. (1982). Multiaxial problem-oriented system for sexual dysfunctions. *Archives of General Psychiatry, 39,* 614–619.

Schreiner-Engel, P. (1987). Developmental psychology. *Infertility.*

Schreiner-Engel, P., & Schiavi, R. (1986). Lifetime psychopathology in individuals with low sexual desire. *Journal of Nervous and Mental Disease,* November, *174*(11), 646–651.

Schreiner-Engel, P., Schiavi, R., Vietorisa, D., & Smith, H. (1987). The differential impact of diabetes type on female sexuality. *Journal of Psychosomatic Research, 31*(1), 23–33.

Scott, J. R. (1986). Spontaneous abortion. In D. Danforth & J. R. Scott (eds.), *Obstetrics and gynecology,* 5th ed. Philadelphia: J. B. Lippincott.

Segraves, K. A., Segraves, R. T., & Schoenberg, H. (1987). Use of sexual history to differentiate organic from psychogenic impotence. *Archives of Sexual Behavior, 16*(2), 125–137.

Semans, J. H. (1956). Premature ejaculation: A new approach. *Southern Medical Journal, 49,* 353–357.

Sevely, J. L., & Bennett, J. W. (1978). Concerning female ejaculation and the female prostate. *Journal of Sex Research, 14,* 1–20.

Shahani, S. K., & Hattikudur, N. S. (1981). Immunological consequences of vasectomy. *Archives of Andrology, 7*(2), 193–199.

Sherfey, M. J. (1973). *The nature and evolution of female sexuality.* New York: Vintage.

Sherwin, B., Gelfand, M., & Brender, W. (1985). Androgen enhances sexual motivation in females: A prospective crossover study in steroid administration in surgical menopause. *Psychosomatic Medicine, 47*(4), 339–351.

Shettles, L. B. (1972). Predetermining children's sex. *Medical Aspects of Human Sexuality, 6,* 172ff.

Shilts, R. (1988). Promising treatment for AIDS. *San Francisco Chronicle,* p. A1–A4.

Shilts, R. (1987). *And the band played on.* New York: St. Martin's Press.

Short, R. V. (1980). The origins of human sexuality. In C. R. Austin & R. V. Short (eds.), *Human sexuality.* Cambridge, MA: Cambridge University Press.

Short, R. V. (1979). The development of human reproduction. Regulation de la fecondité, INSERM, *83,* 355–366.

Silber, S. J. (1981). *The male.* New York: Scribner's.

Singer, I., & Singer, J. (1972). Types of female orgasm. *Journal of Sex Research, 8*(11), 255–267.

Skaakeback, N. E., Bancroft, J., Davidson, D. W., & Androgen, W. P. (1981). Replacement with oral testosterone undecaonate in hypogonadal men. *Clinical Endocrinology, 14,* 49–67.

Small, M. (1987). Semirigid and malleable penile implants. *Urological Clinical of North America,* February, *14*(1), 187–201.

Smith, P. J., & Talbert, R. L. (1986). Sexual dysfunction with antihypertensive and antipsychotic agents. *Clinical Pharmacy,* May, *5*(5), 373–384.

Solomon, S., & Cappa, K. (1987). Impotence and bicycling. A seldom-reported association. *Postgrad-Med,* January, *81*(1), 99–100.

Sondheimer, S. J. (1985). Hormonal changes in premenstrual syndrome. *Psychosomatics,* October, 803–809.

Sparks, R. A., Purrier, B. G., Watt, P. J., & Elstein, M. (1981). Bacteriological colonisation of uterine cavity: Role of tailed intrauterine contraceptive device. *British Medical Journal of Clinical Research, 282*(6271), 1189–1891.

Sparling, P. F. (1988). Sexually transmitted diseases. In J. B. Wyngaarden & L. H. Smith (eds.), *Cecil textbook of medicine,* 18th ed., pp. 1701–1706. Philadelphia: W.B. Saunders.

Speert, H. (1982). "Historical highlights." In Danforth, D. N. (ed.), *Obstetrics and gynecology,* 4th ed. Philadelphia: Harper & Row.

Spitz, R. A., & Wolf, K. D. (1947). Anaclitic depression. An inquiry into the genesis of psychiatric conditions in early childhood, II. *The Psychoanalytic Study of the Child,* II.

Stamm, W. E., & Holmes, K. (1984). Chlamydia trachomatis infections of the adult. In K. Holmes, P. Mardh, P. F. Sparling, & P. Wiesner, *Sexually transmitted diseases.* New York: McGraw-Hill.

Steege, J., Stout, A., & Carson, C. (1986). Patient satisfaction in Scott and Small-Carrion penile implant recipients: A study of 52 patients. *Archives of Sexual Behavior,* October, *15*(5), 393–399.

Steinberger, A., & Steinberger, E. (1976). Secretion of FSH-inhibiting factor by cultured Sertoli cells. *Endocrinology, 99,* 918–921.

Stern, C. (1973). *Principles of human genetics,* 3rd ed. San Francisco: Freeman.

Stipp, D. (1987, April 4). Research on impotence upsets idea that it is usually psychological. *Wall Street Journal,* p. 1.

Stone, K. M., Grimes, D. A., & Magder, L. S. (1986). Primary prevention of sexually transmitted diseases: A primer for clinicians. *Journal of the American Medical Association, 255,* 1763–1766.

Sturup, G. K. (1979). Castration: The total treatment. In H. L. P. Resnick & M. E. Wolfgang (eds.), *Sexual behavior: Social and legal aspects,* pp. 361–382. Boston: Little, Brown.

Suitters, B. (1967). *The history of contraceptives.* London: International Planned Parenthood Federation.

Sullivan, W. (1987, December 29). Sexual potency saved after surgery to remove prostate. *New York Times,* p. 16.

Sullivan, W. (1986, October 31). Scientists developing a new drug that blocks and terminates pregnancy. *New York Times,* p. 13.

Symons, D. (1979). *The evolution of human sexuality.* Oxford: Oxford University Press.

Taba, A. H. (1979). Female circumcision. *World Health.* Geneva, Switzerland: World Health Organization.

Tanner, J. M. (1984). Physical growth and development. In J. O. Forfar & G. C. Arneil (eds.), *Textbook of pediat-*

rics, 3d ed., Vol. 1, pp. 278–330. Edinburgh: Churchill Livingstone.

Tanner, J. M. (1978). *Fetus to man.* Cambridge, MA: Harvard University Press.

Tarabulcy, E. (1972). Sexual function in the normal and in paraplegia. *Paraplegia, 10,* 202–204.

Tatum, H. J. (1987). Contraception and family planning. In M. Pernoll & R. Benson (eds.), *Current obstetric and gynecologic diagnosis and treatment 1987.* Los Altos, CA: Appleton & Lange.

Taylor, C., & Pernoll, M. (1987). Normal pregnancy and prenatal care. In M. Pernoll & R. Benson (eds.), *Current obstetric and gynecologic diagnosis and treatment 1987.* Los Altos, CA: Appleton & Lange.

Terzian, H., & Dale-Ore, G. (1955). Syndrome of Kluver and Bucy reproduced in man by bilateral removal of temporal lobes. *Neurology, 5,* 373–380.

Thompson, R. F. (1985). *The brain.* New York: W.H. Freeman.

Tietze, C. (1983). *Induced abortion: A world review,* 5th ed. New York: Population Council.

Tinbergen, N. (1951). *The study of instinct.* Oxford: Clarendon.

Tolor, A., & DiGrazia, P. V. (1976). Sexual attitudes and behavior patterns during and following pregnancy. *Archives of Sexual Behavior, 5,* 539–551.

Trapp, J. D. (1987). Pharmacologic erection program for the treatment of male impotence. *Southern Medical Journal,* April, *80*(4), 426–427.

Trussel, J., & Kost, K. (1987). Contraceptive failure in the United States: A critical review of the literature. *Studies in Family Planning, 18*(5), 237–283.

Tutin, C. E. G. (1980). Reproductive behavior of wild chimpanzees in the Gombe National Park, Tanzania. In R. V. Short & B. J. Weir (eds.), *The great apes of Africa.* Reproduction and Fertility Suppl. *28,* pp. 43–57.

Udry, J. R., & Morris, N. M. (1968). Distribution of coitus in the menstrual cycle. *Nature, 220,* 593–596.

Ulene, A. (1987). *Safe sex in a dangerous world.* New York: Vintage Books.

Vance, E. B., & Wagner, N. N. (1976). Written descriptions of orgasms: A study of sex differences. *Archives of Sexual Behavior, 5,* 87–98.

Vandereycken, W. (1986). Towards a better delineation of ejaculatory disorders. *Acta Psychiatrica Belgica,* January–February, *86*(1), 57–63.

Veevers, J. E. (1974). The life style of voluntarily childless couples. In L. Larson (ed.), *The Canadian family in comparative perspective.* Toronto: Prentice-Hall.

Veith, I. (1965). *Hysteria: The history of a disease*. Chicago: University of Chicago Press.

Wahl, P. et al. (1983). Effect of estrogen/progestin potency on lipid/lipoprotein cholesterol. *The New England Journal of Medicine, 308*(15), 862–867.

Wallis, C. (1984, December 9). Children having children. *Time*, pp. 78–87.

Walsh, P. (1985). Diseases of the prostate. In J. B. Wyngaarden & L. H. Smith (eds.), *Cecil textbook of medicine*, 17th ed., pp. 1375–1379. Philadelphia: W. B. Saunders.

Washington, A. E., Cates, W., Jr., & Zaidi, A. A. (1984). Hospitalizations for pelvic inflammatory disease: Epidemiology and trends in the United States, 1975–1981. *Journal of the American Medical Association, 251*, 2529–2533.

Waxenburg, S. E., Drellich, M. G., & Sutherland, A. M. (1959). The role of hormones in human behavior, I: Changes in female sexuality after adrenalectomy. *Journal of Clinical Endocrinology, 19*, 193–202.

Wein, A. J., Fishkin, R., Carpiniello, V. L., & Malloy, T. R. (1981). Expansion without significant rigidity during nocturnal penile tumescence testing: A potential source of misinterpretation. *The Journal of Urology, 126*, 343–344.

West, J. B. (ed.). (1985). *Best and Taylor's physiological basis of medical practice*, 11th ed. Baltimore, MD: Williams & Wilkins.

Westoff, C. F., & Jones, E. F. (1977). The secularization of U.S. Catholic birth control practices. *Family Planning Perspectives, 9*, 203–207.

Westoff, C. F., & Rindfuss, R. R. (1974). Sex preselection in the United States: Some implications. *Science, 184*, 633–636.

Wickett, W. H., Jr. (1982). *Herpes: Cause and control*. New York: Pinnacle.

Wigfall-Williams, W. (1987). *Hysterectomy: Learning the facts, coping with feelings, facing the future*. City: Michael Kesend.

Wilkins, L., Blizzard, R., & Migeon, C. (1965). *The diagnosis and treatment of endocrine disorders in childhood and adolescence*. Springfield, IL: Charles C. Thomas.

Williams, J. H. (1987). *Psychology of women. Behavior in a biosocial context*. New York: W. W. Norton.

Williams, P. L., & Warwick, R. (1980). *Gray's anatomy*, 36th ed. Philadelphia: Saunders.

Wilmore, J. H. (1977). *Athletic training and physical fitness*. Boston: Allyn & Bacon.

Wilmore, J. H. (1975). Inferiority of female athletes: Myth or reality. *Journal of Sports Medicine, 3*(1), 1–6.

Wilson, E. O. (1978). *On human nature*. New York: Bantam.

Wilson, E. O. (1975). *Sociobiology: The new synthesis*. Cambridge, MA: The Belknap Press of Harvard University Press.

Wilson, G. (1983). *Sexual positions: A photographic guide to pleasure and love*. New York: Arlington.

Wilson, G. (1982). *The Coolidge effect*. New York: William Morrow.

Wilson, G. T., & Lawson, D. M. (1978a). Effects of alcohol on sexual arousal in women. *Journal of Abnormal Psychology, 87*, 609–616.

Wilson, G. T., & Lawson, D. M. (1978b). Effects of alcohol on sexual arousal in male alcoholics. *Journal of Abnormal Psychology, 87*, 609–616.

Winkelstein, W., Jr., Lyman, D., Padian, N., Grant, R., Samuel, M., Wiley, J., Anderson, R., Lang, W., Riggs, J., & Levy, J. (1987). Sexual practices and risk of infection by the human immunodeficiency virus. The San Francisco Men's Health Study. *Journal of the American Medical Association*, January 16, *257*(3), 321–325.

Witkin, H. A. et al. (1976). Criminality in XYY and XXY men. *Science, 193*, 547–555.

Wolpe, J., & Lazarus, A. A. (1966). *Behavior therapy techniques*. New York: Pergamon.

Yalom, I. D. et al. (1968). Postpartum blues syndrome. *Archives of General Psychiatry, 18*, 16–27.

Yen, S. (1986). Endocrine physiology of pregnancy. In D. Danforth & J. Scott (eds.), *Obstetrics and gynecology*, 5th ed. Philadelphia: J. B. Lippincott.

Ying, S., Becker, A., Ling, N., Ueno, N., & Guillemin, R. (1986). Inhibin and Beta type transforming growth factor have opposite modulating effects on the follicle stimulating hormone-induced aromatase activity of cultured rat granulosa cells. *Biochemical and Biophysical Research Communications, 136*(3), 969–975.

Yupze, A. (1982). Postcoital contraception. *International Journal of Gynecology and Obstetrics, 16*, 497–501.

Zacharias, L., Rand, W. M., & Wurtman, R. J. (1976). A prospective study of sexual development and growth in American girls: The statistics of menarche. *Obstetrics, gynecological survey* (suppl.), *31*, 325–337.

Zelnik, M. (1979). Sex education and knowledge of pregnancy risk among United States teenage women. *Family Planning Perspectives, 11*, 335.

Zelnik, M., & Kantner, J. F. (1980). Sexual activity, contraceptive use and pregnancy among metropolitan area teenagers: 1971–1979. *Family Planning Perspectives, 12*(5), 230–231, 233–237.

Zilbergeld, B. (1978). *Male sexuality: A guide to sexual fulfillment*. Boston: Little, Brown.

Zilbergeld, B., & Evans, M. (1980). The inadequacy of Masters and Johnson. *Psychology Today, 14*.

Subject Index

Name Index

Messe, M.R., 229n
Meyer, A.W., 144n
Michael, R.P., 50n, 94n
Miller, W.B., 177n
Mishell, D.R., 183n, 191n
Mitamura, T., 99n
Moghissi, 161n, 165n, 166n
Money, J., 69n, 73n, 92n, 98n
Monga, T., 214n
Moore, K.L., 38n, 41n, 44n, 45n, 138n, 141n
Moos, R., 86n, 100n
Morganthau, T., 132n
Morgenthau, T., 124n
Morris, D., 7n, 49n
Morrow, P.A., 20
Mosher, B.A., 75n
Munjack, D.J., 212n
Murad, F., 74n, 180n
Murphy, F.K., 118n
Murphy, J., 180n
Musaph, H., 208n

Nadelson, C.C., 201n
Nadler, R.D., 97n
Nathan, S., 205n
Nefzawi, S., 23n, 32n, 208n
Neiberg, P., 161n
Netter, F.H., 45n
Newman, H.F., 54n
Nilsson, L., 45n
Noonan, J.T., Jr., 170n
Nory, M.J., 158n
Novak, E.R., 109n

Oakley, G., 163n
O'Farrell, T., 215n
Olds, J., 70n
Oriel, J.D., 122n
Orkin, M., 119n
Ortner, A., 90n
Ory, H.W., 170n, 174n, 196n
Osser, S., 182n

Page, D.C., 41n
Paige, K.E., 25n, 34n, 88n, 105n
Pear, R., 128n
Pepe, F., 150n
Perkins, R.P., 150n
Pernoll, M., 142n, 162n
Perry, J.D., 59n
Persky, H., 100n
Peterman, T.A., 124n
Pfeiffer, E., 101n
Pincus, G., 178n
Pohlman, E.G., 138n
Pomeroy, W.B., 18–19, 19n
Price, W., 129n
Pritchard, J.A., 142n, 168n, 196n

Qualls, C.B., 230n
Quinn, T.C., 124n, 127n

Rabkin, C., 134n
Raisman, G., 95n
Ramasharma, K., 77n
Raven, P.H., 23n, 125n
Reichlin, S., 92n
Reid, R.L., 89n
Reinhold, R., 123n
Remy, J., 25n
Richart, R., 109n
Rioux, J., 193n
Robbins, M.B., 53n
Robinson, P., 19n
Rodgers, B., 25n, 32n
Ross, C., 188n
Rousseau, J.J., 18
Rowan, R., 119n, 136n
Rowell, T.E., 6n, 97n
Rubenstein, 126n
Rubin, P., 107n
Rubin, Z., 88n
Rubinow, D.R., 88n, 89n
Ruble, D., 88n
Russell, B., 10n
Russell, K.P., 152n
Russell, M.J., 94n
Rutledge, F., 107n

Saah, A.J., 131n
Sadler, T.W.L., 38n, 41n, 45n, 138n
Sadock, B.J., 150n
Salzman, L., 62n
Samuel, T., 191n
Sandberg, G., 211n
Sanger, M., 170
Sarrel, P.J., 20n
Savage, D.C.L., 79n
Scarpinato, L., 132n
Scharfman, M.A., 158n
Schmeck, H.M., Jr., 122n
Schover, L., 206n
Schreiner-Engel, P., 166n, 214n
Scott, J.R., 162n, 196n
Segraves, K.A., 215n
Semans, J.H., 224n
Sevely, J.L., 59n
Shahani, S.K., 193n
Sherfey, M.J., 8, 62n
Sherwin, B., 101n
Shettles, L.B., 140n
Shilts, R., 123n, 126n, 131n
Short, R.V., 97n, 190n
Silber, S.S., 107n, 111n, 136n
Simon, W., 13n
Simons, R.C., 159n
Singer, I., 62n
Skaakeback, N.E., 98n
Small, M., 229n

Smith, P.J., 215n
Solomon, S., 214n
Sondheimer, S.J., 88n
Sparks, R.A., 182n
Sparling, P.F., 112n, 114n, 116n, 118n
Speert, H., 152n, 156n
Spitz, R.A., 48n
Stamm, W.E., 114n
Steege, J., 230n
Steinberger, A., 75n
Steptoe, P., 167
Stern, C., 41n
Stipp, D., 228n
Stone, K.M., 112n, 120n
Sturup, G.K., 98n
Suitters, B., 171n
Sullivan, W., 181n, 213n
Symons, D., 6n, 7n, 8n, 16n, 49n, 50n

Taba, A.H., 25n
Tanner, J.M., 77n, 83n
Tarabulcy, E., 68n
Tatum, H.J., 170n, 193n
Taylor, C., 143n, 161n
Terzian, H., 70n
Thompson, J.F., 162n
Thompson, R.F., 69n, 71n, 95n
Tietze, C., 163n, 196n, 197n
Tinbergen, N., 9n
Tolor, A., 150n
Trapp, J.D., 216n
Trussell, J., 139n, 164n, 177n, 192n
Tutin, C.E.G., 97n
Tyler, E.B., 10

Udry, J.R., 100n
Ulene, A., 133n, 134n, 135n, 136n

Vance, E.B., 61n
Vandereycken, W., 211n
Vatsyayana, 48n
Veevers, J.E., 171n
Veith, I., 28n

Wahl, P., 180n
Wallis, C., 140n, 165n
Walsh, P., 109n, 110n
Washington, A.E., 114n, 115n
Watson, J.B., 10
Waxenburg, S.E., 101n
Wein, A.J., 55n
West, J.B., 73n
Westoff, C.F., 140n, 175n
Wigfall-Williams, W., 213n
Wilkins, L., 92n
Williams, J.H., 13n
Williams, P.L., 23n